# Smart Innovation, Systems and Technologies

Volume 218

The Smart Innovation, Systems and Technologies book series encompasses the topics of knowledge, intelligence, innovation and sustainability. The aim of the series is to make available a platform for the publication of books on all aspects of single and multi-disciplinary research on these themes in order to make the latest results available in a readily-accessible form. Volumes on interdisciplinary research combining two or more of these areas is particularly sought.

The series covers systems and paradigms that employ knowledge and intelligence in a broad sense. Its scope is systems having embedded knowledge and intelligence, which may be applied to the solution of world problems in industry, the environment and the community. It also focusses on the knowledge-transfer methodologies and innovation strategies employed to make this happen effectively. The combination of intelligent systems tools and a broad range of applications introduces a need for a synergy of disciplines from science, technology, business and the humanities. The series will include conference proceedings, edited collections, monographs, handbooks, reference books, and other relevant types of book in areas of science and technology where smart systems and technologies can offer innovative solutions.

High quality content is an essential feature for all book proposals accepted for the series. It is expected that editors of all accepted volumes will ensure that contributions are subjected to an appropriate level of reviewing process and adhere to KES quality principles.

Indexed by SCOPUS, EI Compendex, INSPEC, WTI Frankfurt eG, zbMATH, Japanese Science and Technology Agency (JST), SCImago, DBLP.

All books published in the series are submitted for consideration in Web of Science.

More information about this series at http://www.springer.com/series/8767

Victor Chang · Muthu Ramachandran ·
Víctor Méndez Muñoz
Editors

# Modern Industrial IoT, Big Data and Supply Chain

## Proceedings of the IIoTBDSC 2020

Springer

*Editors*
Victor Chang
Artificial Intelligence and Information Systems Research Group, School of Computing, Engineering and Digital Technologies, Teesside University
Middlesbrough, UK

Víctor Méndez Muñoz
Faculty of Computer Science, Multimedia and Telecommunications
Universitat Oberta de Catalunya (UOC)
Barcelona, Spain

Muthu Ramachandran
School of Built Environment, Engineering, and Computing Software Engineering Technologies, and Emerging Practices (SETEP)
Leeds Beckett University
Leeds, UK

ISSN 2190-3018 ISSN 2190-3026 (electronic)
Smart Innovation, Systems and Technologies
ISBN 978-981-33-6140-9 ISBN 978-981-33-6141-6 (eBook)
https://doi.org/10.1007/978-981-33-6141-6



This Springer imprint is published by the registered company Springer Nature Singapore Pte Ltd.
The registered company address is: 152 Beach Road, #21-01/04 Gateway East, Singapore 189721, Singapore

# Preface

We are pleased to publish this proceedings of the First International Conference on Industrial IoT, Big Data and Supply Chain (IIoTBDSC). The novelty in the scope is attracting contributions to some crucial aspects of our interconnected world. People and devices are defining high-tech services with open problems to face in reliability, efficiency, speed and accuracy.

We have received 120 submissions. After a peer-review process and paper improvement, 46 final papers were selected.

The conference proceeding scope is to demonstrate that today's challenges can be faced with crossover knowledge fields. Industrial Internet of Things (IIoT) allows the network of IoT devices to communicate, analyze data and process information collaboratively. Big Data (BD) infrastructure problem domains require to deal with the 5V's: volume, velocity, variety and veracity. Supply chain requires modern advancement and support from IIoT and big data to make all services efficient, fast, accurate and reliable. These three can combine and work together to produce more significant impacts and contributions as follows.

The maturity of IoT technologies can grow and become part of our everyday lives. BD 5Vs is a must to provide affordable BD analytics using machine learning, AI, statistical and other advanced techniques, models and methods, which can create values for people and organizations adopting it. Suppliers can know the updates on their stocks and demands. Manufacturers and transport companies know the workloads, destination and resource distributions. Owners and customers are also stakeholders receiving benefits from these approaches. At the end of the day, the supply chain can reach a greater sustainable ecosystem with the help of IIoT and big data.

We are grateful for the opportunities to serve the community. We will organize IIoTBDSC 2021 again and will publicize it in due course.

We wish you a happy reading of our selected papers.

Yours sincerely,
Prof. Victor Chang
Dr. Muthu Ramachandran
Prof. Víctor Méndez Muñoz

Middlesbrough, UK
Leeds, UK
Barcelona, Spain
November 2020

# Contents

# About the Editors

**Prof. Victor Chang** is currently a Full Professor of Data Science and Information Systems at the School of Computing, Engineering and Digital Technologies, Teesside University, Middlesbrough, UK, since September 2019. He is a Research Group Leader for Artificial Intelligence and Information Systems Research Group at Teesside University. He was Senior Associate Professor, Director of Ph.D. (June 2016–May 2018) and Director of MRes (September 2017–February 2019) at International Business School Suzhou (IBSS), Xi'an Jiaotong-Liverpool University (XJTLU), Suzhou, China, between June 2016 and August 2019. He was also a very active and contributing key member at Research Institute of Big Data Analytics (RIBDA), XJTLU. He was Honorary Associate Professor at the University of Liverpool. Previously, he was Senior Lecturer at Leeds Beckett University, UK, between September 2012 and May 2016. Within 4 years, he completed Ph.D. (CS, Southampton) and PGCert (Higher Education, Fellow, Greenwich) while working for several projects at the same time. Before becoming an academic, he has achieved 97% on average in 27 IT certifications. He won a European Award on Cloud Migration in 2011, IEEE Outstanding Service Award in 2015, best papers in 2012, 2015 and 2018, the 2016 European special award and Outstanding Young Scientist 2017. He is a visiting scholar/Ph.D. examiner at several universities, Editor-in-Chief of IJOCI and OJBD journals, Former Editor of FGCS, Associate Editor of TII and Information Fusion, Founding Chair of two international workshops and Founding Conference Chair of IoTBDS and COMPLEXIS since the year 2016. He is Founding Conference Chair for FEMIB since the year 2019. He published 3 books as sole authors and Editor of 2 books on cloud computing and related technologies. He gave 22 keynotes at international conferences. He is widely regarded as one of the most active and influential young scientists and experts in IoT/Data Science/Cloud/Security/AI/IS, as he has experiences to develop 10 different services for multiple disciplines.

**Dr. Muthu Ramachandran** is Principle Lecturer (Associate Professor) at the School of Built Environment, Engineering and Computing at Leeds Beckett University, UK. Muthu has extensive research experience coupled with a teaching background and experiences on software and systems engineering methods and lifecycle, software development, agile software engineering, project management skills, process improvement skills, Internet technology, mobile, networks and distributed computing, realtime and embedded systems, cloud computing, service-oriented architecture and IT systems development for the past 25 years. He was a research scientist at India Space Research Labs where he worked on real-time systems development projects. Muthu is an author of two books: *Software Components: Guidelines and Applications* (Nova Publishers, NY, USA, 2008) and *Software Security Engineering: Design and Applications* (Nova Publishers, NY, USA, 2011). He is also an edited co-author of a book, *Handbook of Research in Software Engineering* (IGI, 2010), and has edited the book KE for SDLC (2011). He has widely authored published journal articles, book chapters and conferences materials on various advanced topics in software engineering and education. He received his master's degree from the Indian Institute of Technology, Madras, and from Madurai Kamaraj University, Madurai, India. He is a member of various professional organizations and computer societies: IEEE, ACM, Fellow of BCS and Fellow of HEA. He was a speaker to the 5th international symposium on SOA Cloud 2012, London, COMPLEXIS 2016 and FEMIB 2019.

**Prof. Víctor Méndez Muñoz** is Consultant Professor at UOC, Spain. He has a background in several fields of big data, cloud computing and IoT, both in academia and in the industry. Currently, he is contributing to the IoT ecosystem for microbiological control at IUL, which is an innovative company in the biotechnology. We are building new models and architectures taking full advantage of the state-of-the-art opportunities in IoT microcomputers, microservices and edge computing, to ensure genuine progress. Furthermore, he fosters bridging the gap between theory and practice, as Collaborating Editor in scientific journals and conferences, training postgraduates at UOC and partaking in innovation programs with Universities.

# Part I
# Area 1: Industrial Internet of Things (IIOT) Fundamentals and Applications

# Chapter 1
# A Modelling Framework for CPS-Based Industry 4.0: Application to Manufacturing Systems

**Zakaria Benzadri, Takieddine Bouheroum, Youcef Ouassim Cheloufi, Mohamed Nadir Hassani, and Faiza Belala**

**Abstract** Recently, digital transformation, known as the Fourth Industrial Revolution (Industry 4.0), has become a promising technological framework used to integrate and extend manufacturing processes at the intra and inter-organizational levels of smart factories. Among the most important technologies in Industry 4.0, cyber-physical systems (CPS) are recent complex systems, subject to distributed control, cooperation, influence, cascading effects, and emerging behaviours. Nowadays, few research attempts and industrial companies are interested in integrating CPSs to study, design, and implement more intelligent manufacturing systems. The main objective of this paper is to bring the migration to CPS-based Industry 4.0 and its new features while proposing a cross-layers and generic architecture (I4.0-CPS) validated through a realistic case study of the manufacturing process of a grader in an Algerian company (ENMTP-Somatel-Liebherr).

## 1.1 Introduction

The industrial revolution is a concept that has fundamentally changed our society and economy. It marks the passage from a system of artisanal production based on coal, textiles, and steam engines (First Industrial Revolution—mechanization), through a

---

Z. Benzadri (✉) · T. Bouheroum · Y. O. Cheloufi · F. Belala
LIRE Laboratory, University of Constantine, 2-Abdelhamid Mehri, Constantine, Algeria
e-mail: zakaria.benzadri@univ-constantine2.dz

T. Bouheroum
e-mail: takieddine.bouheroum@univ-constantine2.dz

Y. O. Cheloufi
e-mail: youcef.cheloufi@univ-constantine2.dz

F. Belala
e-mail: faiza.belala@univ-constantine2.dz

M. N. Hassani
ENMTP-SOMATEL, Constantine, Algeria

 
V. Chang et al. (eds.), *Modern Industrial IoT, Big Data and Supply Chain*,
Smart Innovation, Systems and Technologies 218,
https://doi.org/10.1007/978-981-33-6141-6_1

system that takes advantage of the fields of electricity, oil and chemistry (Second Industrial Revolution—Electrification), and to a system whose dynamics come from electronics, computers, audio–visual, and nuclear power (Third Industrial Revolution—automation). Recently, digital transformation, known as the Fourth Industrial Revolution, is a promising technological framework used to integrate and extend manufacturing processes at the intra- and inter-organizational levels of smart factories [1]. This emerging digital transformation claims to be a fusion of the virtual world, computer design, management (operations, finance, and marketing) and real-world products, and objects. It refers to the use of digital technology to make manufacturing operations more agile, flexible, and customer-focused.

Today, the evolution of information and communication technologies has led to the birth of the new industrial revolution. As a result, several strategic initiatives have been launched around the world: "smart manufacturing" in the USA, "Internet+" in China (see [2]), "industry of the future" in France, and "Industry 4.0" in Germany; the term Industry 4.0 seems to be gaining international acceptance. Companies in the world including Algerian ones are facing the Industry 4.0 revolution which imposes new challenges and new modes of production, such as exploitation and massive management of data, interconnection of machines, dematerialization of communication and distribution channels, the restructuring of the company for flexible and personalized production, etc. In this view, we identify the following key challenges, requiring each factory to act for the transformation to Industry 4.0: **(C1)** standardization of rules and processes that permits us to identify a set of findings to move towards Industry 4.0; **(C2)** connectivity and change tracking of hundreds of heterogeneous services, machines, or factories distributed over different sites; **(C3)** transition from re-engineering software systems and manufacturing processes to modular and maintainable solutions.

### *1.1.1 Context and Problematic*

The Industry 4.0 is defined as a collective term for technologies and concepts of company's value chain [1, 3–5]. Cyber-physical systems, Internet of things, big data, cloud computing, edge computing, and fog computing are among the core technologies for Industry 4.0. In this context, cyber-physical systems (CPSs) are recent complex systems, subject to distributed control, cooperation, influence, cascading effects, and emerging behaviours. They may be considered as sensors and actuators that monitor physical processes and create a virtual copy of the physical world in a plant [6]. In general, CPSs are a new class of engineered systems that offer close interaction between cyber and physical components; they are expected to play a major role in the design and development of future systems. However, nowadays, few research attempts and industrial companies are interested in integrating CPSs to study, design, and implement more intelligent manufacturing systems, in the context of Industry 4.0. CPSs can play a major role in smart manufacturing and production processes, thus offering significant advantages in terms of time, resources, and costs compared to conventional production systems. Indeed, current modelling approaches of this type of systems are based on classical notations and languages assuming that are monolithic, based on central control, global visibility, hierarchical structures, and coordinated activities.

Within the software engineering framework, model-based solutions may hold promise for the development cycle of CPS-based production systems. These solutions focus on a more abstract concern than conventional programming, taking into account the seamless integration of heterogeneous CPS components and their coordination. In this work, we identify, in an explicit manner, all the stages of designing a distributed and intelligent production system, depending on the above CPS and Industry 4.0 principles [7], from abstraction to architecture, and from model to realization (in response to the identified challenges: C1; C2; and C3).

**Related Work**
In the literature, few research works exist for modelling CPS; most are driven by a generic methodology without reference to an explicit design process, focusing instead on isolated stages such as simulation, development, or verification. On the other hand, the used model notations (BPMN, Rdps, UML, etc.) for this type of systems are often inappropriate, emphasizing only one aspect while neglecting others. Some architectures, deduced from published works, have been identified and summarized in the table below, and their authors, each according to its particular area of interest, take into consideration only significant layers of these architectures, which can possibly host the Industry 4.0 components, but not leading to a model abstracting the details of the corresponding production systems and facilitating their design.

Through Table 1.1, we notice that almost of the existing approaches represent good attempts to model their corresponding manufacturing systems. Their established architectures are structured on several hierarchical layers: the network, the plant, the system, the cell, and the station. We also note that only the RAMI4.0 reference architecture makes it possible to describe the production system according to several views. In this case, a complete three-dimensional description has been defined. Concerning the third column of Table 1.1, it indicates how to define the individual components of a production system; obviously the architecture 5C inherits the CPS principle to distinguish between the physical elements of a production system, its computation components, and its connection elements (network). So, more importance must therefore be given to their integration. The genericity criterion depicts that the last two works in Table 1.1 are intended for the description of particular systems developed in a very particular context. Examples of technologies and components are considered.

In this paper, we propose a multi-view, cross-layer, and generic architecture (called I4.0-CPS) to respond to our major concerns, namely the abstraction of static and behavioural aspects, the distribution and heterogeneity of its entities, and the intelligent handling of data. The proposed approach is illustrated through a realistic case study of the manufacturing process of a grader in an Algerian factory (ENMTP-Somatel-Liebherr).

**Table 1.1** Related work traits

| Architecture | Cross-layer | Multi-views | CPS-based | Generic | Underlying model |
|---|---|---|---|---|---|
| RAMI 4.0 [8] | + | + | – | + | – |
| Architecture 5C [9] | + | – | + | + | – |
| PERFoRM [10] | + | – | – | – | – |
| FC-CPMTS [11] | + | – | – | – | – |

### *1.1.2 Paper Organization*

The remainder of the paper is organized as follows. Section 1.2 is dedicated to the presentation of the proposed cross-layers CPS-based architecture for Industry 4.0 (I4.0-CPS). Section 1.3 describes the case study to be used throughout the paper to illustrate our approach. Finally, Sect. 1.4 concludes the paper by giving some perspectives.

## 1.2 A Cross-Layers Architecture for CPS-Based Industry 4.0

Knowing that the architectural reference model of Industry 4.0 (RAMI4.0) consists of a three-dimensional coordination system, it describes all the essential aspects of Industry 4.0, namely [8] a layer dimension, a life cycle dimension, and lastly a dimension specifying the hierarchy. In the same thought, we give in this section a multi-layered architecture based on CPS, which makes it possible to understand and model an intelligent factory while integrating new concepts related to the decentralization and intelligence of production processes in Industry 4.0. This lies in the way to make the computational components of Industry 4.0 collaborate with the physical world in a distributed way using data access and processing services available on the Internet. The proposed architecture, illustrated in Fig. 1.1, offers two different views to describe a production system in an intelligent factory: **(1)** ***the functional view*** consists of four levels (see Fig. 1.1) which permit to specify the physical (resources, products, machines, routers, sensors, etc.) or computational (servers, computers, web application, etc.) elements of a production system that are described according to their position in the different layers (or levels); **(2)** ***the network view*** consists of three levels: edge, fog, and cloud computing, associating and specifying the means intended for the corresponding functional level. In what follows, we describe each of the functional view levels by referring to the associated network view levels:

- **Manufacturing asset level (MA).** In this level, different robots, machines, and applications are available for the manufacturing process. The machines can detect products and retrieve the process information they need. As a result, some corresponding functionalities are identified (summarized on the right in Fig. 1.1).

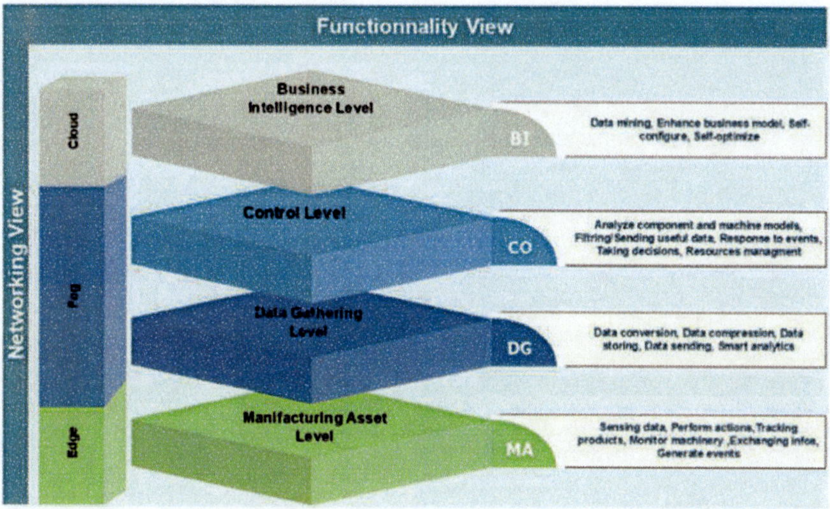


**Fig. 1.1** Multi-layered architecture I4.0-CPS

- **Gathering data level (GD).** Significant information must be inferred from the data collected. The IoT should offer promising transformational solutions for this, which can also be realized at the fog level.
- **Control level (CO).** This level plays the pivotal role of the proposed architecture. Information is transmitted to it from every machine, process, or system connected to form the plant network. After collecting massive amounts of information, specific analyses must be used to extract useful information to better understand the state of individual components.
- **Business intelligence level (BI).** This level acts as a supervisory control to make the machines self-configurable and self-adaptive.

## 1.3 Case Study: A Grader Manufacturing System

Algerian companies are facing a new industrial revolution that imposes new challenges, particularly with regard to new production and communication models. We contribute, through this work to bring a solution to a modelling problem posed by production systems in the context of Industry 4.0. ENMTP-Somatel-Liebherr, one of the Algerian national companies, is active in the production of hydraulic excavators, wheel loaders, self-propelled cranes, and graders. The proposed architecture dealing with the modelling of intelligent production systems is illustrated and projected on an example of the manufacture of the "SOMATEL grader 5410". This allows us to study in depth the manufacturing process of a grader in this company and to identify a set of findings to move this factory towards Industry 4.0, such as the appropriation

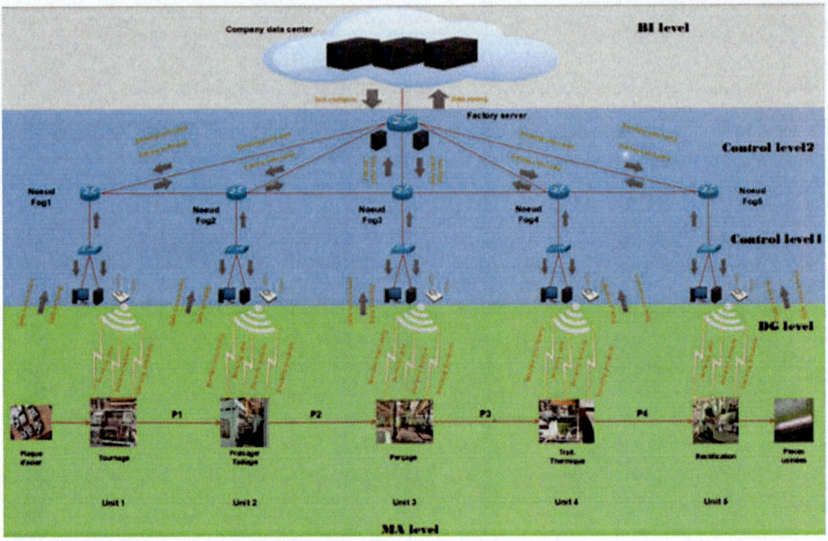


**Fig. 1.2** Architecture instance of I4.0-CPS for the machining process

of new technologies, the control and the sharing of data across the network, organizational restructuring of the company, and the development of a digital plan. In Somatel-Liebherr, several micro-processes constitute the producing grader process: Debitage, Mecano welding, machining, pre-assembly, and final assembly. In this work, we are interested in the machining process, which begins with the blank steel wafer and ends with factory parts, having the desired shapes and characteristics. According to our formalization approach, Fig. 1.2 shows the application of the I4.0-CPS architecture for our case study. We deduce the following generic and cross-layers architecture for describing the spatial organization for our manufacturing process. We focus on the collaboration that occurs during the machining of parts used in, product manufacturing, synchronization, deployment, and decision-making. This collaboration is governed by a well-defined organization of the different elements constituting this I4.0-CPS architecture. Obviously, the physical and software components of the corresponding CPS are deeply intertwined, each operating at different spatial scales, represented by the different levels of the I4.0-CPS architecture: MA, GD, CO, and BI, and interacting with each other through fog or cloud computing, to describe the functioning of the intelligent machining process in SOMATEL as shown in Fig. 1.2. Two aspects are considered in this model:

**Static aspect**. We specify the physical elements (machines, resources, and manufactured products) at the MA level as well as some sensors useful to collect certain information. We have chosen to host each operation (turning, milling, drilling, heat treatment, and grinding) of the machining process in a separate unit. The next level in the architecture (GD level) contains the data conversion tools for exploitation. Level CO (control), distributed on two levels ($CO_1$ and $CO_2$) in our case study, contains the supervisory computers or servers, which collect the information from the lower levels and provide useful data or actions for other layers. We note that at level $CO_1$, the functionality associated with each operation in the machining process is performed by a separate fog node. $CO_2$ level is provided by two different servers: one for hosting the used applications and the other is reserved for data storage. At the BI level, we combine cloud servers to study, analyse, and make decisions in order to improve the functioning of the production process in question. Eventually, an optimization or reconfiguration can be considered.

**Interaction aspect**. The essential contribution of our approach is that it allows seamless integration between heterogeneous systems consisting of computing devices and/or distributed sensors and actuators. Sensors and actuators provide an interface between the physical and cyber worlds. The proposed I4.0-CPS architecture takes into account this aspect and pays particular attention to the interactions that may exist between the different components of this machining process in a well-defined organizational context, and they are schematized by arrows in Fig. 1.2. We thus distinguish between the inter-layer interactions linking the elements constituting the CPS, belonging to different levels of the I4.0-CPS architecture, and the intra-layer interactions governing the functional links between the elements of the same layer.

## 1.4 Conclusion and Perspectives

In this paper, we tackled the challenges in managing the life cycle development of production systems in smart manufacturing. We relied on CPSs to give an efficient framework for modelling this type of complex systems. Especially, we have defined a multi-level architecture I4.0-CPS dealing with structure and behaviour aspects.

Future research will aim at (1) the elaboration of the corresponding service-oriented architecture in order to describe typically the possible interactions between the physical and logical entities of CPS-based systems; (2) a transcription to a component-based model to take advantage of its abstraction and reuse aspects during the production system development life cycle: design, implementation, and deployment.

## References


1. Qin, J., Liu, Y., Grosvenor, R.: A categorical framework of manufacturing for industry 4.0 and beyond. Procedia Cirp **52**, 173–178 (2016)
2. Chang, V., Xu, Y.K., Zhang, J., Xu, Q.: Research on intelligent manufacturing development approach for China's local valve industry. Smart Sustain. Built Environ. (2020)
3. Chang, V.I., Lin, W.: How Big Data transforms manufacturing industry: a review paper. Int. J. Strateg. Eng. (IJoSE) **2**(1), 39–51 (2019)
4. Benzadri, Z., Bouanaka, C., Belala, F.: Big-CAF: a bigraphical-generic cloud architecture framework. Int. J. Grid Util. Comput. **8**(3), 222–240 (2017)

5. Bouheroum, A., Benzadri, Z., Belala, F.: Towards a formal approach based on bigraphs for fog security: case of oil and gas refinery plant. In: 2019 7th International Conference on Future Internet of Things and Cloud (FiCloud), pp. 64–71. IEEE (2019).
6. Khaitan, S.K., McCalley, J.D.: Design techniques and applications of cyberphysical systems: a survey. IEEE Syst. J. **9**(2), 350–365 (2014)
7. Xu, L.D., Xu, E.L., Li, L.: Industry 4.0: state of the art and future trends. Int. J. Prod. Res. **56**(8), 2941–2962 (2018).
8. Hankel, M., Rexroth, B.: The reference architectural model industry 4.0 (RAMI 4.0). ZVEI, April, 410. IBM. (2020). IBM Rational Software Architect Designer (2015). https://www.ibm.com/developerworks/downloads/r/architect/index.html.
9. Lee, J., Bagheri, B., Kao, H.-A.: A cyber-physical systems architecture for industry 4.0-based manufacturing systems. Manuf. Lett. **3**, 18–23 (2015)
10. Leitão, P., Barbosa, J., Pereira, A., Barata, J., Colombo, A.W.: Specification of the perform architecture for the seamless production system reconfiguration. In: Iecon 2016–42nd annual conference of the IEEE industrial electronics society, pp. 5729–5734 (2016).
11. Zhou, Z., Hu, J., Liu, Q., Lou, P., Yan, J., Li, W.: Fog computing-based cyber-physical machine tool system. IEEE Access **6**, 44580–44590 (2018)

# Chapter 2
# A Deep Classifier for Crowdsourcing User Requests

**Feifei Niu, Chuanyi Li, and Bin Luo**

**Abstract** Software feature requests are proposed by users online to ask for new features of software covering diverse aspects of software properties, such as usability, security, and performance. They are valuable sources of software requirements. Early detection of request categories enables them to be stored structurally and incorporated into requirements specifications. But manually analyzing and labeling the category of user requests is a labor-intensive and time-consuming task. In this paper, we propose a deep learning-based approach to automatically classify user requests, where both statistical and semantical text features are adopted. The main contributions of this work are comparing the effectiveness of different deep learning algorithms in feature request classification and exploring the proper way to apply deep learning to solve the requests classification issue. Three research questions are answered by experiments to illustrate the contributions. The experimental results derived from the dataset collected from Sourceforge.net show that the deep classifier works properly on utilizing different types of features, classifying user requests, and combining with active learning.

## 2.1 Introduction

As an early phase of software development, requirement classification is crucial. It can help to distinguish between types of requirements and help to detect early aspects in software development. For example, security is a significant aspect that counts for much [3] and usually has a high priority amount all requirements [2].

User requests are proposed by the crowd on open forums to ask for new features and are valuable resources of requirements. However, the number of user requests is growing rapidly and covers all aspects, making it a challenge to manage and utilize them. Besides, user requests are usually unstructured and colloquial, which turns out to be an obstacle to manually analyze and label their category.

F. Niu · C. Li (✉) · B. Luo
Software Institute, Nanjing University, Nanjing 210093, China
e-mail: lcy@nju.edu.cn

 
V. Chang et al. (eds.), *Modern Industrial IoT, Big Data and Supply Chain*,
Smart Innovation, Systems and Technologies 218,
https://doi.org/10.1007/978-981-33-6141-6_2

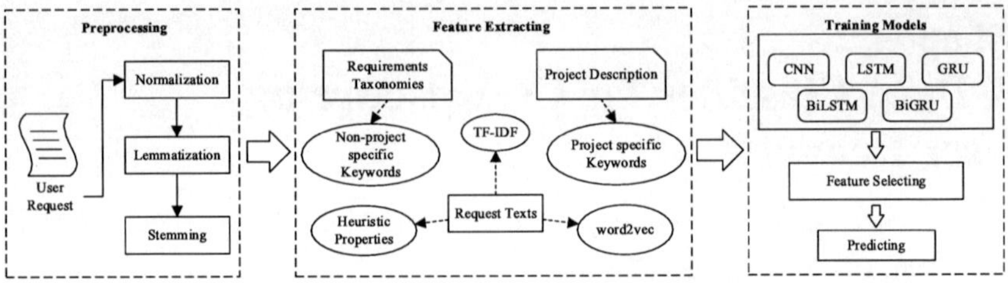


**Fig. 2.1** Framework of the proposed deep learning based user feature requests classification approach

Recently, the rise of artificial intelligence has provided better solutions to many problems [16]. In this paper, we propose deep learning (DL) classifier to classify user requests automatically. We employ both semantic and statistical text features for training the classifiers including Word2vec [15], TF-IDF [12], keyword frequency (KF) [13], and heuristic properties (HP) [13]. The DL classifiers are convolutional neural network (CNN) [11], long short-term memory network (LSTM) [9], BiLSTM [17], gated recurrent units (GRU) [4], and BiGRU [17]. We proposed more HPs and evaluate the effectiveness of different HPs to select an optimized set of HPs for different projects. We carried out experiments on 3000 user requests collected from three projects on Sourceforge.net. The results are evaluated on accuracy, precision, recall, and $F$-measure. Most importantly, we designed two strategies based on cross-prediction and active learning to employ the classifier into actual scenarios. The framework of our research can be found in Fig. 2.1.

The remainder of this paper is organized as follows. Section 2.2 surveys requirement classification and related techniques. Section 2.3 introduces the approach. Experimental processes and results are illustrated in Sect. 2.4. Threats to validity are discussed in Sect. 2.5. Section 2.6 concludes the paper with a discussion of future work.

## 2.2 Background

### *2.2.1 Software Requirement Classification*

Requirement classification has been studied for years. Cleland-Huang et al. [5] proposed an information retrieval-based approach to classify non-functional requirements (NFRs) with weighted indicator terms. Winkler and Vogelsang [19] have proposed an automatic classification of requirements employing convolutional neural networks. Kurtanovic and Maalej [10] automatically classify requirements into functional and non-functional. Baker et al. [1] leverage artificial neural networks and convolutional neural networks to classify non-functional requirements into five categories: maintainability, operability, performance, security, and usability. Mohamad et al. [14] identified security requirements from 3003 requirements with the random forest classifier. However, requirements are confidential assets for companies and are rarely available for domain experts [7]. The recent study for requirements engineering has transferred from requirements specifications to users' online reviews. Li et al. [13] defined taxonomy for user requests' classification and classified user requests into seven categories including security (SE), reliability (RE), performance (PE), life cycle (LI), usability (US), capability (CA), and system interface (SI). In this paper, we follow Li et al.'s [13] definitions of the seven categories.

### *2.2.2 Related Techniques*

Recently, the blossoming of DL models allows us to extract semantic and syntactic information of texts automatically and obtain better and more comprehensive features [15]. Convolutional neural networks (CNNs) can retain word order information and learn sentence patterns composed of word sequences that span multiple words [8]. Recurrent neural networks (RNNs) have been widely used in processing sequential data and have achieved pretty good results. LSTM [9] and GRU [4] are the most commonly used RNNs. Besides, Schuster and Paliwal [17] proposed BiRNN models, which connect two opposite direction layers to the same output and can increase the input information to the network. In this paper, we evaluate these DL models, i.e., CNN, LSTM, BiLSTM, GRU, and BiGRU to classify user requests.

First, we employ feature extraction techniques to convert natural language texts into machine understandable representations including:

**Term Frequency-Inverse Document Frequency (TF-IDF)**. TF-IDF is commonly used to calculate the importance of a word in a document [12]. It increases proportionally to the number of times a word appears in a user request but is offset by the number of documents in the corpus that contains the word.

**Keyword Frequency (KF)**. KF calculates the frequency of each type of keywords that appeared in each user request, where the keywords are predefined and categorized into the target seven classes, which are defined by Li et al. [13].

**Heuristic Properties (HPs)**. Li et al. [13] have defined HPs as certain parts that strongly imply the possible category of the request.

**Word2vec** [15]. Word embedding maps words or phrases into vectors of numbers and is used to extract semantic information from texts.

## 2.3 Approach

The proposed approach, as depicted in Fig. 2.1, consists of three phases: preprocessing, extracting features, and training classifiers. Firstly, the user requests are preprocessed using common NLP techniques: normalization, lemmatization, and stemming. Then, both semantic and statistical features are extracted from feature

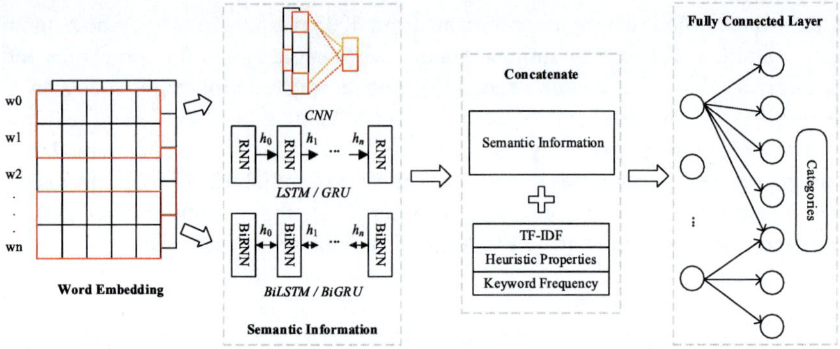


**Fig. 2.2** Structure of CNN, LSTM, BiLSTM, GRU and BiGRU models

**Table 2.1** Proportion of each type of user request in each project (1000 for each project)

| Project | SE (%) | RE (%) | PE (%) | LI (%) | US (%) | CA (%) | SI (%) |
|---|---|---|---|---|---|---|---|
| KeePass | 10.1 | 6.8 | 1.9 | 3.9 | 34.3 | 38.1 | 4.9 |
| Mumble | 2.4 | 6.3 | 3.5 | 12.3 | 33.2 | 33.7 | 8.6 |
| WinMerge | 0.9 | 7.4 | 2 | 5.7 | 41 | 40 | 3 |

texts. Semantic features are extracted via DL models. All features are concatenated as the input of the training model. After several fully connected layers, a softmax layer outputs a seven-dimensional vector that stands for the possibility of classifying to each category. The highest possibility is thought to be the final classification result. The model is shown in Fig. 2.2.

Our dataset consists of 3000 user requests of three projects on Sourceforge.net. The three projects are KeePass, Mumble, and WinMerge. The percentage of each category in each project is shown in Table 2.1.

To evaluate fairly, we use fivefold cross-validation experiments on each project. Each time, we use one fold for testing, one for validation, and one for training.

We use average accuracy, precision, recall, and $F1$ of fivefolds to evaluate the performance of the classifiers, where accuracy is the percentage of records classified correctly in the test set.

## 2.4 Experiment and Results

### *2.4.1 Experiment*

Our experiments revolve around solving three research questions. The questions, research methods, and corresponding results are as follows:

**RQ1** Are all the HPs helpful to identify the category of user requests? Is there an optimal set of HPs that is most useful for classification?

*Method*: To examine the effectiveness, we test on each HP to get accuracy variation with the number of added HPs changes. Firstly, we define an empty set $S_{\mathrm{HP}}$ and a set $S$ that contains all eighteen HPs. $S_{\mathrm{HP}}$ is the set of HPs used as the input of models. For each experiment, choose one HP from $S$ and add to $S_{\mathrm{HP}}$, train the model with features extracted from $S_{\mathrm{HP}}$, which is concatenated with Word2vec, TF-IDF, and KF. In this step, we test on all the eighteen HPs one by one and pick out the one that gains the highest result. The high accuracy indicates that this HP is more indicative. Move this HP from $S$ to $S_{\mathrm{HP}}$. Then, repeat this step, until $S$ is empty. We can obtain the accuracy curve that depicts the accuracy varies with the added number of HPs increases. The algorithm that we employ is LSTM, which can learn long-term dependency information.

*Results*: We test on three projects, and the accuracy curves are shown in Fig. 2.3. The $X$-axis stands for the number of added HPs, and the $Y$-axis is the accuracy result. From the result, we can see that the accuracy curve increases as HPs added at first, but it reaches a peak and then begins to decline. The optimal HP set is obtained when the curve reaches pear. The accuracy improvement can be one percent with the optimal HP set compared with all HPs in three projects. The accuracy improvement of the optimal HP set is nearly 2 percent higher than models without HPs. However, the optimal HP set is different in three projects. In KeePass, the optimal HP set is 9, 13, 1, 5, 14, 4, 3, 7. In Mumble, the optimal HP set is 9, 13, 1, 12, 5, 7, 15, 8, 17, 3, and in Winmerge, the optimal HP set is 9, 1, 4, 10, 16, 8, 5, 17, 6. The optimal set is different between different projects. The reason may be that the semantic is different in different projects, so does the HPs' effectiveness. During the experiment process, we find that the ninth HP is most effective in both three projects.

**RQ2** How well can we automatically classify user requests with DL classifiers? Do DL classifiers perform better than traditional machine learning classifiers?

*Methods*: Firstly, the word vector model is trained with the training set. Then, we build DL classifiers with word vectors as inputs. To inspect the accuracy improvement when combined with TF-IDF, KF, and HPs, we concatenate TF-IDF, KF, and HPs to the word vectors. We experiment on different DL classifiers to explore the distinction. To

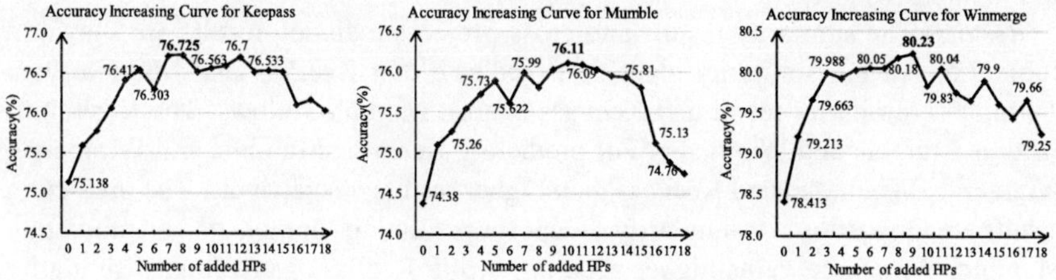


**Fig. 2.3** Accuracy curve of projects with the number of input HPs increases

compare the performance of every algorithm, we apply one-way analysis of variance (one-way ANOVA). The hypothesis is that there is no significant difference between different DL classifiers.

*Results*: Table 2.2 presents the experimental results of different combinations of text features and models. The highest average accuracy of Word2vec is achieved by LSTM, which is 47.83, 39.93, and 52.20, respectively. The accuracy improvement of combining TF-IDF with Word2vec is 16–25%. When adding KF, the improvement is 8–11%. When the optimal set of HPs is added in, the improvement is 0.9–2.3%. The results show that TF-IDF, KF, HPs are all effective in improving accuracy results. When using all the statistical and semantic text features, the classifiers can obtain the best results. Our average results are higher than the previous results achieved by traditional machine learning algorithms by Li et al. [13] apart from the CNN classifier. One-way ANOVA test result is shown in Table 2.3. The results indicate that there is no significance between the five algorithms for KeePass. LSTM and BiLSTM gain the best result in Mumble, with CNN gaining the worst result. In Winmerge, there is no significance among LSTM, BiLSTM, and GRU, which are both better than BiGRU and CNN. In general, LSTM and BiLSTM perform better than GRU and BiGRU, and they are all better than CNN in all the three projects. The precision, recall, and $F1$-measure results of the LSTM classifier are shown in Table 2.4. $F1$-measure of all categories has been improved than that in Li et al. [13].

**RQ3** In the actual scenarios, when users start a new project, all the user requests are unlabeled at first, and there is no available training set. How can we predict these user requests with DL algorithms?

*Method*: To label new user requests, there are two strategies. The first solution is to predict these requests with classifiers trained by other similar projects, which is cross-project prediction. We study the relationship between project similarity and cross-prediction accuracy. Firstly, we calculate project similarity through user requests with different algorithms like cosine similarity, Euclidean distance, Manhattan distance, and Jaccard index. Then, we train the classifier on one project to predict another project. The second solution is by means of active learning strategy. Active learning is a semi-supervised learning algorithm in the context of classification to reduce the efforts involved in labeling [6]. The chosen algorithm is LSTM and all the extracted features. The original dataset is split into training set and testing set with a rate of 0.2. The chosen sampling strategies are least confident (LC), small margin (SM), and random (R) [18].

*Results:* The similarity results and cross-project prediction results are shown in Table 2.5. All the similarity algorithms indicate that KeePass and Mumble gain the highest similarity result and KeePass against Mumble follows. The least similarity is Mumble and Winmerge. For predicting KeePass, Mumble, and Winmerge, Winmerge, Winmerge and KeePass gain higher results, respectively. The predicting results are proportional to similarity except for predicting Mumble. When predicting Mumble, Winmerge gains higher accuracy while KeePass gains higher similarity

**Table 2.2** Accuracy of deep classifiers with different text features

| Accuracy (%) | KeePass | | | | Mumble | | | | Winmerge | | | |
|---|---|---|---|---|---|---|---|---|---|---|---|---|
| | W | WT | WTK | WTKH | W | WT | WTK | WTKH | W | WT | WTK | WTKH |
| CNN | 44.70 | 65.97 | 75.20 | 76.83 | 35.60 | 61.63 | 74.33 | 75.23 | 45.67 | 67.53 | 77.63 | 79.00 |
| LSTM | **47.83** | 65.43 | 75.18 | 77.00 | 39.93 | 63.20 | **75.43** | 76.53 | **52.20** | 68.53 | **78.55** | **80.45** |
| BiLSTM | 46.50 | **66.95** | **75.55** | 77.05 | 37.95 | **63.30** | 74.95 | **76.58** | 46.90 | 68.48 | 78.50 | 80.25 |
| GRU | 42.13 | 66.28 | 75.20 | **77.13** | 39.48 | 63.08 | 74.93 | 76.20 | 47.15 | 68.73 | 78.50 | 80.20 |
| BiGRU | 44.30 | 65.78 | 74.78 | 77.05 | 38.53 | 63.05 | 74.48 | 76.03 | 47.80 | **68.98** | 77.83 | 79.78 |

K—for keyword frequency, H—for heuristic property, T—for TF-IDF, and W—for Word2vec

**Table 2.3** One-way ANOVA Test of CNN, LSTM, BiLSTM, GRU, and BiGRU

| DL Algorithm | KeePass | | | | Mumble | | | | Winmerge | | | |
|---|---|---|---|---|---|---|---|---|---|---|---|---|
| | LSTM | BiLSTM | GRU | BiGRU | LSTM | BiLSTM | GRU | BiGRU | LSTM | BiLSTM | GRU | BiGRU |
| CNN | 0.996 | 0.983 | 0.931 | 0.836 | 0.000 | 0.000 | 0.004 | 0.022 | 0.000 | 0.000 | 0.000 | 0.018 |
| LSTM | – | 1.000 | 0.996 | 0.928 | – | 0.856 | 0.043 | 0.008 | – | 0.965 | 0.481 | 0.004 |
| BiLSTM | – | – | 1.000 | 0.995 | – | – | 0.064 | 0.013 | – | – | 0.454 | 0.003 |
| GRU | – | – | – | 1.000 | – | – | – | 0.492 | – | – | – | 0.023 |

**Table 2.4** Precision (*P*), recall (*R*), and *F*-measure (*F*) of each category when applying all features on LSTM

| Projects | SE (%) | | RE (%) | | PE (%) | | LC (%) | | US (%) | | CA (%) | | SI (%) | |
|---|---|---|---|---|---|---|---|---|---|---|---|---|---|---|
| | *P* | *R* | *P* | *R* | *P* | *R* | *P* | *R* | *P* | *R* | *P* | *R* | *P* | *R* |
| KeePass | 81.1 | 72.3 | 69.8 | 64.7 | 81.8 | 47.4 | 100 | 87.2 | 77.6 | 78.7 | 76.5 | 81.9 | 78.3 | 73.5 |
| Mumble | 87 | 83.3 | 80.4 | 71.4 | 79.4 | 77.1 | 78 | 74.8 | 77.1 | 79.2 | 75 | 77.5 | 81.3 | 75.6 |
| WinMerge | 85.7 | 66.7 | 66.7 | 59.5 | 93.8 | 75 | 86.7 | 68.4 | 84.2 | 83.4 | 79.5 | 86 | 88.9 | 80 |
| Average | 84.6 | 74.1 | 72.3 | 65.2 | 85 | 66.5 | 88.2 | 76.8 | 79.6 | 80.4 | 77.0 | 81.8 | 82.8 | 76.4 |
| *F*1 | 79.0↑ | 68.57↑ | 74.62↑ | 82.11↑ | 80↑ | 79.3↑ | 79.47↑ | | | | | | | |

**Table 2.5** Similarity results between projects and cross-project prediction results

| | Cosine | Euclidean | Manhattan | Jaccard | Prediction between Projects (%) | |
|---|---|---|---|---|---|---|
| $(K, M)$ | 0.987 | 0.861 | 0.444 | 0.363 | $K \rightarrow M$: 63.55 | $M \rightarrow K$: 58.49 |
| $(K, W)$ | 0.992 | 0.868 | 0.512 | 0.41 | $K \rightarrow W$: 69.76 | $W \rightarrow K$: 66.05 |
| $(M, W)$ | 0.978 | 0.77 | 0.365 | 0.346 | $M \rightarrow W$: 69.48 | $W \rightarrow M$: 65.63 |

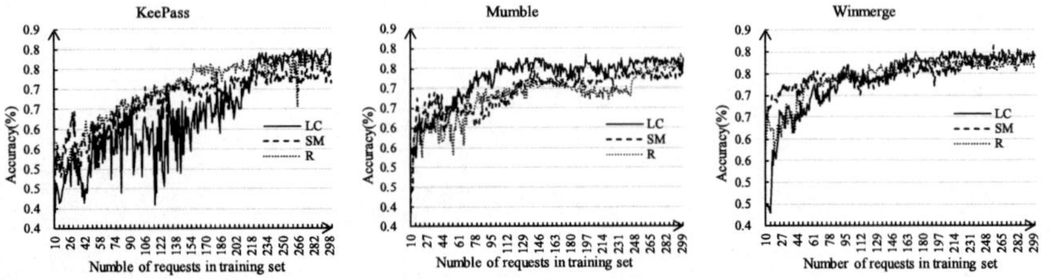


**Fig. 2.4** Accuracy curve of active learning employing different strategies: **a** KeePass, **b** Mumble, and **c** Winmerge

results. The result shows that a higher semantic similarity means higher classification accuracy to a certain extent. When predicting a new project, a more similar project is more operational.

For solution 2, as the manual labeled requests expanding gradually, we can gain the accuracy curve of the classifiers on the test set. The curves of the three strategies on different projects are shown in Fig. 2.4. The accuracy curves converge when the number of requests in the training set is about 250, which means that labeling 250 user requests manually is enough to obtain quite high accuracy. In all three projects, the LC strategy gains higher results when converges. The active learning results are much higher than the cross-project prediction.

## 2.4.2 *Discussion*

In this paper, we propose a DL-based method to automatically classify user requests and study the effectiveness of text features. Results show that text features—KF, HPs, TF-IDF, and Word2vec—can promote the classification accuracy to a large extent. When adding TF-IDF to Word2vec as inputs, the accuracy can improve by 16–25%. With both project-specific keyword frequency and non-project-specific keyword frequency, the improvement can be 8–11%. The improvement of adding HPs is 0.9–2.3%. All the text features show a positive effect on classification accuracy, and the combination of all features gains the best result.

DL classifiers show excellent performance compared to traditional machine learning classifiers. In general, LSTM and BiLSTM gain better results than other algorithms. Compared with the results shown by Li et al. [13], there is an improvement in precision and recall of small samples and all the seven $F1$-measure. The conclusion is that deep classifiers can extract more text features even though the rare appearance and can help to distinguish small samples.

We propose several HPs to help extract features relevant to categories. They show positive effects on the classification result. We pick out the optimal HPs set for each project, which can achieve the highest accuracy result on average. The most effective is the ninth HP: "The core functions of the system that the owner of the request wants to be implemented," which is powerful in extracting capability features.

In actual scenarios, active learning works even with small annotated data.

## 2.5 Threats to Validity

The quality of user requests is not as high as requirements specifications and will greatly affect accuracy. To mitigate this threat, we employ common effective natural language process techniques to preprocess our dataset. The scale of the dataset can influence the final accuracy result. Besides, the dataset is unbalanced. To reduce this threat, we employ fivefold cross-validation technique and evaluation metric based on retrieval to evaluate the results. The distribution of data in the training set and test set is the same, and the order is shuffled.

## 2.6 Conclusion and Future Work

In this paper, we proposed an automatic user request classification model based on DL. Compared with traditional machine learning algorithms, DL classifiers improve by 2%. Among all the classifiers, LSTM and BiLSMT gain the best performance. We propose several HPs and select the optimal set of HPs for each project experimentally. The promotion was 1–4%. In actual scenarios, active learning can achieve better performance than cross-prediction with quite little annotated data.

In this study, we use KF, HPs, TF-IDF, and word embedding as text features and evaluate their effectiveness. However, we believe that there must exist other features that can improve the result, for example, *Role of the Author*. Besides, more sentence patterns may indicate the category. In future, we will try to mine more helpful features that not only can be used in classification, but also other fields. In our work, keywords lists are extracted manually. In future, we will come up with solutions to obtain text features automatically.

## References


1. Baker, C, Deng, L, Chakraborty, S, Dehlinger, J.: Automatic multi-class non-functional software requirements classification using neural networks. In: 2019 IEEE 43rd Annual Computer Software and Applications Conference (COMPSAC), pp. 610–615 (2019)
2. Chang, V., Ramachandran, M.: Towards achieving data security with the cloud computing adoption framework. IEEE Trans. Serv. Comput. **9**(1), 138–151 (2015)
3. Chauhan, R., Kaur, H., Chang, V.: An optimized integrated framework of Big Data analytics managing security and privacy in healthcare data. Wirel. Personal Commun. 1–22 (2020)
4. Cho, K., Van Merriënboer, B., Gulcehre, C., Bahdanau, D., Bougares, F., Schwenk, H., Bengio, Y.: Learning phrase representations using RNN encoder-decoder for statistical machine translation. arXiv preprint arXiv:1406.1078 (2014)
5. Cleland-Huang, J., Settimi, R., Zou, X., Solc, P.: Automated classification of non-functional requirements. Requir. Eng. **12**(2), 103–120 (2007)
6. Dhinakaran, V.T., Pulle, R., Ajmeri, N., Murukannaiah, P.K.: App review analysis via active learning. In: International Requirements Engineering Conference, 2018
7. Ferrari, A.: Natural language requirements processing: from research to practice. In: Proceedings of the 40th International Conference on Software Engineering: Companion Proceedings, pp. 536–537 (2018)
8. Graves, A., Mohamed, A.R., Hinton, G.: Speech recognition with deep recurrent neural networks. In: 2013 IEEE International Conference on Acoustics, Speech and Signal Processing, pp. 6645–6649 (2013)
9. Hochreiter, S., Schmidhuber, J.: Long short-term memory. Neural Comput. **9**(8), 1735–1780 (1997)
10. Kurtanović, Z., Maalej, W.: Automatically classifying functional and non-functional requirements using supervised machine learning. In: 2017 IEEE 25th International Requirements Engineering Conference (RE), pp. 490–495 (2017)
11. LeCun, Y., Bottou, L., Bengio, Y., Haffner, P., others: Gradient-based learning applied to document recognition. Proc. IEEE **86**(11): 2278–2324 (1998)
12. Leskovec, J., Rajaraman, A., Ullman, J.D.: Mining of Massive Datasets, pp. 7–9. Cambridge University Press (2014)
13. Li, C., Huang, L., Ge, J., Luo, B., Ng, V.: Automatically classifying user requests in crowdsourcing requirements engineering. J. Syst. Softw. **138**, 108–123 (2018)
14. Mohamad M, Steghöfer JP, Scandariato, R.: Cross-project classification of security-related requirements. arXiv preprint arXiv:2003.14134 (2020)
15. Mikolov, T., Chen, K., Corrado, G., Dean, J.: Efficient estimation of word representations in vector space. arXiv preprint arXiv:1301.3781 (2013)
16. Roy, P.P., Kumar, P., Chang, V.: A hybrid classifier combination for home automation using EEG signals. Neural Comput. Appl. 1–13 (2020)
17. Schuster, M., Paliwal, K.K.: Bidirectional recurrent neural networks. IEEE Trans. Signal Process. **45**(11), 2673–2681 (1997)
18. Settles, B., Craven, M.: An analysis of active learning strategies for sequence labeling tasks. In: Proceedings of the Conference on Empirical Methods in Natural Language Processing, pp. 1070–1079 (2008)
19. Winkler, J., Vogelsang, A.: Automatic classification of requirements based on convolutional neural networks. In: 2016 IEEE 24th International Requirements Engineering Conference Workshops (REW), pp. 39–45 (2016)

# Chapter 3
# PS-AENN-Based Prediction Model of Billet Temperature in Industrial Steelmaking Process

**Jiangnan Gong, Hong Yu, and Cheng Huang**

**Abstract** In the process of industrial steelmaking, the existing predictive models and control models are not good enough for the recognition and extraction of important processes. How to use these data to extract important processes in the process industry to achieve high-quality automated production and more accurate and rapid prediction effect is also a problem worth challenging at present. Therefore, this paper proposes a parameter selection auto-encoder neural network predictor (PS-AENN) based on the interpretability of the fully connected neural network. Combining the multi-dimensional characteristics of process industry data, the neural network interpretability scheme is first introduced to identify important process parameters, then the feature extraction is carried out by the auto-encoder, and finally to make prediction. This method of deep learning and process selection provides new solutions for important process extraction and prediction results.

## 3.1 Introduction

Tapping temperature is the temperature when the billet leaves the heating furnace after a series of heating procedures. The optimum value of the temperature of the billet in the hot rolling process of the steel industry is 1180–1250 °C. Controlling the temperature in this range is beneficial to improve the quality of steel products, reduce energy consumption, and improve the productivity of plants. However, due to the high-temperature environment of the heating furnace and the thickness of the

---

J. Gong · H. Yu (✉)
Chongqing Key Laboratory of Computational Intelligence, Chongqing University of Posts and Telecommunications, Chongqing 400065, China
e-mail: yuhong@cqupt.edu.cn

J. Gong
e-mail: gongjiangnan1108@qq.com

C. Huang
Shenzhen Tuoxingyuan Automation Technology Tech. Co., LTD., Shenzhen, China
e-mail: hc116@sina.com

 
V. Chang et al. (eds.), *Modern Industrial IoT, Big Data and Supply Chain*,
Smart Innovation, Systems and Technologies 218,
https://doi.org/10.1007/978-981-33-6141-6_3

billet, the surface temperature may be inconsistent with the temperature of the core. Therefore, real-time online measurement of the tapping temperature of the billet is a difficult problem to solve.

The process industry is an extremely important category of industry, and it has obvious process steps. Generally, it is through *n* orderly process steps to get products, so the prediction and control of the process industry are also important works. Gerardo et al. [2] find out some physical parameters that characterize the quality changes during the deposition process, select these parameters through online tracking, and propose a method to monitor and predict the quality of the electrolyte in the copper electroplating process. Wang et al. [6] propose an economic benefit index to evaluate the performance of CDQ system, and adopt deep neural network based on SAE carrying out CDQ operation prediction. Zhang et al. [9] also propose an industrial-state classification and equilibrium method, and a heuristic method of state transition algorithm to optimize unknown model parameters, and use the data-based ANNs-ARIMA model to compensate for the error and establish a comprehensive prediction model of heavy metal iron concentration.

The combination of traditional statistical models and neural networks dramatically improves the prediction effect and accuracy. Undey et al. [5] developed an integrated online multivariate statistical process monitoring (MSPM) method for mass production processes in order to achieve online process performance monitoring and real-time product quality prediction. Ge et al. [1] divided the data into multiple categories according to different data distributions and combined with principal component analysis(PCA) and Gaussian process regression(GPR) to propose CLGPR for the quality prediction of polypropylene production process. Yang et al. [7] proposed a new soft-sensor method based on Gaussian process regression (GPR) model. This method simultaneously strengthens integrated learning and real-time learning to ensure the accuracy and robustness of the measurement. Hu et al. [3] first performs clustering to classify the data, and then uses LS-SVM support vector machines with different kernel functions for different operating conditions, and finally uses the PLSR method as an integration strategy to form an MMEP model to predict the carbon efficiency of iron ore sintering.

Another group of researchers focused on the multi-dimensional characteristics of the process industry data, developing feature extraction and parameter selection methods. Lu and gao [4] proposed a quality prediction scheme for PLS in the batch process stage to extract the stages that have a key impact on the relevant quality, and make predictions based on those selected stages. You et al. [8] propose to use neural network to reduce dimensionality and then use Gaussian process regression to predict crop yield in space and time. However, this kind of dimensionality reduction method is to extract the features of some selected process parameters, which cannot be used for automatic identification and extraction of important parameters in the process industry.

## 3.2 Preliminary Work

### 3.2.1 Problem Description

During the production process of the steel industry, a lot of process parameters will be generated by sensors. The temperature of the billet can reflect the production status of the furnace and whether the furnace is in the best production environment. It is necessary to sample and analyze the upper, middle, and lower parts of the billet during the production process to measure the tapping temperature. However, the cost of this kind of temperature measurement is too high and the billet temperature cannot be obtained in time. In addition, in the process industry under the big data environment, there are many process parameters in the steel industry; if we take all these parameters into account, it will cause huge time and calculation cost and is hard to achieve rapid prediction. Considering the problems of quality prediction and extraction of important process parameters in process industry, it is very important to put forward a method of extraction of important process parameters and accurate prediction of steel temperature in the process of industrial steelmaking.

### 3.2.2 Parameter Selector

In this paper, the main principle of the neural network parameter selector is to find a path with the most important weight through the weight information in the condition of accurate prediction. This path plays the largest contribution to the prediction of the final result. By finding this path, the weight sequence of input process parameters is determined, and then the effect of important parameter identification is achieved.

As shown in Fig. 3.1, $a_p^l$ represents the $p$-th node in the $l$-th layer of the neural network, and $w_{p,q}^l$ represents the weight between the $p$-th node in the $l$-th layer and $q$-th node in the next layer. The input and output data in first layer and final layer

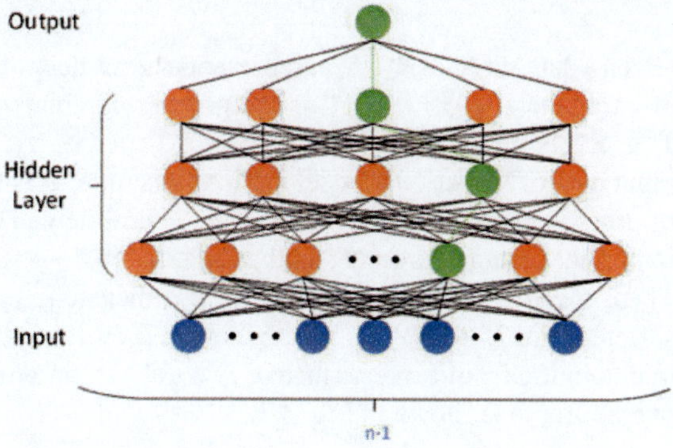


**Fig. 3.1** Schematic diagram of parameter selection process

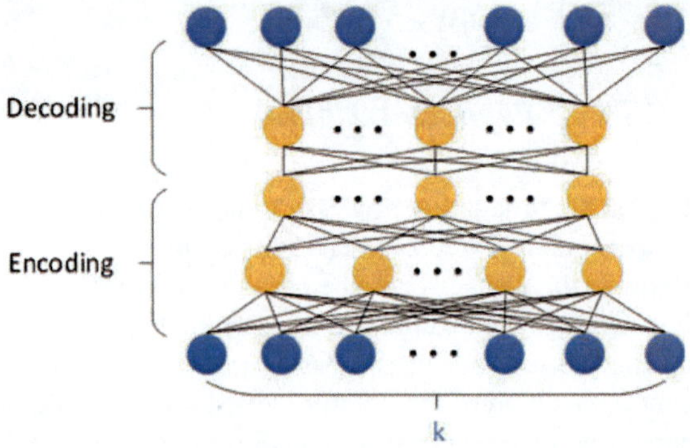


**Fig. 3.2** Auto-encoder structure diagram used in this paper

are process parameters and quality parameter, respectively. As shown in Fig. 3.2, the connection between the green nodes represents the path with the largest weight. A node in the hidden layer closest to the input layer is determined through the weight path. Then, we only need to compare the weight of the input parameters, which connected to this node in order to achieve the purpose of sorting and identifying important parameters.

### 3.2.3 *Auto-Encoder*

Auto-encoder is composed of encoder and decoder. It is a branch of neural network model. It can learn the hidden features of input data and do the feature extraction of data, which is called encoding. At the same time, it uses the learned features to reconstruct the original input data, which is called decoding. The structure of auto-encoder used in this paper is shown in Fig. 3.2.

### 3.2.4 *Model Building*

Given a data set $X \in R^{m \times n}$, $m$ represents the number of data, $n$ represents the number of parameters, $T \in R^{m \times n-1}$ is the process parameter, where $T = \{t_1, t_2, \ldots, t_m\}$. $Y \in R^m$ is the result parameter, where $Y = \{y_1, y_2, \ldots, y_m\}$, and $X = TY$. The input data of PS-AENN model is $T$. At this time, $T$ is not selected for the extraction of process parameters. First, using the neural network parameter selector to select important parameters for $T$, so that $T \in R^{m \times n-1} \to T' \in R^{m \times k}$, where $k \leq n - 1$. Then, the $n - 1$ dimensions of the $k$ dimensions process parameters have been selected, and $T' \in R^{m \times k}$ is sent to the five layers auto-encoder for process extraction to obtain the feature vector matrix $F \in R^{m \times \frac{1}{3}k}$, and finally to make prediction. The overall frame is shown in Fig. 3.3.

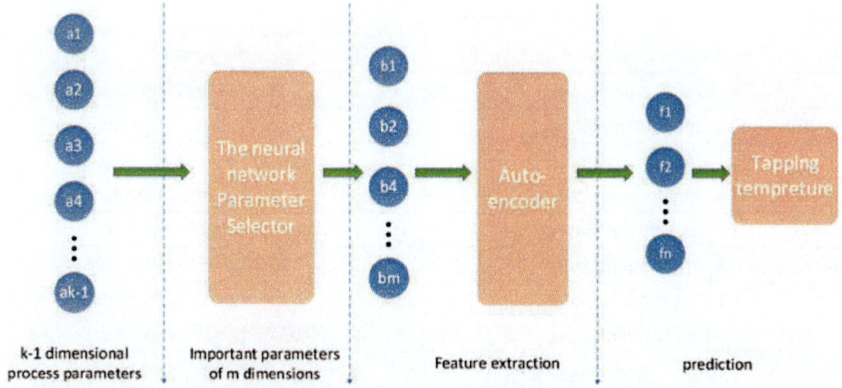


**Fig. 3.3** Parameter selection auto-encoder neural network model

## 3.3 Industry Case Study

### *3.3.1 Industry Case and Data Description*

In this section, we applied parameter selection auto-encoder neural network (PS-AENN) to a specific steel industry heating furnace hot rolling process. The experimental data is real-time captured by a high-temperature sensor during daily production by a steel plant data. The process of industrial steelmaking and hot rolling belongs to the traditional process industry. The billet is sent from the entrance of the heating furnace and is transmitted through the conveyor belt. The upper and lower surfaces of the billet are heated, respectively, in four heating sections, and then the target temperature is reached after the rolling process is carried out.

The entire heating furnace process has more than 30 process parameters. The data source is composed of high-temperature sensors at different positions in the heating furnace and the residence time of the billet in each heating section. The process parameters are shown in Table 3.1.

### *3.3.2 Experiment*

This paper proposes a prediction method for the temperature of the steel billet in the hot rolling process of industrial steelmaking. Since the temperature is a continuous attribute, the evaluation indicators predicted by the model use the mean square error, the root mean square error, the absolute value error, and the average absolute percentage error.

In Sect. 3.3.1, it can be seen that the range of tapping temperature is 1150–1250 °C so the small fluctuation of predicted value has little influence on the final billet quality

**Table 3.1** Heating furnace process parameters

| No | Parameter names | No | Parameter names |
|---|---|---|---|
| 1 | 1-Upper temperature | 16 | 4-Max Lower temperature |
| 2 | 2-Upper temperature | 17 | 1-Min Upper temperature |
| 3 | 3-Upper temperature | 18 | 2-Min Upper temperature |
| 4 | 4-Upper temperature | 19 | 3-Min Upper temperature |
| 5 | 1-Lower temperature | 20 | 4-Min Upper temperature |
| 6 | 3-Lower temperature | 21 | 1-Min Lower temperature |
| 7 | 4-Lower temperature | 22 | 2-Min Lower temperature |
| 8 | 4-Lower temperature | 23 | 3-Min Lower temperature |
| 9 | 1-Max Upper temperature | 24 | 4-Min Lower temperature |
| 10 | 2-Max Upper temperature | 25 | 1-Heating time |
| 11 | 3-Max Upper temperature | 26 | 2-Heating time |
| 12 | 4-Max Upper temperature | 27 | 3-Heating time |
| 13 | 1-Max Lower temperature | 28 | 4-Heating time |
| 14 | 2-Max Lower temperature | 29 | Total heating time |
| 15 | 3-Max Lower temperature | 30 | Tapping temperature |

and can be accepted. During model training, process parameters 1–29 in Table 3.1 are taken as input, and the output steel temperature is taken as the prediction target. Before the training, the data generated by more than 20,000 heating furnace sensors in actual production were screened and more than 8000 pieces of data were selected for training. 80% of the data is training data, and the remaining 20% is test data.

First of all, we make a comparative experiment on the number of important process parameters. Each column in the first row represents the proportion of important parameters selected by the model, and the results are predicted by the model for the test data.

As shown in Table 3.2, when we select 1/2 important parameters, the prediction effect of the model is the best, and the average error is less than 10 °C.

However, the number of training epochs is also a key factor in deep learning. Too few training epochs will make the model unable to learn the potential characteristics of the data, and too many epochs will make the model fall into the overfitting situation. Both of them will reduce the prediction effect. Therefore, in order to make the model

**Table 3.2** Parameter selection ratio comparison

| | 1/4 | 1/3 | 1/2 | 2/3 |
|---|---|---|---|---|
| MSE | 763.24 | 201.51 | **95.84** | 135.71 |
| RMSE | 27.63 | 14.19 | **9.79** | 11.65 |
| MAE | 18.87 | 10.42 | **6.97** | 7.84 |
| MAPE | 1.5% | 0.8% | **0.53%** | 0.65% |

**Table 3.3** Prediction result of different epochs

| | | 50 epochs | 100 epochs | 200 epochs | 500 epochs |
|---|---|---|---|---|---|
| PS-AENN Training Data | MSE | 172.17 | **47.54** | 123.89 | 235.08 |
| | RMSE | 13.12 | **6.89** | 11.13 | 15.33 |
| | MAE | 10.18 | **5.37** | 7.85 | 11.89 |
| | MAPE | 0.6% | **0.3%** | 0.4% | 0.6% |
| PS-AENN Test Data | MSE | 196.58 | **82.06** | 158.01 | 299.07 |
| | RMSE | 14.02 | **9.06** | 12.57 | 17.29 |
| | MAE | 10.40 | **5.97** | 8.22 | 12.69 |
| | MAPE | 0.87% | **0.5%** | 0.6% | 1.2% |

have a better prediction effect, we have different iterations of the model. Each epoch is repeated 5 times and the average value of the evaluation index is calculated for judgment. The results are shown in Table 3.3.

In addition, after learning the internal characteristics of the data, the different predictors will also lead to the quality of the results. Next, we use different predictors for experiments to compare the advantages and disadvantages of the prediction effect.

As can be seen from Table 3.4, since we are using the weight information of the fully connected neural network for the selection of important parameters, the predictor is also fully connected neural network. The remaining parameters are relatively important parameters for network prediction, which leads the result to the best (Table 3.5).

As shown in Fig. 3.4, it can be seen that PS-AENN is significantly better than other deep learning prediction models in four indicators, and PS-AENN has higher accuracy and faster training speed under the same number of training iterations, and it is found that the accuracy of prediction after selecting important parameters is significantly improved compared with Table 3.4. In addition, we can clearly see the fitting between the predicted data and the real data through the line chart. Due to the large amount and dense distribution of data, we select some of them for display.

**Table 3.4** Prediction result of different predictor

| | | CNN | SAE | BPNN |
|---|---|---|---|---|
| PS-AENN Training Data | MSE | 861.34 | 130.71 | **50.81** |
| | RMSE | 29.35 | 11.43 | **7.13** |
| | MAE | 23.37 | 9.19 | **5.36** |
| | MAPE | 1.3% | 0.5% | **0.3%** |
| PS-AENN Test Data | MSE | 868.85 | 188.61 | **72.19** |
| | RMSE | 29.47 | 13.73 | **8.51** |
| | MAE | 23.44 | 10.39 | **5.76** |
| | MAPE | 1.9% | 0.8% | **0.4%** |

**Table 3.5** Prediction result of different model

| | | CNN | SAE | BPNN |
|---|---|---|---|---|
| PS-AENN Training Data | MSE | 50.59 | 86.03 | **48.47** |
| | RMSE | 7.11 | 9.27 | **6.94** |
| | MAE | 5.05 | 7.47 | **5.06** |
| | MAPE | 0.3% | 0.4% | **0.3%** |
| PS-AENN Test Data | MSE | 81.91 | 109.86 | **76.07** |
| | RMSE | 9.05 | 10.48 | **8.72** |
| | MAE | 7.68 | 7.68 | **5.48** |
| | MAPE | 0.4% | 0.6% | **0.4%** |

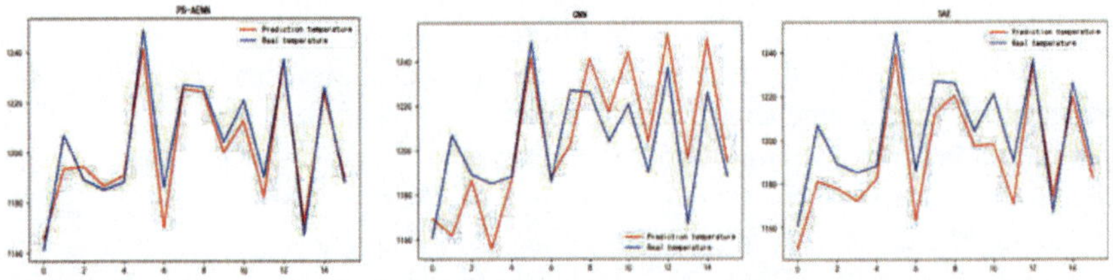

**Fig. 3.4** PS-AENN, CNN, SAE prediction fitting

The blue line indicates the actual temperature, and the red line indicates the predicted temperature. It can be seen that under the same number of rounds of training, PS-AENN has a better fitting effect than CNN and SAE.

### *3.3.3 Summary*

In this paper, a parameter selection auto-encoder neural network model (PS-AENN) based on deep learning is proposed to predict the tapping temperature accurately and quickly. Different from the other prediction models used in this field, PS-AENN model uses a new method to automatically select important parameters, which can not only save time and cost but also improve prediction accuracy. At the same time, we use the real data of a steel plant to verify the feasibility of the model and the accuracy of prediction, effectively solve the prediction problem in process industry. In the future work, we will further explore the control problems of the process industry based on the prediction results of this model, and achieve automatic, efficient, and high-yield control of the process industry.

## References


1. Ge, Z., Tao, C., Song, Z.: Quality prediction for polypropylene production process based on clgpr model. Control. Eng. Pract. **19**(5), 423–432 (2011)
2. Granados, G.E., Lacroix, L., Medjaher, K.: Condition monitoring and prediction of solution quality during a copper electroplating process. J. Intell. Manuf. (2018).
3. Hu, J., Wu, M., Chen, X., Cao, W., Pedrycz, W.: Multi-model ensemble prediction model for carbon efficiency with application to iron ore sintering process. Control Eng. Pract. **88**, 141–151 (2019).
4. Lu, N., Gao, F.: Stage-based process analysis and quality prediction for batch processes. Ind. Eng. Chem. Res. **44**(10), 3547–3555 (2005).
5. Undey, C., Ertunc, S., Cinar, A.: Online batch/fed-batch process performance monitoring, quality prediction, and variable-contribution analysis for diagnosis. Ind. Eng. Chem. Res. **42**(20), 4645–4658 (2003)
6. Wang, J.G., Wang, Y., Yao, Y., Yang, B.H., Ma, S.W.: Stacked autoencoder for operation prediction of coke dry quenching process. Control Eng. Pract. **88**, 110–118 (2019).
7. Yang, K., Jin, H., Chen, X., Dai, J., Wang, L., Zhang, D.: Soft sensor development for online quality prediction of industrial batch rubber mixing process using ensemble just-in-time Gaussian process regression models. Chemom. Intell. Lab. Syst. **155**(155), 170–182 (2016)
8. You, J., Li, X., Low, M., Lobell, D.B., Ermon, S.: Deep gaussian process for crop yield prediction based on remote sensing data, pp. 4559–4566 (2017).
9. Zhang, F., Yang, C., Zhu, H., Li, Y., Gui, W.: An integrated prediction model of heavy metal ion concentration for iron electrocoagulation process. Chem. Eng. J. **391**, 123628 (2020). https://doi.org/10.1016/j.cej.2019.123628

# Chapter 4
# Research for Field Strength Measuring Experience of Low-Frequency Time-Code Timing Signal

**Xin Wang, Luxi Huang, Shuanglin Li, Fan Zhao, Ping Feng, and Xiaohui Li**

**Abstract** The field strength of low-frequency time-code timing signal can describe the strength information of the signal in space directly and is the most important parameter of low-frequency time-code timing system coverage area and communication effectiveness assessment also. Therefore, the high precision of field strength measurement becomes premise and necessary condition of signal monitoring technology development and performance evaluation research of low-frequency time-code timing system. The paper is optimizing of the existing low-frequency time-code timing signal measurement theory, researching the measurement method of low-frequency time-code timing signal, setting up the low-frequency time-code timing signal measurement system for the high precision measurement of field strength, processing the measurement data and analysis the measurement result, providing actual testing data to support the quality of signal service and analysis assessment research for the low-frequency time-code timing system. The result shows that the mean value of field strength measurement error is 0.33 dBμv/m, meets the actual needs of the engineering completely.

## 4.1 Introduction

As the only technology encouraged by the International Telecommunication Union (ITU) strongly, low-frequency time-code timing technology was developed by the United States, Japan, Britain, Germany, and other countries 50 years ago, restarted a new round development after the 2000 year. Characterized by the low-frequency

X. Wang (✉) · L. Huang · S. Li · F. Zhao · P. Feng · X. Li
National Time Service Centre, Chinese Academy of Science, Xi'an 710600, China
e-mail: wangx@ntsc.ac.cn

X. Wang · L. Huang · F. Zhao · P. Feng · X. Li
Fengkai Low Frequency Time-Code Time Service Station, Zhaoqing 526500, China

X. Wang · L. Huang · P. Feng · X. Li
University of Chinese Academy of Science, Beijing 100049, China

 
V. Chang et al. (eds.), *Modern Industrial IoT, Big Data and Supply Chain*,
Smart Innovation, Systems and Technologies 218,
https://doi.org/10.1007/978-981-33-6141-6_4

time-code timing signal is transmission stability and has a large transmission area, it can meet the medium time service accuracy (hundreds of microseconds to milliseconds level) which are users need. The construction of low-frequency time-code timing system has solved the time demand within the coverage range. However, the correctness, performance, quality, and safety of transmission of low-frequency time-code timing signals need to be measured, assessed, and guaranteed by corresponding technical means.

As an important part of low-frequency time-code signal monitoring research, field strength measuring of the signal can reflect the strength of the signal in space information directly and truly and is the most important parameter of low-frequency time-code timing system coverage area and communication effectiveness assessment. Therefore, the high precision of field strength measurement becomes premise and necessary condition of signal monitoring technology development and performance evaluation research of low-frequency time-code timing system.

ITU-R were put forward a series topics about the application of the low-frequency time-code timing technology and signal measurement, among them, the new issue in 245/7 (the interference run by 20–90 kHz between standard frequency and time service) mentioned, "the signal field strength measurement method and standard instruments have not been developed well, or this kind of method and instrument is rarely." However, up to now, the application of low-frequency time-code timing technology in China, in the research of low-frequency signal, the appropriate test theory and method are still not completely solved, which will become an important constraint in the study of low-frequency time-code monitoring technology.

The paper is optimizing of the existing low-frequency time-code timing signal measurement theory, researching the measurement method of low-frequency time-code timing signal, setting up the low-frequency time-code timing signal measurement system for the high precision measurement of field strength, processing the measurement data and analysis the measurement result, providing actual testing data to support the quality of signal service, and analysis assessment research for the low-frequency time-code timing system. The result shows that the error of field strength measurement is less than 0.33 dBμv/m, meet the actual needs of the engineering.

## 4.2 The Foundation of Theory

### *4.2.1 Definition of Signal Field Strength*

The signal field strength is the electromagnetic field intensity of the electromagnetic wave in a certain place of space which is emitted by the high-power antenna [1]. In practical engineering, dB (dB) is often expressed in terms of 1 (dB/m), which is regarded as dBμv/m.

$$\mathrm{dB}(\mu \mathrm{v/m}) = 20 \lg \frac{E(\mu \mathrm{v/m})}{1(\mu \mathrm{v/m})} \tag{4.1}$$

### 4.2.2 Principle of Field Strength Measurement

The field strength measurement mainly uses antenna with simple shape such as ring magnetic antenna and whip electric antenna [2]. When the receiving antenna is placed in the electromagnetic field with the electric field intensity component of $E$, the relation between the induced voltage $U$ on the antenna and the signal field strength is equation [3], where, $H$ is the effective height of the receiving antenna, and $H$ can be calculated by the antenna theory.

$$U = H * E \tag{4.2}$$

In actual engineering, receiving antenna should be polarization matching with measured signal and impedance matching with the measuring instrument. The maximum induced signal can be achieved through rotating the receiving antenna in measurement. The signal level value is obtained by the measuring instrument through cable. At the same time, considering the receiving antenna and cable loss of transmission, the field strength measured value calculation is as follows [4]:

$$E_r = E + K + L \tag{4.3}$$

where $E_r$ is the signal field strength value, $K$ is the receiving antenna factor, and $L$ is the attenuation value of the transmission cable and other accessories.

### 4.2.3 Field Strength Propagation Mechanism

**Field Strength Propagation of Ground Wave**

For low-frequency time-code timing signal, the field strength propagation of ground wave can be expressed as:

$$E(\mathrm{dB}) = 109.54 + A - 20 \lg d + 10 \lg P_\Sigma \tag{4.4}$$

where $d$ is the spatial path of wave propagation (great circle distance), $A$ is the attenuation factor of ground wave (dB), and $P_\Sigma$ is the radiation power of transmitting antenna. The basic field is $E_{\mathrm{basic}} = 109.54 + A - 20 \lg d$, that is the field strength at the receiving point when $P_\Sigma$ is 1 kw.

**Field Strength Propagation of Sky Wave**

For low-frequency time-code timing signal, the field strength propagation of sky wave can be expressed as:

$$E = \frac{0.6p_t^{\frac{1}{2}} R_i^m R_g^{m-1} D_m D_g^{m-1} F_t F_r (\cos^2 \psi)^m}{L} \quad (4.5)$$

where type of $p_t$ is the antenna radiation power, $R_i$ is the ionosphere reflection coefficient, $R_g$ is ground emission coefficient, $D_m$ is the ionosphere is focused on the factors, $D_g$ is spherical divergence factor, $F_t$ is transmitting antenna factor, $F_r$ is receiving antenna factor, $\psi$ is launch angle and arrival angle of the sky wave on ground, which is relative to the horizon, $L$ is sky wave propagation length of ray, and $m$ is the wave-step number of sky wave.

## 4.3 Design of Measurement Test

### *4.3.1 Signal Field Strength Measurement System*

There is a period of time in the low-frequency time-code timing signal every second that the signal amplitude will decrease, and the time length of the amplitude will change with the difference of the code information. Therefore, in order to reflect the amplitude of the measurement site correctly, it is actually necessary to measure the amplitude that the signal amplitude has not decreased. Due to the code rate of low-frequency time-code timing signal is 1 baud/s, when the pulse repetition rate of 1 Hz, detector to obtain the RSL value is corresponding to the range of the carrier. It is important to note that the detection method has three kinds, mean value detector (AV), peak detection (PK), and quasi-peak detector (QP), as measured in the low-frequency time-code timing signal, we use quasi-peak detector (QP) because in this mode with continuous spectrum signal strength testing capabilities. Meanwhile, the receiving antenna factor and transmission cable loss should be considered in the measurement, and the field strength of the low-frequency time-code timing signal can be calculated comprehensively.

Based on the above ideas, the signal field strength measurement system is built in this experiment, and the block diagram is shown in Fig. 4.1. The system is divided into antenna and receiving module and signal processing and display module. Through to the receiving antenna input signal is processed by frequency conversion to intermediate frequency signal, and amplifying, adopting QP detection after filtering operation, signal processing and display module test and calculation the signal amplitude after digital processing. The field strength of low-frequency time-code timing signal can be obtained by eliminating outliers and interference data through the system select proper threshold and transmit the data to computer display and storage.

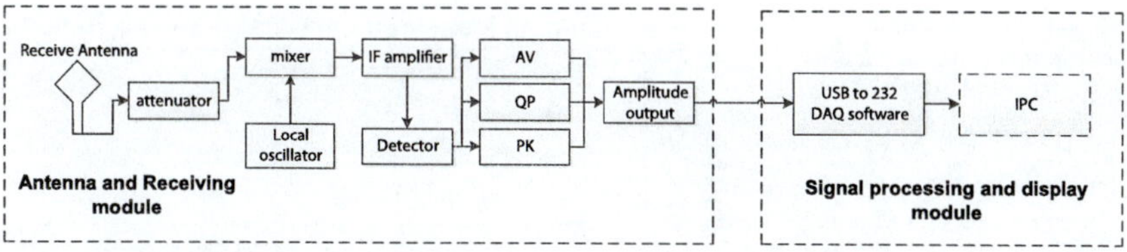


**Fig. 4.1** Block of signal field strength measurement system

### 4.3.2 *Design of Test Plan*

The low-frequency time-code timing signals are transmitted over long distances in the concentric spherical shells formed by the earth surface and the low ionosphere [5] (D layer during the day and the bottom of *E* layer at night). It is important to note that, in different areas, along the surface of ground wave diffraction signals and the reflection of the sky wave along the conductive walls with different occupies the proportion of different transmission distance. Typically, 0–300 km range with ground wave is given priority to, more than 300 km sky wave stepped in, and with the increase of the distance, the sky wave influence is also growing [6].

Based on the study of the propagation theory of low-frequency time-code timing signal, the test plan selected 50–300 km (from Shangqiu to Huai'an) and 300–800 km (from Huai'an to Qidong), respectively, to conduct signal field strength measurement. The detailed test route is shown in Fig. 4.2.

**Fig. 4.2** Route of signal field strength measurement test

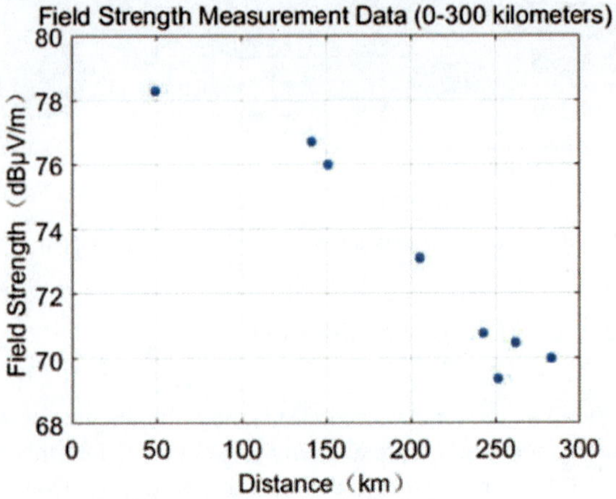


**Fig. 4.3** Field strength measurement data at each test point of 50–300 km

## 4.4 Data Processing Analysis

### *4.4.1 Ground Wave Zone (50–300 km)*

Figure 4.3 shows the signal strength measured data at each test point of 50–300 km. It can be seen from the figure and table that the coverage of ground wave area of BPC signal is good, and most of the measured results of field strength are over 70 dBμV/m. Because the ground wave signal is dominant, it can be seen that the change of the signal field strength with the distance basically conforms to the trend of monotone decreasing.

### *4.4.2 Ground and Sky Wave Interference Zone (300–800 km)*

Figure 4.4 shows the signal strength measured data at each test point of 300–800 km. Through data analysis, it can be seen that the coverage of signal in the interference area of ground waves and sky waves is good, and the measured results of field strength is greater than 55 dBμV/m. As the sky wave begins to intervene, the signal presents the superposition of ground wave and sky wave. With the increase of distance, the field strength distribution shows an oscillating attenuation trend.

### *4.4.3 The Value Compared with the Theoretical Value*

According to the theoretical model of low-frequency radio signal propagation, the theoretical value of the field strength distribution of low-frequency time-code timing

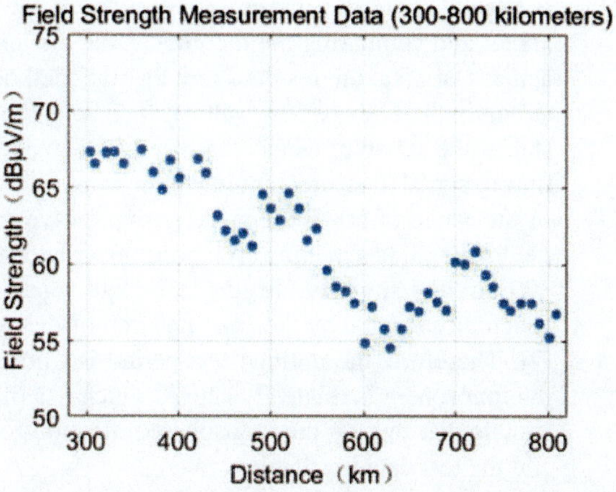


**Fig. 4.4** Field strength measurement data of each test point of 300–800 km

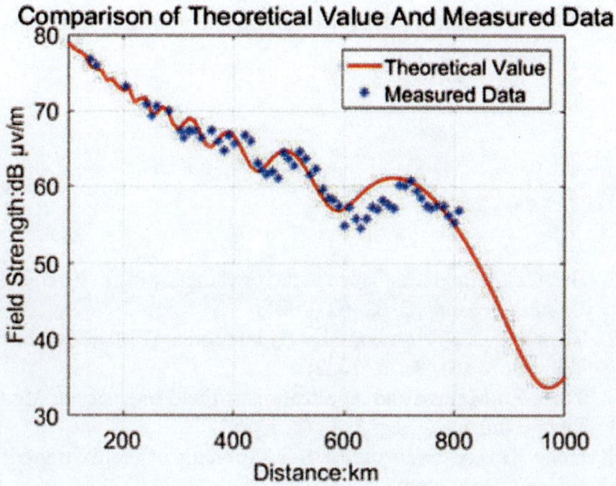


**Fig. 4.5** The comparison graph of the theoretical value and the measured data of the field strength of each test point

signal can be simulated. The result of comparison between the measured value and the theoretical value is shown in Fig. 4.5.

As can be seen from Fig. 4.5, the actual measured value of field intensity is in good agreement with the theoretical value curve of low-frequency time-code timing signal, and the mean error is 0.33 dBμv/m.

## 4.5 Conclusions

From the above analysis of the experimental data, we can preliminary to the following conclusion:

(1) By establishing the transmission model of low-frequency time-code timing signal and comparing the theoretical calculation value of the model with the actual test data, the results show that the method is used to measure the field strength, the average error value is 0.33 dBμv/m, and the test data is relatively stable and reliable, which can meet the demand of low-frequency time-code timing signal measurement in engineering.

(2) In the range of 50–300 km, the ground wave transmission is stable, and the field strength has no direct relationship with the test time. In the range of 300–800 km, the sky wave begins to be intervened, and the signal field strength is actually formed by the superposition of the sky wave and the ground wave [7]. Therefore, the optimal test period is from 11:30 to 14:00. At this time, the ionosphere is relatively stable, which has little influence on the sky wave signals that rely on the ionosphere reflection for long-distance transmission, and the test data is reliable.

(3) In order to avoid the introduction of electromagnetic interference or other error terms, the signal field strength measurement system should use battery instead of alternating current during the actual test to ensure the accuracy and reliability of the test data.

## References


1. Bai, C.: Method to enhance veracity of measuring radio signal's electric field intensity. Commun. Countermeasures **3**, 62–64 (2007)
2. Tianming, X.: Television Signal Field Strength Measurement and Air Management. China Cable Television. **10**, 48–50 (2002)
3. Yuan, Z.: Method and Application of Field Intensity of Medium and Short Wave Broadcasting Signal. Inf. Syst. Eng. **2**, 88 (2019)
4. Dong Xin. Analysis on the Key Problems of Field Strength Measurement of FM TV.Wireless Internet Technology. 2016.4:3–5
5. Liu Jun, The theory and the technique of engineering on low-frequency time-code in time service system. Graduate University of Chinese Academy of Sciences,2002
6. Wang Na, Study on Field Strength and Delay of Low frequency Ground-wave. Graduate University of Chinese Academy of Sciences,2012
7. Noboru Wakai, Noriyuki Kurihara, Atsushi Otsuka, Kuniyasu Imamura, and Yukio Takahashi, Wintertime survey of LF field strengths in Japan, RADIO SCIENCE, 2006, VOL. 41, RS5S13.

# Chapter 5
# Smart Medical and Its Ethical Problems

**Victor Chang, Rui Ying, and Qianwen Ariel Xu**

**Abstract** With the rapid development of smart devices, big data and the Internet of Things, the concept of "smart city" has been raised and developed. Over the past decade, the concept of smart medical has received extensive attention. Currently, some simple smart medical technologies have begun to be applied, such as online appointment registration, electronic medical records, etc. This report aims to introduce the vision of "smart medical." Based on the analytics of this study, some potential ethical problems are discovered, including privacy problems, data ownership problems, security and liability problems, and unemployment problems.

## 5.1 Introduction and Background

In 2009, Samuel Palmisano, the CEO of IBM, put forward the vision of "smart city," which led to a smart city trend in a large number of countries and multinational corporations [1]. A smart city is an urban area that collects various data by applying sensors to offer information and services efficiently. The smart city concept includes smart power, smart medical, smart transportation, smart supply chain, and smart banking. This report will introduce the functions and explore the ethics problems of smart medical.

Generally, smart medical refers to the use of information technology such as the Internet of Things and cloud computing to connect physical, information, social, and commercial infrastructure related to medical and health construction through perceptual, instrumented, and intelligent ways. The goal of smart medical is to meet the

V. Chang (✉) · Q. A. Xu
Artificial Intelligence and Information Systems Research Group, School of Computing, Engineering and Digital Technologies, Teesside University, Middlesbrough, UK

Q. A. Xu
e-mail: iamarielxu@163.com

R. Ying · Q. A. Xu
International Business School of Suzhou, Xi'an Jiaotong-Liverpool University, Suzhou, China
e-mail: 541255071@qq.com

 
V. Chang et al. (eds.), *Modern Industrial IoT, Big Data and Supply Chain*,
Smart Innovation, Systems and Technologies 218,
https://doi.org/10.1007/978-981-33-6141-6_5

requirement of the corresponding healthcare ecosystem intelligently. Smart medical is superior to ordinary digital medical services in terms of information system integration, information sharing, and intelligent processing [2]. Smart medical is usually composed of smart hospitals, regional health, and family health. Section 5.2 will introduce smart hospital information and the application of big data and IoT in smart medicine. Section 5.3 will explore smart medical issues from privacy, data ownership, security, and labor disruption. Section 5.4 will conclude this report, and Sect. 5.5 will give some suggestions.

## 5.2 The Vision of Smart Medical

### *5.2.1 Interaction in a Future Smart Hospital*

A smart hospital is a highly interactive, intelligent environment with a large number of computing devices and types of equipment, consisting of handheld computers which are able to generate and gain access to a limited amount of data, PCs that can be employed at fixed locations for long-term and semi-public displays that are easy to locate, allowing users or patients to share and discuss information [3]. Smart hospital staff can use a specific set of services to interact with the environment, interpreting contextual information.

For a patient who enters the smart hospital for help, semi-public displays or even intelligent robot can give them corresponding guide services, including registration appointments, department selection guides, and map location queries. They can then register and login to the hospital system via their smartphone or other micro wearable devices, which has a guiding role in their further medical treatment.

Hospital members or staff has to be very familiar with the hospital environment and context to generate the data required for their work. As the environment and personnel of smart hospitals change frequently, this learning process is time-consuming and high-cost. The context-aware application is developed and integrated into smart hospitals to save the time of the staff by helping them to be informed about the location and status of medical device, patients, or other hospital staff, as well as applying a real-time digital map or lists of artifacts and people which is automatically updated [4].

In order to deliver germane data to hospital staff, smart hospitals will consider contextual information, including the user's basic information, the time of diagnosis and treatment, and the status of clinical information (e.g., availability of test outcomes). For instance, when a doctor arrives near the patient's bed, his tablet or another mobile device will display the patient's data and situation, reminding the doctor of some essential information that is of great importance to the whole process of medical treatment. The smart hospital also takes into account the user's personalized context. For instance, physical access to a public device or workstation only shows the physician's patient data, personnel scheduling, and recommendations, which are significant for him to integrate his further working project.

Stanford [5] stated that the developed technology makes a significant contribution to advancing the efficiency of medical staff collaboration, such as consulting with experts or requesting help in transferring patients. For instance, based on the video call technology, the teleconference can help the hospital in a remote area to find a better method to address their severe problems. Furthermore, the 3D simulation technology can be used by physicians in different regions to solve complex surgical case problems, reducing the time cost of solving complex problems.

### 5.2.2 *Smart Medical Based on Big Data and the Internet of Things*

With the technologies of big data and the Internet of Things becoming more advanced, their applications are becoming more widespread. Utilizing IoT, big data, cloud computing and other related information technologies, smart medical connects patients with medical staff, medical equipment, and medical institutions to promote a comprehensive information model and improve service efficiency.

A large number of medical archive information resources generated in real-time and at high speed are the basis of smart medical care. In the context of big data, Chen [6] stated that various types of medical treatment systems related to medical diagnosis, and treatment generate a large amount of electronic data almost every moment. Due to the diversity and rapid growth of medical data, traditional analytical techniques and storage relationship methods are difficult to cope with, making it more challenging to meet the needs of clinical applications. Fortunately, the technology of cloud computing can deal with this problem. With further breakthroughs in this technology, big data can be universally used in smart medical in the near future.

For medical institutions, smart health care means the rational allocation of medical resources. In China, large hospitals are often full of patients. Many of them are only infected with common diseases that primary health care institutions can solve, and primary healthcare institutions are often idle resources. Through big data and cloud computing, smart medical systems can reasonably distribute patient traffic, suggesting that patients infected with common diseases go to nearby primary hospitals for treatment. Simultaneously, smart medical can guarantee the safety and high quality of medical services in primary hospitals. Doctors can access patients' health records from databases and quickly retrieve relevant information such as their medical history and medication status. Utilize this auxiliary information, doctors make more accurate judgments and avoid medical accidents such as repeated medications and drug allergies [7]. Through big data, smart medical care can break the independent state of traditional hospitals, realize information sharing for patients, and rationally utilize medical resources, thereby improving work efficiency and service quality.

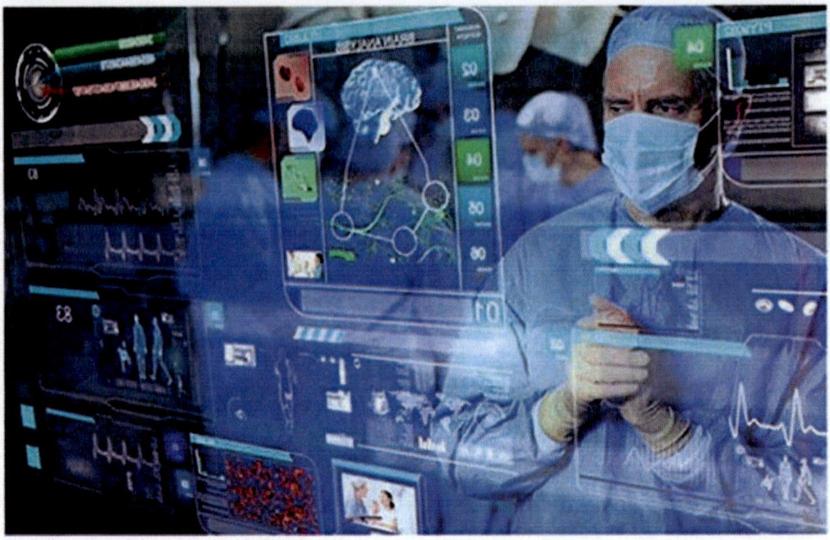

**Fig. 5.1** Diagnose rare diseases

Artificial intelligence, based on big data analysis and cloud computing, can significantly help doctors diagnose and treat patients. For example, if there is a patient with a rare disease that the doctor cannot diagnose, without artificial intelligence, the patients can only try a better hospital. In this case, artificial intelligence can use artificial neural networks to identify and sift diseases similar to or related to the disease from a database with huge data volumes. Then, search for past solutions and the latest research results, and summarize this information to give the best solution based on their algorithms. Moreover, big data analysis can provide data support for diagnosis and treatment decision-making. Currently, the majority of doctors or surgeons only refer to their past experience when developing a medical plan. In some cases, for lack of a large sample of data, they have to try solutions that may be effective, causing great pain to patients. With data support of previous medical experience from big data, they can access an efficient solution. To be specific, for instance, if a doctor wants to diagnose whether a patient has lung cancer, most of the current methods are based on the doctor's experience and consultations to reach a corresponding conclusion. When machine learning and data mining technologies can be applied more maturely in the medical industry, researchers can use the classification and cluster analysis methods to train the corresponding models based on historical data. These models show that case data that meet some characteristics have a higher probability of getting lung cancer. These characteristics can include several aspects of a person's situation, such as smoking, eating habits, location pollution, income status, genetic history, etc. This model provides a more reliable data reference when doctors diagnose patients. Simultaneously, this technology of artificial intelligence can also play

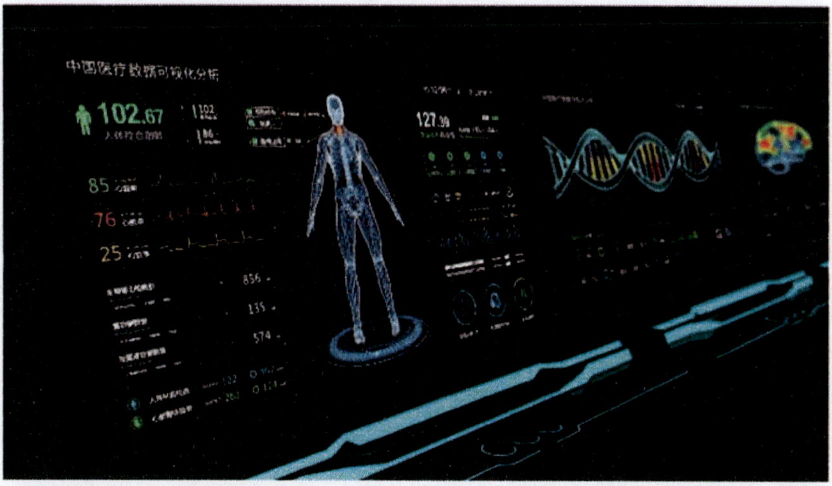


**Fig. 5.2** Data reference for medical decision

a more significant role in improving efficiency and saving labor costs (Fig. 5.1 and Fig. 5.2).

Shang [8] shows that data mining and data analysis of massive medical data through big data analysis systems can improve the scientific decision-making support of health-related departments. Smart data based on big data often means that medical departments can make more timely emergency measures. Through establishing an intelligent emergency response system, medical departments are linked to systems such as medical institution disease control centers, enabling timely and effective supervision and control. In the face of sudden public health incidents, they can also respond rapidly to improve the flexibility and timely response of the regulatory authorities in handling incidents.

In Sect. 5.2.1, some functional implementations in the smart hospital employ the technology of IoT. Specifically, established on the IoT, the structure of a smart hospital is often separated into three parts: perception layer, network layer, and application layer [9]. As is shown in Fig. 5.3, the perception layer contains the data acquisition layer and access layer. The data collection layer is used to sense and acquire relevant data from hospitals and around the hospital. The access layer is used to convey data obtained from the data acquisition layer and input it into the global object connection network. The network layer includes a network transmission platform and an application platform [10]. The former ensures that the data obtained by the perception layer is transmitted in real-time and efficiently. The latter is the integration of various data so that third parties can develop a variety of applications on this platform for the relevant personnel. The application layer is mainly used in the hospital's informatization and advanced applications in management and decision-making (Fig. 5.4 and Fig. 5.5).

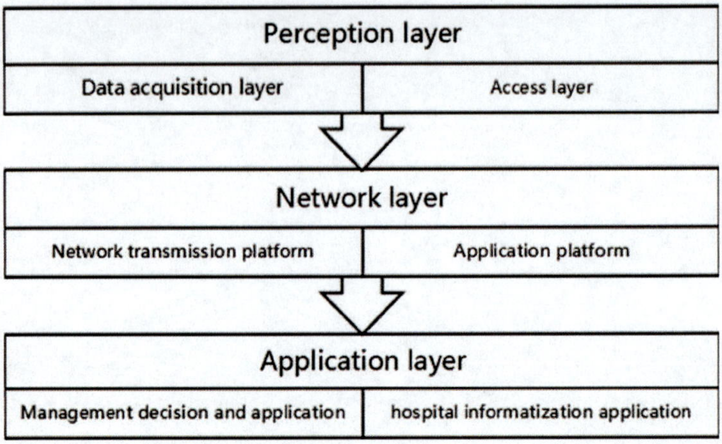


**Fig. 5.3** The architecture of the smart hospital

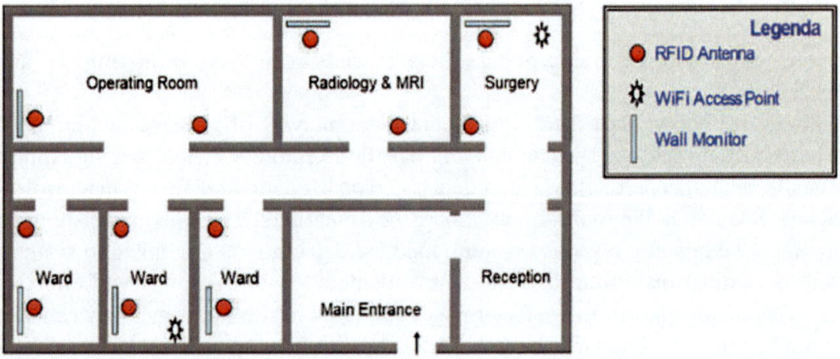


**Fig. 5.4** Hospital layout [11]

Outside the hospital, the Internet of Things will also have a wide range of smart medical applications. In terms of residents' health management, utilizing IoT technology, relevant medical institutions can conduct physical examinations, preventive reminders, and knowledge popularization. It is essential to use artificial intelligence technology to quantify the patient's health trajectory to customize unique health guidance for different patient individuals, thereby improving residents' health and disease prevention awareness. Additionally, intelligent wearable devices based on the Internet of Things technology, such as Apple Watch, can monitor the user's physical indicators in real time. The user is alerted when it is determined that the user is at risk of illness. In a critical moment, this device can even automatically call for help [2]. This will significantly increase the probability of surviving users in the event of an emergency.

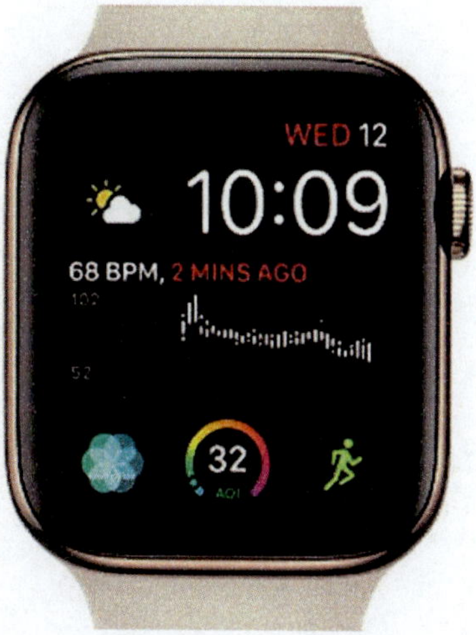


**Fig. 5.5** Smart wearable device [12]

Nowadays, medical resources are very tight, and hospital wards are often not enough for patients who need to be hospitalized. Using IoT technology, it provides remote consultation for patients in critical situations and enables continuous monitoring. Utilizing IoT technology, advanced medical technology is no longer limited by time and space and can greatly alleviate medical problems in backward areas (Fig. 5.6 and Fig. 5.7).

In terms of drug management, by providing RFID tags for each drug, both medical personnel and patients can quickly learn about drug information. Dong, Wang, and

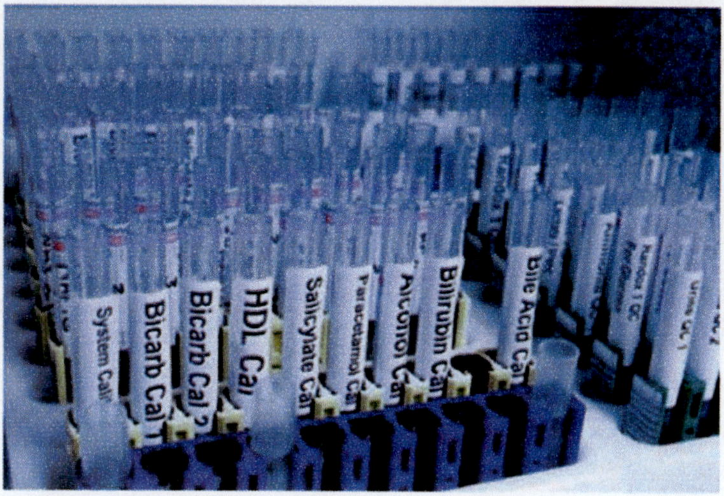


**Fig. 5.6** Drug storage environment monitoring

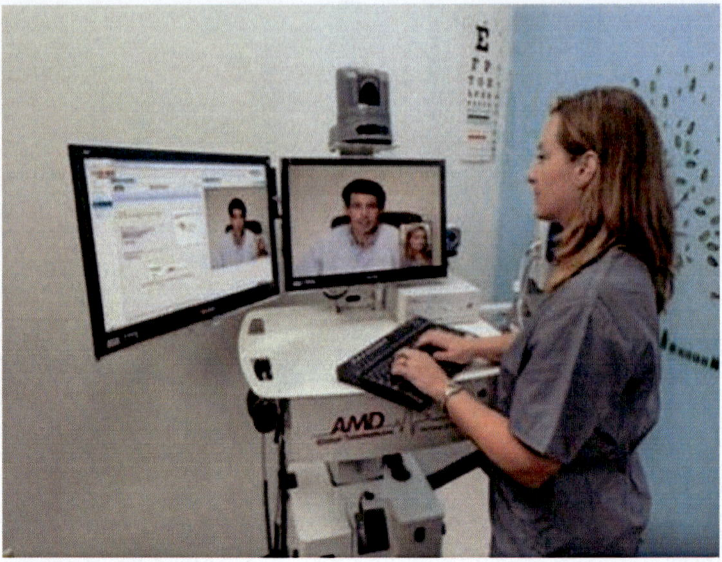


**Fig. 5.7** Integration of telemedicine

Zou [2] underlined that environmental factors such as temperature and humidity could be collected using IoT equipment. Once the environment is found to be unsuitable for the storage of certain types of drugs, a warning message can be given in time to prompt the management personnel to adjust the storage environment or replace the storage location, thereby effectively managing drugs. Poor management of medical waste can cause great damage to environmental safety. IoT technology can effectively detect whether the disposal location and method of waste are reasonable. When unreasonable conditions occur, it can quickly issue an alarm, which helps achieve reasonable supervision of medical waste.

## 5.3 Potential Ethics Problems

Based on analytics of the application of smart medical technologies, this study has identified a few potential ethical issues, including privacy problems, data ownership problems, security and liability problems, and risk of unemployment.

### *5.3.1 Privacy Problems*

Privacy refers to personal information that the parties do not want others to know or are inconvenient to know. By applying big data and IoT in the area of smart healthcare, privacy protection will become a huge problem. If the privacy data is leaked, it will cause considerable damage to patients and doctors. The majority of privacy problems are caused by poor data management, cyber-attacks, and patient's weaker privacy awareness.

For medical research, the data are usually anonymous or de-identified to obtain the consent of the data subject and to comply with data protection regulations, which is essential for data management [13]. However, for some databases of the medical institution, the data anonymization is not insufficient. The reason for the poor insufficient anonymization lies in the fact that some medical workers understand the deviation of privacy information. Ou and Lan [14] discovered that about 46.9% of medical staff believe that certain medical information of patients is not in the scope of privacy protection. Based on this understanding, they naturally do not pay attention to the security of such knowledge of patients. Moreover, healthcare workers know very little about the technical characteristics and operational methods of smart medical care, causing poor data management. According to the survey of Ou and Lan [14], only 31% of healthcare workers understand the smart health system, which means that 69% of medical workers do not understand the technical characteristics and operation of smart health care. In this situation, even if all healthcare workers have an awareness of patients' privacy, they still cannot protect patients' privacy when they use and manage medical data.

The patient's lack of privacy awareness is also one of the causes of data leakage. The survey found that many people do not realize that their medical information is in the privacy category. For example, when asked if they are worried about others seeing their medical information, which may affect their job opportunities and professional status, 11% of respondents expressed great concern, and 45% of them expressed some concern. In other words, from the patient's perspective, 44% of respondents are not aware of protecting their medical information [14]. Besides, some people, such as the elderly and children, cannot understand the importance of medical data privacy. They are also potential victims of data breaches.

The data leakage caused by cyber-attacks is really dangerous because hackers may utilize the data for fraud or other criminal activities. In the era of smart medical care, social media, mobile APP, telemedicine, and other medical applications are widely used between doctors and patients, which may lead to that patients' diagnoses, habits, and other private data are likely to leave traces on the Internet. This increases the risk of personal data leakage. Since the current privacy protection techniques are primarily established on static data sets and big medical data has a dynamic characteristic, it may increase the risk of private medical data leakage [15]. Referring to the statistics from the United States Department's Health Civil Rights Office, in the first quarter of 2015 alone, 87 data breaches were incurred in the US, involving over 500 medical facilities and a total of 92.23 million private data leakage [16]. The secondary use of big medical data is able to create priceless value. However, once it is leaked, the harm is usually unpredictable. According to a Ponemon Institute study [17], the average expense of every global medical information breach caused by loss or theft is $380, which is a substantial economic loss.

### 5.3.2 Data Ownership Problems

In the era of big data, the data should be owned by whom it is always a problem. This is the issue of interest coordination among data users, collectors, and producers [18]. After comprehending the scope, manner and consequences of data utilization, the data subject should have the right to make decisions on how to utilize private data. However, many research institutions collect and develop their private data without achieving explicit consent from the data subjects, contrary to business ethics. In 2013, Snowden's exposure to the "PEISM" event caused great concern. In this incident, the National Security Agency collected a large amount of user privacy data from multiple Internet companies and communication companies for political purposes, reflecting the ethical issues of data abuse. Therefore, it is essential for the data subject to have the right to know who owns the data and decides how to use the data.

### 5.3.3 Security and Liability Problems

In Sect. 5.2.2, it is mentioned that big data analysis has a reference function for medical decision-making. However, when a physician's experience is contrary to the big data analysis results or the artificial intelligence algorithm and its recommendations, which one should be chosen?

The application of big medical data attempts to use data to present all human activities, including vital signs, disease information, living environment, education level, eating habits, etc. Nevertheless, the application of big medical information brings a problem; in the interaction between people and numbers, data often exists away from the subject, and the subject is replaced by data [19]. This phenomenon may have a negative impact on both doctors and patients. On the one hand, if medical workers rely too much on data to diagnose and treat diseases, they are likely to limit their thinking ability and lose their traditional clinical experience, which will make it difficult for doctors to form a complete impression on patients. On the other hand, for the patient, as they are labeled with different kinds of data and become digital symbols without thinking abilities, consciousness and emotion, their pain becomes a bunch of numbers and images, and then their human subjectivity is lost [6]. At the same time, the patients will be confused and have a feeling that they are not truly listened and understood, which may undermine the contact and human care between doctors and patients, thereby reducing the effect of treatment [16].

In the future, artificial intelligence will be more universally used in smart healthcare, which may generate medical liability issues. If a doctor utilizes the data analysis results or AI recommendations, rather than believe his own experience. Still, he fails to heal the sick, who will be responsible for the medical accident? The AI, the doctor, or the data model researcher? Therefore, doctors' over-reliance on data analysis and AI may generate responsibility problems, which is one of the ethical issues that need to be solved in smart medical.

### *5.3.4 Risk of Unemployment*

In 2016, the AI robot AlphaGo defeated the Go game world champion, which means that the robot will surpass humanity in more and more fields. Since the invention of steam engines, machine replacement of human labor had always been a serious ethical issue. As mentioned in Sect. 5.2.2, the application of AI contributes to improving efficiency and saving labor costs. With the gradual popularization of smart health care, AI and mechanical automation may replace some medical workers' work. First of all, the work tasks of some positions are relatively simple, such as the production and packaging of drugs. These mechanical and physical work are easily replaced by artificial intelligence, which leads to the unemployment of workers. Furthermore, doctors who are mostly engaged in mental work will also be affected. For example, AI can help doctors improve their efficiency when they perform diagnostic and therapeutic work, which means that more medical problems can be solved in less time. The workload of workers will be significantly reduced [20]. In this case, the hospital may dismiss some employees to save costs.

It is worth mentioning that artificial intelligence is a conditional alternative to human labor. First of all, the primary condition for its replacement of human labor is that AI can accomplish higher-quality tasks than laborers. Typically, machines cannot automate the entire job. They can only automate part of specific tasks in the job [21]. Moreover, AI systems need to be superior to human labor in terms of cost.

## 5.4 Recommendations

According to the analytics of problems and conclusions in this study, we have some recommendations for each identified issue.

For privacy problems, the authorities of countries worldwide are supposed to improve the corresponding regulations to protect data subjects' privacy. Medical workers must improve their understanding of smart medical to protect data subjects' privacy better as well. Meanwhile, data subjects need to pay attention to the importance of data protection. Additionally, more advanced databases and firewall technologies will significantly contribute to resisting network attacks in the future.

In terms of data ownership problems, the institution or company that can make the best use of it deserves to own data. Specifically, a party that can protect data and uses data to benefit all sides can own the data. However, if the institution wants to use the data, it must ask for the consent and tell the data subject how the data will be used [22].

From my perspective, the data analysis is just a tool or a reference, and the final judgments have to be made by doctors. Therefore, doctors still have to bear the primary responsibility of a medical accident.

Piero Scaruffi [23] believes that future AI society will create new jobs that we cannot even imagine today for the unemployment problems. Although the application of artificial intelligence will replace part of the work, it will also create numerous jobs. Medical workers are supposed to improve their knowledge structure and actively understand AI applications, which may help them utilize AI technology to assist their work. Additionally, developers should realize that AI is designed to do things that people cannot do, rather than replacing them.

## 5.5 Conclusion

In conclusion, based on big data, IoT, and artificial intelligence technologies, smart medical is going to change the means of interaction between patients and doctors. Especially under circumstances like the COVID-19 pandemic, people must go through a quarantine period, and the approach of face-to-face counseling cannot be used. This is not convenient for the people with health requirements, the patients who need regular consultation in particular. In such a case, the existence of smart medical care, such as teleconsultation, will help solve this kind of problem [24][25]. However, although smart health care will improve the patient's treatment experience and the doctor's efficiency, it will also bring many ethical issues, including privacy problems, data ownership problems, security and liability problems, and the risk of unemployment. In the future, governments, medical institutions, medical workers, and patients are supposed to pay attention to these issues and try to avoid them.

**Acknowledgements** This work is supported by VC Research (VCR 0000081).

## References

1. Jin, P., Meng, T., Guo, K., Shao, F.: Research of routing algorithm based on smart city and the design of multi-network integration. In: 2015 Chinese Automation Congress (CAC). Chinese Automation Congress (CAC), vol. 2015, 1289 (2015). https://doi.org/10.1109/CAC.2015.7382698
2. Dong, Y., Wang, M., Zou, B.: Application of Internet of Things technology in the field of smart medical. Logist. Mater. Handl. (7), p. 107 (2017). Available at: https://ez.xjtlu.edu.cn/

login?url=https://search.ebscohost.com/login.aspx?direct=true&db=edscoj&AN=edscoj.wljsyyy201707017&site=eds-live&scope=site. Accessed 2 Dec 2018

3. Sánchez, D., Tentori, M., Favela, J.: Activity recognition for the smart hospital. IEEE Intell. Syst. **23**(2), 50 (2008). https://doi.org/10.1109/MIS.2008.18
4. Moran, E.B., et al.: Mobility in hospital work: towards a pervasive computing hospital environment. Int. J. Electron. Healthc. **3**(1), 72–89 (2006)
5. Stanford, V.: Beam Me Up, Doctor Mc-Coy. IEEE Pervasive Comput. **2**(3), 13–18 (2003)
6. Chen, H., Compton, S., Hsiao, O.: DiabeticLink: A health Big Data system for patient empowerment and personalized healthcare. In: Zeng, D. et al. (eds.) Smart Health. ICSH 2013, Lecture Notes in Computer Science, vol 8040. Springer, Berlin, Heidelberg (2013). https://doi.org/10.1007/978-3-642-39844-5_10
7. Dighriri, M., Lee, G.M., Baker, T.: Big Data environment for smart healthcare applications over 5G mobile network. In: Alani, M., Tawfik, H., Saeed, M., Anya, O. (eds.) Applications of Big Data Analytics. Springer, Cham (2018). https://doi.org/10.1007/978-3-319-76472-6_1
8. Shang, Y.: Research of present situation of application of smart healthcare on the background of Big Data. Sci. Technol. Ind. (10), p. 19 (2016). Available at: https://ez.xjtlu.edu.cn/login?url=https://search.ebscohost.com/login.aspx?direct=true&db=edscoj&AN=edscoj.kjhcy201610005&site=eds-live&scope=site. Accessed 2 Dec 2018
9. Qin, X., Ying, W.: Design of explosive production information and managing system based on Internet of Things. In: 2015 International Conference on Control, Automation & Robotics, Singapore, pp. 173–176 (2015). https://doi.org/10.1109/ICCAR.2015.7166025
10. Yu, L., Lu, Y., Zhu, X.: Smart hospital based on internet of things. J. Netw. **7**(10), 1654–1661 (2012)
11. Coronato, A., Esposito, M.: Towards an implementation of smart hospital: a localization system for mobile users and devices. In: Sixth Annual IEEE International Conference on 2008 Sixth Annual IEEE International Conference on Pervasive Computing and Communications (PerCom), Pervasive Computing and Communications, 2008. PerCom 2008, p. 715. https://doi.org/10.1109/PERCOM.2008.79
12. Pathak, K.: The 10 Best Apple Watch Series 4 Features [Online image] (2018). Available at: https://www.iphonehacks.com/2018/09/best-apple-watch-series-4-features.html. Accessed: 3 Dec 2018
13. Mittelstadt, B., Floridi, L.: The ethics of Big Data: current and foreseeable issues in biomedical contexts. Sci. Eng. Ethics **22**(2), 303–341 (2016). https://doi.org/10.1007/s11948-015-9652-2
14. Ou, S., Lan, X.: On the Informal Institution Gaps and Addendum in the Protection of Patients´ Right of Privacy. J. Zhejiang Wanli Univ. **2016**(5), 38–42 (2016). Available at: https://ez.xjtlu.edu.cn/login?url=https://search.ebscohost.com/login.aspx?direct=true&db=edscoj&AN=edscoj.zjwlxyxb201605006&site=eds-live&scope=site. Accessed 3 Dec 2018
15. Mu, L., Lei, Z.: Big data processing technology research and application prospects. In: 2014 Fourth International Conference on Instrumentation and Measurement, Computer, Communication and Control, Harbin, vol. 2014, pp. 269–273 (2014). https://doi.org/10.1109/IMCCC.2014.63.
16. Wang, L., Hao, M.: Ethical reflection on characteristics of medical Big Data and its application. Med. Philos.: Humanist. Soc. Med. Edition (4), 32 (2017). Available at: https://ez.xjtlu.edu.cn/login?url=https://search.ebscohost.com/login.aspx?direct=true&db=edscqv&AN=edscqv.671782863&site=eds-live&scope=site. Accessed 2 Dec 2018.
17. Ponemon Institute: 2017 Cost of Data Breach Study, IBM Security [Online] (2017). Available from: https://info.resilientsystems.com/hubfs/IBM_Resilient_Branded_Content/White_Papers/2017_Global_CODB_Report_Final.pdf
18. Zhang, N., et al.: Ethical issues of medical data sharing under the background of Big Data. Chin. J. Inf. Traditional Chin. Med. **25**(8), 9–11 (2018).
19. Shao, F., Yang, Y.: Reconstruction of the "Body": diagnosis of the crisis of modernity in medicine. Med. Philos. **9**, 21 (2016). https://doi.org/10.12014/j.issn.1002-0772.2016.05a.06
20. Pan, W.: The multiple impacts of artificial intelligence technology development on employment. Huxiang Forum **31**(4), 145–153 (2018)

21. Chui, M., Manyika, J. Miremadi, M.: Four fundamentals of workplace automation. McKinsey Q. (1), pp. 50–57 (2016). Available at: https://ez.xjtlu.edu.cn/login?url=https://search.ebscohost.com/login.aspx?direct=true&db=bth&AN=116464173&site=eds-live&scope=site. Accessed 2 Dec 2018
22. Li, X., Wang, P.: Informed consent in Big Data era. Med. Philos. **9**, 9 (2016). https://doi.org/10.12014/j.issn.1002-0772.2016.05a.03
23. Scaruffi, P.: Intelligence is not artificial. [Online] (2014). Available from: https://www.scaruffi.com/singular/download.pdf (Accessed: 2 December 2018)
24. Pandey, N., Srivastava, R.M., Kumar, G., Katiyar, V., Agrawal, S.: Teleconsultation at a tertiary care government medical university during COVID-19 lockdown in India—a pilot study. Indian J. Ophthalmol. **68**(7), 1381–1384 (2020). https://doi.org/10.4103/ijo.IJO_1658_20
25. Keshvardoost, S., Bahaadinbeigy, K., Fatehi, F.: Telemedicine and e-Health: role of telehealth in the management of COVID-19: lessons learned from previous SARS, MERS, and Ebola outbreaks. Telemed. E-health. **26**(7), 850–852 (2020) https://doi.org/10.1089/tmj.2020.0105

# Chapter 6
# Network-Aware Distributed Machine Learning Over Wide Area Network

**Pan Zhou, Gang Sun, Hongfang Yu, and Victor Chang**

**Abstract** Machine learning requires accessing all dataset to train a high-quality model. Due to the data regulations and privacy concerns, the dataset of different data centers cannot be collected into one data center. It is unavoidable to conduct distributed training across multiple data centers. However, state-of-the-art distributed learning algorithms suffer from high communication cost due to the low-speed, highly heterogeneous wide area network connecting the data centers. In this paper, we propose a novel network-aware decentralized distributed training algorithm, namely NAD-PSGD, to overcome the problem. NAD-PSGD can enable worker nodes to mainly use high-speed links to exchange information and thus significantly reduce communication cost. Through our experiment on Amazon clouds and testbed cluster, NAD-PSGD can reduce the convergence training time by up to 42.8 and 66.9%, in comparison with advanced algorithms AD-PSGD and Allreduce-SGD, respectively.

## 6.1 Introduction

Machine learning has made great success in fields like image classification, nature language processing, and recommendation systems. Many large companies, such as Google, Facebook, and Microsoft, have deployed machine learning models in their productions [3]. Training a high-quality model requires accessing all dataset from data centers in various regions or countries [1, 5]. However, these dataset cannot be transferred among data centers due to data regulations and privacy concerns [8].

P. Zhou · G. Sun · H. Yu (✉)
University of Electronic Science and Technology of China, Chengdu, China
e-mail: yuhf@uestc.edu.cn

G. Sun
e-mail: gangsun@uestc.edu.cn

V. Chang
Artificial Intelligence and Information Systems Research Group, School of Computing, Engineering and Digital Technologies, Teesside University, Middlesbrough, UK
e-mail: V.Chang@tees.ac.uk

 
V. Chang et al. (eds.), *Modern Industrial IoT, Big Data and Supply Chain*,
Smart Innovation, Systems and Technologies 218,
https://doi.org/10.1007/978-981-33-6141-6_6

Thus, it is unavoidable to conduct distributed training across multiple data centers in various regions or countries without transferring the raw dataset.

However, the wide area network (WAN) connecting data centers in various regions or countries significantly differs from local area network (LAN) within a data center. On the one hand, the achievable link speed of WAN is very slow and is up to 60$\times$ slower than that of LAN [14]. On the other hand, the speeds of WAN links are highly heterogeneous compared with LAN due to the differences in underlying infrastructures and background applications. The fastest links can be up to $12\times$ faster than the slowest links [5].

Due to the communication intensive nature of distributed training, it is challenge for state-of-the-art distributed training algorithms to conduct training over such low-speed and heterogeneous network. Centralized parallel stochastic gradient decent (SGD) is widely implemented by popular distributed learning frameworks using parameter server, such as MXNET [2] and bytePS [6]. This algorithm requires all the distributed worker nodes to communicate with a central parameter server to synchronize the models of worker nodes. Such a design suffers from the communication bottleneck at the server side. Synchronous decentralized algorithms, such as Allreduce-SGD [7] and D2 [13], enable the worker nodes to communicate in a defined graph (e.g., a ring) to synchronize the models of worker nodes every iteration. This kind of design can evenly distribute the network traffic among worker nodes and thus avoid centralized bottleneck, but the faster nodes have to wait for the slower nodes each iteration. Consequently, its training speed is limited by the slowest node (straggler problem). Asynchronous decentralized algorithms, such as AD-PSGD [10], GoSGD [4], and CHOCO-SGD [9], enable the worker nodes to independently train their model to avoid the straggler problem. For any worker node, at each iteration, it randomly chooses a neighbor node in the graph to pull the model from the neighbor and update its local model. However, the worker nodes frequently choose neighbors with low-speed links. As a result, the worker nodes frequently communicate through low-speed links and suffer from a high communication cost.

In summary, state-of-the-art learning algorithms face at least one of the following three problems: centralized communication bottleneck, straggler problem, and high communication cost caused by low-speed links of WAN. This motivates us to design a **N**etwork-**A**ware **D**ecentralized **P**arallel SGD, namely NAD-PSGD, to address all the mentioned problems. NAD-PSGD has the following three characteristics: (a) the worker nodes train their local model independently and exchange information through peer-to-peer communication in a connected graph; (b) at each iteration, each of the nodes only randomly chooses one neighbor to communicate with; (c) the neighbor with high-speed link can be chosen at a very high frequency and the neighbor with low-speed link can be chosen at a low frequency. Such a communication policy cannot only let the worker nodes mainly use the high-speed links but also detect the variation of other links' speeds and accordingly adjust the frequencies to adaptive to network dynamics.

In conclusion, we make the following original contributions:

- We propose a new decentralized learning algorithm, namely NAD-PSGD, to enable the worker nodes to mainly communicate through high-speed links of WAN, which can significantly reduce the communication cost of the training over WAN.
- We implement the proposed algorithm and conduct the experiments on both Amazon clouds and testbed cluster. Experimental result shows that NAD-PSGD can reduce the convergence training time by up to 42.8 and 66.9\% without losing any accuracy, compare with advanced decentralized algorithms AD-PSGD and Allreduce-SGD, respectively.

The rest of the paper is organized as follows. Section II provides the design of the proposed training algorithm. Section III presents the experimental evaluation of NAD-PSGD. Section IV presents the most related work, and Section V concludes the paper.

## 6.2 Algorithm Design

### *6.2.1 Problem Formulation*

Consider a network of \$N\$ worker nodes. The communication links connecting the nodes can be abstracted as an undirected connected graph $\boldsymbol{G}$ with vertex set $\boldsymbol{V} = \{1, 2, \ldots, N\}$ and edge set $\boldsymbol{\varepsilon} = \boldsymbol{V} \times \boldsymbol{V}$. Each of the nodes trains a copy of a model over its local dataset and communicates with its neighbors in the graph to collaboratively train a high-quality model. We formulate the decentralized training as an optimization problem as follows:

$$\min_{\boldsymbol{x}} F(\boldsymbol{x}) = \sum_{i=1}^{N} \left( E[f(x_i; D_i)] + \sum_{i=1}^{N} x_i - x_j^2 \right) \tag{6.1}$$

where $\boldsymbol{x}$ is the vector of models of worker nodes. $x_i$ and $D_i$ are the model and dataset of worker $i$, respectively. $f(x_i; D_i)$ denotes the loss function of worker $i$. $\|x_i - x_j\|^2$ denotes the model difference between node $i$ and $j$. Formulation Eq. (6.1) is aimed to minimize both the loss functions of worker nodes and the model difference among them. Then models of workers can converge to the same optima.

### *6.2.2 Network-Aware Decentralized Parallel Sgd*

We proposed a new decentralized training algorithm, namely NAD-PSGD, to solve the optimization problem formulated in Eq. (6.1). NAD-PSGD is an asynchronous training algorithm where the training progresses of worker nodes are independent and the slower node could not impact the other nodes' training speed. As shown in Algorithm \ref{alg:nadsgd}, at each iteration, worker node $i$ only pulls model from one randomly chosen neighbor and updates its local model by using the local computed gradient and the pulled model, as shown in *line* 9–10. The neighbor with higher link speed can be chosen at a higher probability, which means the neighbor with higher link speed can be chosen at a higher frequency. This can enable the worker nodes to mainly communicate through high-speed links and thus significantly reduce the communication cost. To adaptive to the network dynamic, in our algorithm design, each node chooses the neighbors with low link speed at very low probabilities to detect the link speed and accordingly update the probabilities. In our implementation of NAD-PSGD, we do not directly measure the link speed since it is resource consuming. Instead, we use the iteration time as a metric of link speed. The reason is that the iteration time is composed of computation time and communication time. The computation time is typically stable and small compared to communication time while communication time varies largely due to the differences in link speeds of neighbor nodes. Thus, a longer iteration time means a slower link speed.

**Algorithm 1** NAD-PSGD at node $i$

**Input:** learning rate $\alpha$, weight $\beta$, iteration number $K$, maximum probability $P_{max}$.
**Output:** Trained model $x_i$

1: Initialize model $x_i^0 \leftarrow x^0$
2: Initialize probability vector $[p_{i,1}, p_{i,2}, \ldots, p_{i,N}] \leftarrow [0]_{1\times N}$
3: Initialize communication time vector $T_i \leftarrow [\infty]_{1\times N}$
4: $N_i \leftarrow$ number of neighbor nodes
5: If $i$ and $j$ are neighbors, then $p_{i,j} \leftarrow 1/N_i, T_i[j] \leftarrow 0$
6: **for** $k \in \{1, 2, 3, \ldots, K\}$ **do**
7:     Randomly select a neighbor $j$ at priobability $p_{i,j}$
8:     Pull node $j$'s model $x_j$
9:     $x_i^k \leftarrow x_i^k - \alpha \nabla f(x_i^k)$ ▷ Update model using local gradient
10:     $x_i^k \leftarrow x_i^k - \beta(x_i^k - x_j^k)$ ▷ Update model using pulled model
11:     $x_i^{k+1} \leftarrow x_i^k$ ▷Store parameter for next iteration
12:     $T_i[j] \leftarrow$ Recording iteration time $t_{i,j}$
13:     UPDATEPROBABILITYVECTOR($T_i$)
14: **end for**
15: **return** $x_i$.
16: **procedure** UPDATEPROBABILITYVECTOR($T_i$)
17:     $m \leftarrow min(T_i).index$ ▷ find index of the smallest item
18:     $p_{i,m} \leftarrow P_{max}$
19:     Get $p_{i,n}, \forall n \neq m$ such that $\sum_{n=1}^{N} p_{i,n} = 1$
20: **end procedure**

## 6.3 Evaluation

### *6.3.1 Experiment Settings*

We deploy experiments on both Amazon EC2 clouds and our testbed cluster.

**Cloud experiment settings**. To evaluate the effectiveness of the proposed NAD-PSGD in the real-world, we conduct the experiment across six Amazon EC2 regions. The geographical distribution of the clouds and bandwidths across the clouds are shown in Fig. 1a, b, respectively. In each cloud, we start an instance of c5.4× large as a worker node. Each node trains a CNN model (i.e., GoogleNet) on the MNIST dataset and communicates with peer nodes at the other clouds through the wide area network.

**Testbed experiment settings**. As renting cloud instance to train deep models are very expensive, we use testbed cluster to conduct our experiment on the more complex dataset (e.g., CIFAR10). We use Docker and Mininet to mimic the WAN across the clouds. In the testbed experiment, each docker container works as a worker node. Each node trains a CNN model (i.e., MobileNet) on the CIFAR10 dataset and communicates with peer nodes in the cluster through the network constructed by Mininet.

**Baseline settings**. We compare NAD-PSGD with state-of-the-art decentralized training algorithms, like AD-PSGD [10] and Allreduce-SGD [7]. We implement these algorithms over Pytorch such that they have the similar runtime environment. For all the algorithms, we set learning rate to 0.01 and set batch-size to 32. For NAD-PSGD, we set the maximum probability $P_{\max}$ to 0.5.

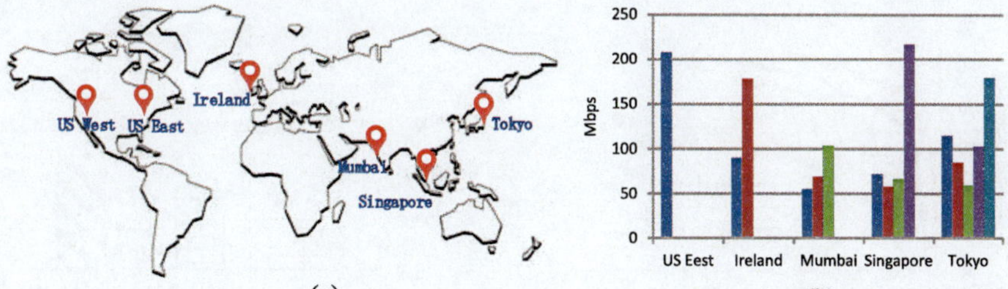


(a) (b)

**Fig. 6.1** Experiment settings across Amazon clouds. **a** Six clouds are chosen to conduct the experiment; **b** the bandwidth cross the chosen clouds

### 6.3.2 *Experiment Settings*

NAD-PSGD can significantly speeds up the training without losing the model accuracy. As shown in Fig. 6.2, when training GoogleNet on MNIST dataset over WAN, NAD-PSGD can reduce the convergence time by up to 40.6 and 49.5%, compared with AD-PSGD and Allreduce-SGD, respectively. As shown in Fig. 6.3, when training MobileNet on CIFAR10 dataset in the testbed cluster, NAD-PSGD can reduce convergence time by up to 42.8 and 66.9%, compared with AD-PSGD and Allreduce-SGD, respectively. The reason is that NAD-PSGD can dynamically detect the link speed and let the worker nodes mainly communicate through high-speed links. As a result, the communication cost of NAD-PSGD is much lower compared with the other algorithms.

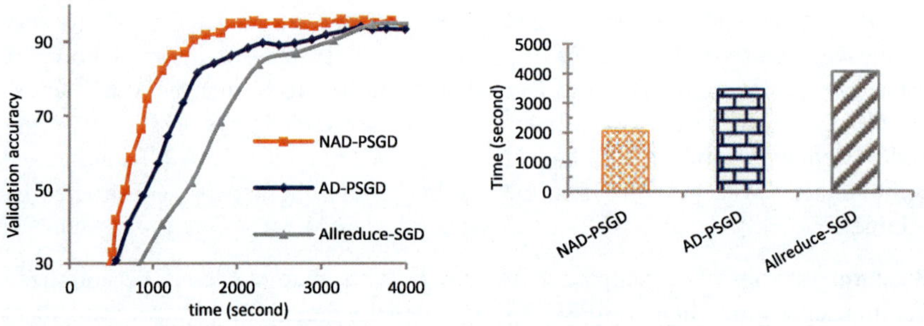


(a) (b)

**Fig. 6.2** Training GoogleNet on MNIST dataset across multiple clouds. **a** Validation accuracy varying with time; **b** convergence time

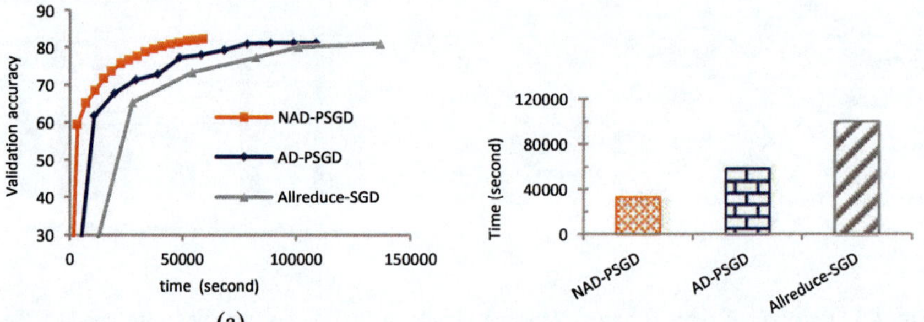


(a) (b)

**Fig. 6.3** Training MobileNet on CIFAR10 dataset in tested cluster. **a** Validation accuracy varying with time; **b** convergence time

## 6.4 Related Work

**Decentralized parallel SGD**. Unlike centralized parallel SGD algorithms implemented by central parameter servers, decentralized algorithms can enable the worker nodes to communicate in a connected graph. This communication mode can evenly distribute the network traffic among worker nodes and thus avoid centralized bottleneck. For example, Allreduce-SGD [7] enables the worker nodes to communicate in a defined graph (e.g., a ring) to average the gradients of all nodes every iteration. The training progress of worker nodes are kept in the same paces, a worker node has to wait for the completion of other nodes' communication. While in algorithms like AD-PSGD [10], GoSGD [4], and CHOCO-SGD [9], the training progresses of workers are independent and each worker communicates with its neighbors without waiting for other workers. However, in these algorithms, the workers frequently communicate with neighbors whose link speeds are very low. Consequently, the training is dragged down by low-speed links.

**Machine learning over heterogeneous networks**. To address heterogeneous resources, Hop [12] and Gaia [5] use bounded staleness for decentralized setting. While when network links experience continuous slowdown (i.e., links in WAN), the whole training would be dragged down by these low-speed links. In addition, Ripples [11] randomly generates a node group at each iteration. And the nodes in this group perform AllReduce algorithm to average models of the group. However, the generated groups usually contain low-speed links, leading to high communication cost.

## 6.5 Related Work

In this paper, we find that the training speed of state-of-the-art machine learning algorithms typically suffer from low-speed links over WAN. Therefore, we propose a novel network-aware learning algorithm, namely NAD-PSGD, to mainly use high-speed links. The experiment shows that NAD-PSGD can significantly speed up machine learning over WAN, compared with advanced learning algorithms.

## Reference

1. Boughorbel, S., Jarray, F., Venugopal, N., Moosa, S., Elhadi, H., Makhlouf, M.: Federated uncertainty-aware learning for distributed hospital ehr data. arXiv preprint arXiv:1910.12191 (2019)
2. Chen, T., Li, M., Li, Y., Lin, M., Wang, N., Wang, M., Xiao, T., Xu, B., Zhang, C., Zhang, Z.: Mxnet: A flexible and efficient machine learning library for heterogeneous distributed systems. arXiv preprint arXiv:1512.01274 (2015)

3. Hazelwood, K., Bird, S., Brooks, D., Chintala, S., Diril, U., Dzhulgakov, D., Fawzy, M., Jia, B., Jia, Y., Kalro, A., et al.: Applied machine learning at facebook: A datacenter infrastructure perspective. In: 2018 IEEE International Symposium on High Performance Computer Architecture (HPCA). pp. 620–629. IEEE (2018)
4. Heged}us, I., Danner, G., Jelasity, M.: Gossip learning as a decentralized alternative to federated learning. In: IFIP International Conference on Distributed Applications and Interoperable Systems. pp. 74–90. Springer (2019)
5. Hsieh, K., Harlap, A., Vijaykumar, N., Konomis, D., Ganger, G.R., Gibbons, P.B., Mutlu, O.: Gaia: Geo-distributed machine learning approaching fLANg speeds. In: 14th fUSENIXg Symposium on Networked Systems Design and Implementation (fNSDIg 17). pp. 629–647 (2017)
6. Inc., B.: Byteps (2019), https://github.com/bytedance/byteps
7. Jia, X., Song, S., He, W., Wang, Y., Rong, H., Zhou, F., Xie, L., Guo, Z., Yang, Y., Yu, L., et al.: Highly scalable deep learning training system with mixed-precision: Training imagenet in four minutes. arXiv preprint arXiv:1807.11205 (2018)
8. Kairouz, P., McMahan, H.B., Avent, B., Bellet, A., Bennis, M., Bhagoji, A.N., Bonawitz, K., Charles, Z., Cormode, G., Cummings, R., et al.: Advances and open problems in federated learning. arXiv preprint arXiv:1912.04977 (2019)
9. Koloskova, A., Stich, S.U., Jaggi, M.: Decentralized stochastic optimization and gossip algorithms with compressed communication. arXiv preprint arXiv:1902.00340 (2019)
10. Lian, X., Zhang, W., Zhang, C., Liu, J.: Asynchronous decentralized parallel stochastic gradient descent. In: International Conference on Machine Learning. pp. 3049–3058 (2018)
11. Luo, Q., He, J., Zhuo, Y., Qian, X.: Heterogeneity-aware asynchronous decentralized training. arXiv preprint arXiv:1909.08029 (2019)
12. Luo, Q., Lin, J., Zhuo, Y., Qian, X.: Hop: Heterogeneity-aware decentralized training. In: Proceed- ings of the Twenty-Fourth International Conference on Architectural Support for Programming Languages and Operating Systems. pp. 893–907. ACM (2019)
13. Tang, H., Lian, X., Yan, M., Zhang, C., Liu, J.: D2: Decentralized training over decentralized data. In: International Conference on Machine Learning. pp. 4855–4863 (2018)
14. Zhou, A.C., Xiao, Y., Gong, Y., He, B., Zhai, J., Mao, R.: Privacy regulation aware process mapping in geo-distributed cloud data centers. IEEE Trans. Parallel Distrib. Syst. **30**(8), 1872–1888 (2019)

# Chapter 7
# Avoidance of Suicidal Bombing Attack in Pakistan Using Machine Learning Models

**Muhammad Umer, Saima Sadiq, Saleem Ullah, Gyu Sang Choi, Arif Mehmood, and Michele Nappi**

**Abstract** Suicide attack of terrorists fear is a standout amongst the most horrendous difficulties for Pakistan as the nation has been the essential and most regular focus of suicide terrorism in the whole world during ongoing years. Suicide is a genuine psychological issue that needs our consideration, and to control and avoid, it is not a simple task. In this paper, we propose a suicidal attack early detection system, for predicting the suicidal acts using data that can automatically analyze the circumstances related to those attacks. Then, we explore tools of machine learning to extract useful information for classification algorithms. Experimental results show that our method for analyzing the suicidal attack data by applying machine learning models verify the effectiveness of performance to control further suicidal attempts.

## 7.1 Introduction

In general, information and statistics can be extracted from data and can be used to analyze, and conclusion can be drawn from it. Information accumulation is a urgent and efficient way to deal with gathering data from an assortment of sources to acquire an exact image of a territory of intrigue. It empowers the analyst to respond to questions, test speculations, and assess results. Strategies for information gathering can differ by field of study, yet exact information gathering is fundamental to guaran-

M. Umer · S. Sadiq · S. Ullah
Department of Computer Science, KFUEIT, Rahim Yar Khan, Pakistan

G. S. Choi
Department of Information and Communication Engineering, Yeungnam University, Gyeongsan 38542, Korea
e-mail: castchoi@ynu.ac.kr

A. Mehmood
Department of Computer Science, The Islamia University of Bahawalpur, Bahawalpur, Pakistan

M. Nappi (✉)
Department of Computer Science, University of Salerno, Fisciano, Italy
e-mail: mnappi@unisa.it

 
V. Chang et al. (eds.), *Modern Industrial IoT, Big Data and Supply Chain*, Smart Innovation, Systems and Technologies 218,
https://doi.org/10.1007/978-981-33-6141-6_7

tee the trustworthiness of research. The objective of information accumulation is to catch quality proof that at that point converts into rich information investigation, permitting the structure of persuading and valid answers to research questions. Information accumulation in the domain of suicide bombarding and related psychological oppression exercises is of crucial significance. Exact also, solid information can help stop psychological oppression exercises and actualize security approaches that can avert the development of fear-based oppressor gatherings. In the course of the most recent 15 years, Pakistan has been a casualty of many suicide bomb assaults, planted bomb assaults, gunfire and murdering occurrences, and numerous other fear-based oppression exercises. It has additionally been a piece of a few automaton strikes directed against the fear mongers in its region. To investigate the example of these episodes and to anticipate the following occurrence, dependable and state-of-the-art information is required. There are a few global associations that are gathering information on psychological oppression and related exercises; in any case, there are a few deficiencies of those sources. The Pakistan Body Count gathers and shares exact and significant information and insights on fear mongering incidents in Pakistan. To stop such events, it is therefore very important to identify victims, and some preventive measures should be taken. The following research questions will be answered in this research:

1. What number of individuals got murdered and harmed every year?
2. Is there any connection with suicide bombarding attacks with important occasions given in the dataset?
3. Is there any relationship between weight of explosive material and number of individuals murdered and harmed?
4. Is there any effect of type of holidays and attacks?
5. Is there any connection between Islamic date and attack day, time and number of victims?
6. Which are top ten targeted. Locations?
7. Would we be able to predict the next suicide bombarding attack?

Applying machine learning strategies for the distinguishing proof of suicide has developed as of late. Semantic Request and Word Count Version 2007 (LIWC2007) assess various words or passionate, psychological and basic English articulations introduced in oral and composed sentences of people. LIWC could be used to recognize the pattern of a passionate post. Ramirez-Esparza et al. [1] investigated on people for depression by comparing with non-depressed . They additionally discovered that discouraged individuals with medical problems. Assessment investigation has been dealt with a Whiz language handling task at numerous degrees of granularity. There has been a wide scope of research done on feeling examination, from standard-based, pack of words ways to deal with AI strategies. Sana Rasheed analyzes suicide bombing with the help of data science [2]. Zeeshan et al. [3–5] explore novel technique to find out pattern in suicidal bombarding. Irfan et al. [6] find out the new targets of suicidal attacks in Pakistan. In this work, we figure closeness of events in informational collection and preparing information. We utilize a few machine learning calculation to fabricate our work. In this specific situation, we center on critical, terrible contemplations and the reality of thinking by suicide.

This paper is organized as follows. Section 7.2 presents the material and methods of this research and visualization of suicidal attacks data. Section 7.3 discusses the experimental results and discussion. Section 7.4 summarizes the conclusion and gives an outlook on future work.

## 7.2 Methodology

Our research work is divided into following parts.

### *7.2.1 Dataset*

This dataset is based on the suicidal bombing attacks in Pakistan between 1995 and 2016. It contains information regarding 475 suicide attacks in which 6982 people killed and 17,624 injured. Columns of dataset are serial no, incident date, Islamic date, time, city, province, location, location based on type and sensitivity, target type and sect, open or close space, number of people killed, number of people injured, and number of suicide bombers [7, 8].

### *7.2.2 Classification Using Machine Learning*

Machine learning algorithms automate the classification process by using different algorithms and by processing on different type of data that is also known as training set. Performance of these machine learning algorithm is evaluated on test data. Supervised as well as un-supervised machine learning techniques are being used in different methods. Supervised learning techniques based on its characteristics can be applied to analyze the data of suicidal attacks. Random forest and logistic regression models are used for that purpose.

#### 7.2.2.1 Random Forest

RF is a tree classifier where each tree is the result of a random vector from the input vector. First, RF makes a forest by creating multiple decision trees on random features. Then, it aggregates voting from all decision trees in order to predict the class labels for the test data. It gives a higher weight to votes from a decision tree with a low error rate, and vice versa. By considering low error rate decision trees, this in turn reduces the chances of a wrong prediction [9, 10].

#### 7.2.2.2 Logistic Regression

LR is a statistical machine learning algorithm that classifies the data by considering the output variables on extreme ends, and it tries to make a logarithmic line that distinguishes between them [11, 12]. Owing to its name, the model is sometimes misunderstood as only used for regression. Instead, being a statistical model, it is also used for classification tasks. It is one of the simplest machine learning algorithms and provides great efficiency and low variance. Logistic regression is also used to extract significant features from the dataset. Logistic regression has the ability to get updated easily with new data using stochastic gradient descent.

### *7.2.3 Suicidal Attacks*

Data is visualized by applying random forest.

**1. What number of individuals got murdered and harmed every year?**

As we can clearly see that in Fig. 7.1, the highest average of people killed was in 2003. Nearly 35% of peoples killed in that year which was really a terrifying situation. The next graph explains that 2005 was the year in which a maximum number of the accident took place. But if we relate injured average ratio with killed people then it shows that most injured people were saved because the highest killed ratio was in 2003.

**2. Is there any connection with suicide bombarding attacks with important occasions given in the dataset?**

In Fig. 7.2, the stats obtained shows that the maximum number of people was killed either on the weekend or on Eid-ul-fitar celebrations. These are the events in which many peoples gather at one place, and this was an easy target for an attacker to create anarchy and to frighten the peoples of that country.

The same scenario was seen in Fig. 7.3. In these two events, the maximum number of peoples gets injured as the thronging of peoples always creates mess.

**3. Is there any relationship between the weight of explosive material and number of individuals murdered and harmed?**

The correlation of Fig. 7.4 shows that in most cases in which many peoples were killed or injured the explosive used was of less weight. The less weight explosives are easy to hide and to carry away in rush places in between the peoples.

**4. Is there any effect of the type of holidays and attacks?**

The beautifully designed count plot of Fig. 7.5 clearly explains to us that the maximum number of blast occurs on a working day.

**5. Is there any connection between Islamic date and attack day, time, and number of victims?**

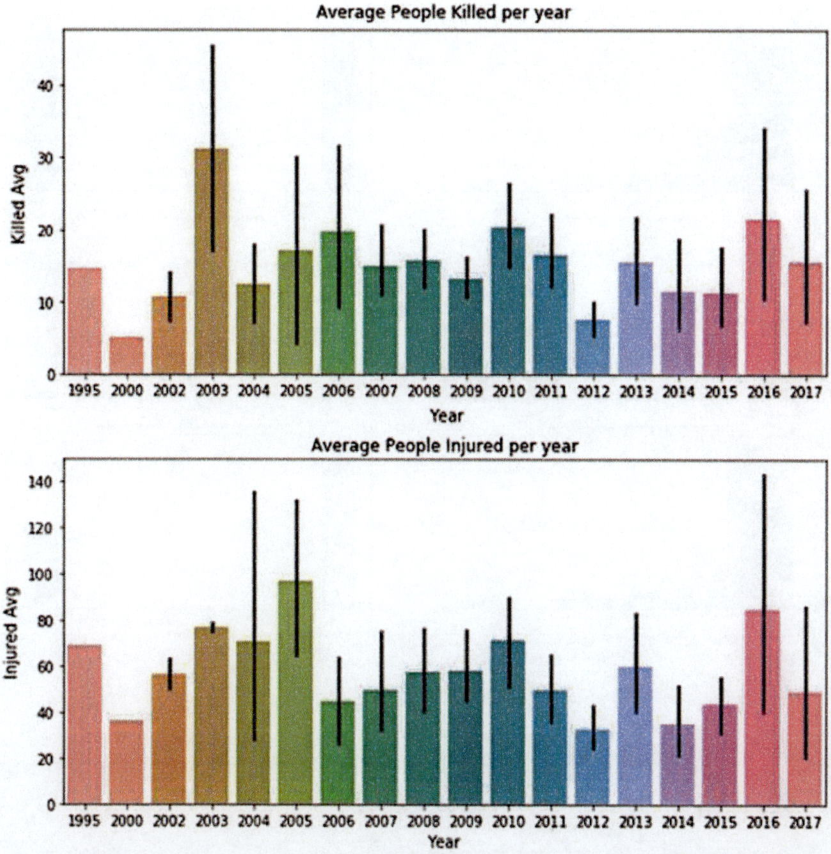


**Fig. 7.1** Average number of people killed and injured every year

Figure 7.6 builds a relation between killed or injured people with Islamic months. If we talk about the direct connection between them, then there is no truth lies behind that. But, if we look at the stats obtained, then Ramadhan is the month in which mostly people killed ratio is obtained. As in this Islamic month, saturation of people will remain at peak. Muslim peoples are busy in their worships as this month considered as the month of blessing for Muslims.

**6. Which are the top ten targeted? Locations?**

We know that terrorist always tries to weaken the defense of any country as much as they can. In Fig. 7.7, the results show that in most cases the target of a terrorist attack was police. 90% of the attack done was on the police department.

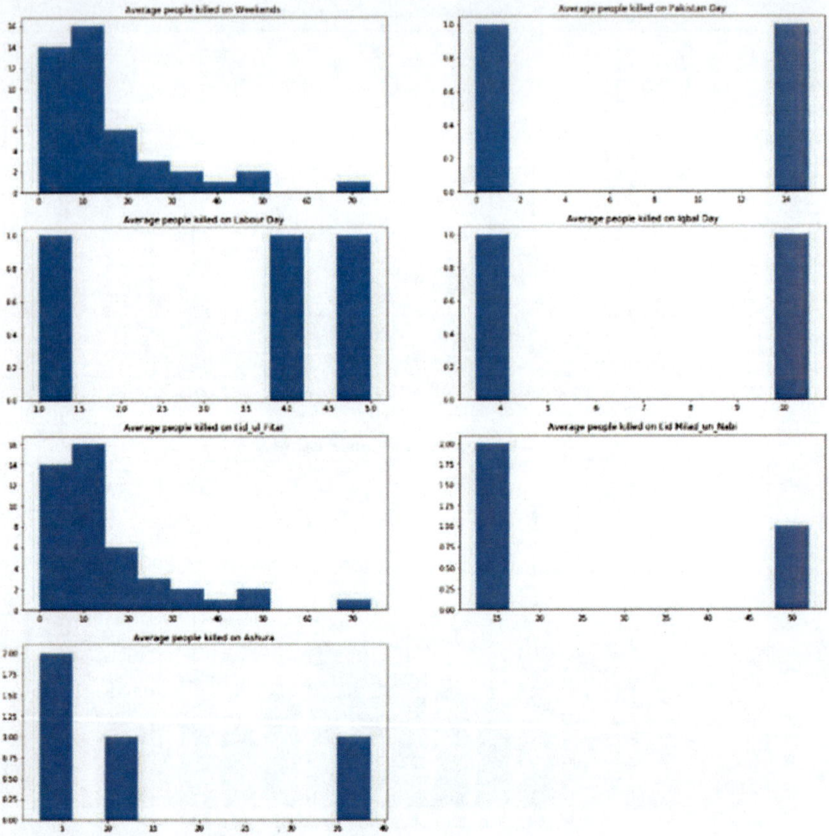

**Fig. 7.2** Correlation between average people killed on important events

## 7.3 Results and Discussion

In Figs. 7.8 and 7.9, the evaluation metric area under the curve shows that the ensemble method random forest classifier outperforms by giving an accuracy of 80.1%. As we know, logistic regression is used in cases when there is a linear relationship between the features and target but ensemble methods with cross-validation perform well in all cases. First, it builds multiple decision trees and then averages to find the best prediction tree. So, in our case, random forest classifier performs well.

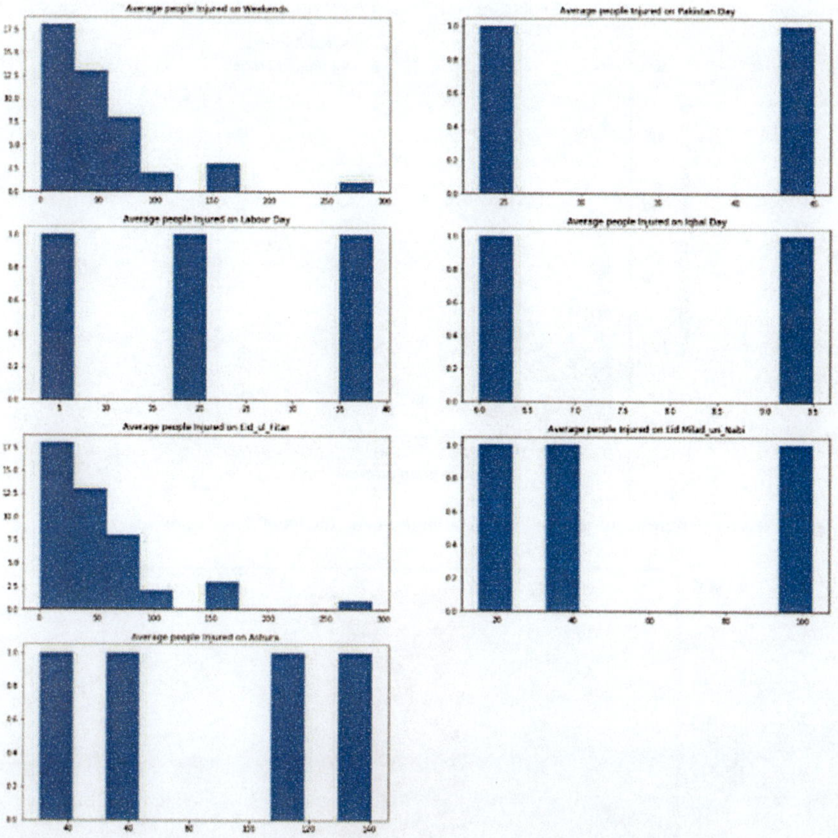


**Fig. 7.3** Correlation between average people injured on important events

## 7.4 Conclusion and Future Work

We analyzed the dataset of suicide bombing based on Data of Pakistan Body Count, an authenticated suicidal bombing data source of Pakistan. In this paper, we analyze the correlation between different variables of dataset such as average people killed and injured per year, suicide bombing attacks on important occasions, relationship between weight of explosive material and number of people killed and injured, type of holidays, important Islamic dates, and top targeted locations. Both machine learning models show good result in terms of accuracy, LR achieved 73%, and random forest achieved 80.1% and gives better results than LR. In future, deep learning techniques can be applied on this scenario.

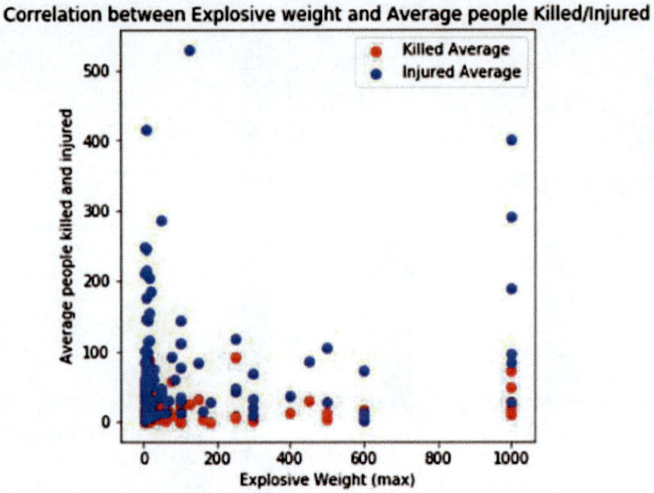


**Fig. 7.4** Average number of people killed and injured and the weight of explosives

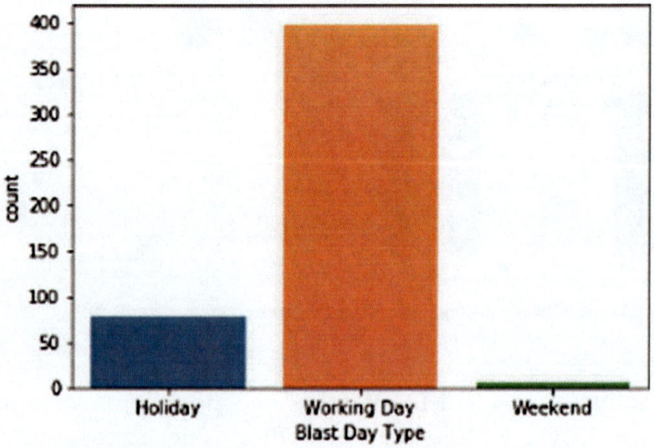


**Fig. 7.5** Blast day type

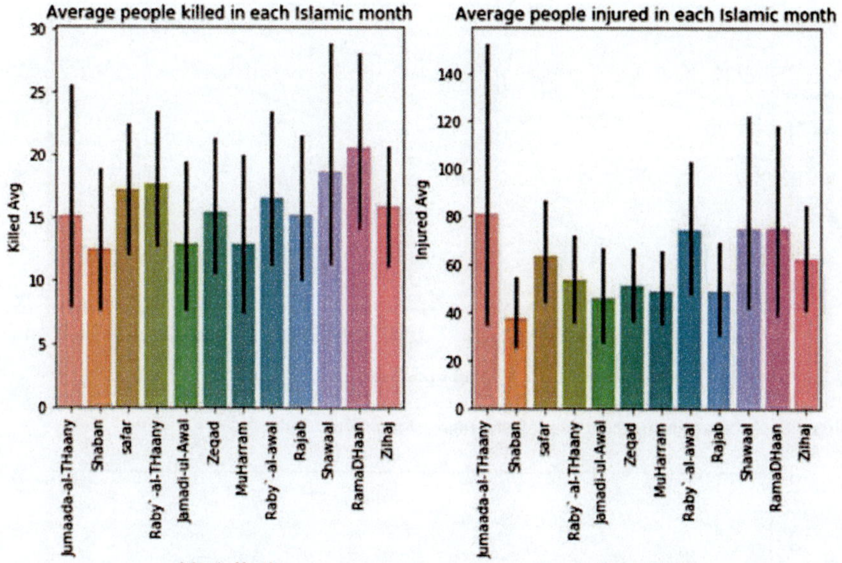


Fig. 7.6 Average people killed and injures in each Islamic month

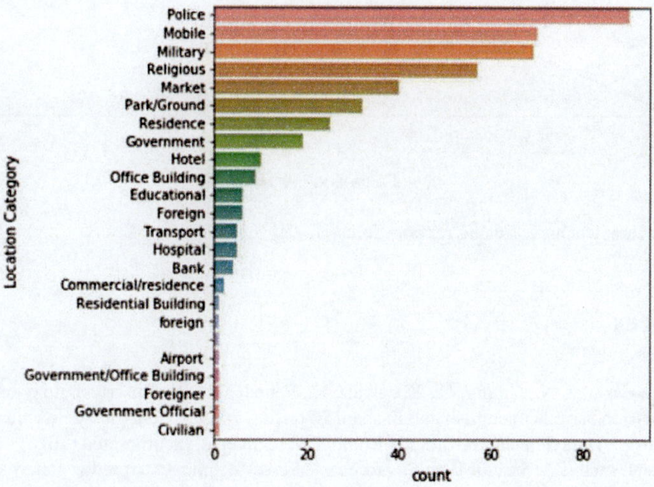


Fig. 7.7 Top locations of the blast

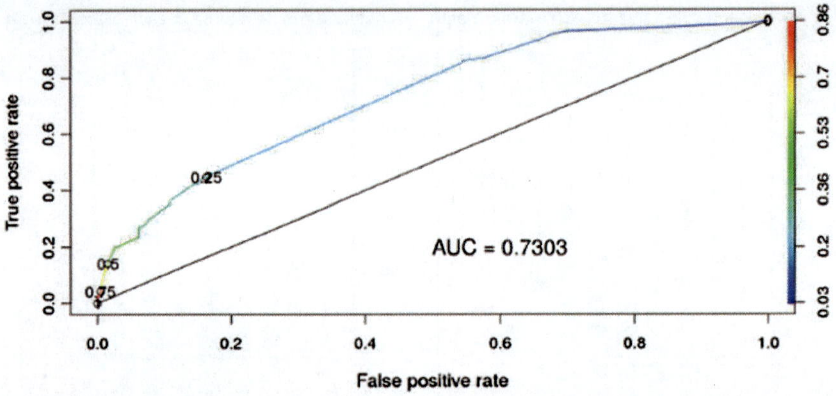


**Fig. 7.8** Accuracy achieved using logistic regression model

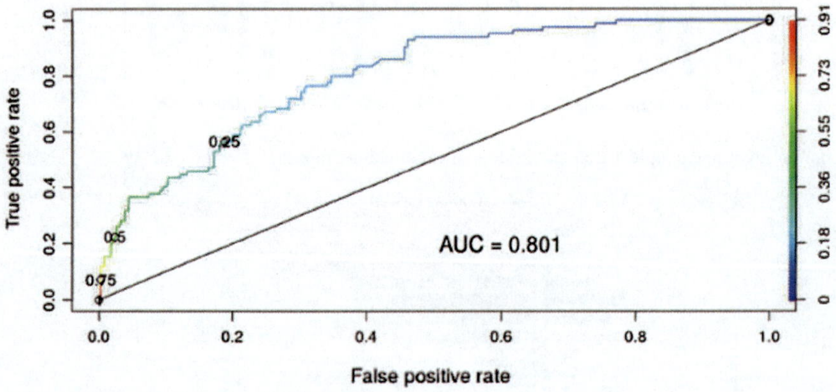


**Fig. 7.9** Accuracy achieved using random forest model

# References


1. Ramírez-Esparza, N., Chung, C., Kacewicz, E., Pennebaker, J.: The psychology of word use in depression forums in english and in spanish: testing two text analytic approaches (2008)
2. Rasheed, S.: Data Science for Suicide Bombings. iUniverse, Incorporated (2016)
3. ul hassan Usmani, Z.: Suicide bombing forecaster novel techniques to predict patterns of suicide bombing in pakistan. In: Autumn Simulation Multi Conference. Springer (2012)
4. Usmani, Z., Kirk, D.R.: Emergency 101 - suicide bombers, crowd formations and blast waves. In: MILCOM 2008–2008 IEEE Military Communications Conference (2008), pp. 1–7
5. Usmani, Z., Imana, E.Y., Kirk, D.: Random walk in extreme conditions–an agent based simulation of suicide bombing. In: IEEE Symposium on Intelligent Agents, 2009, pp. 114–121 (2009)
6. Irfan, M.: Suicide terrorism and its new target—Pakistan (06 2011)
7. Criteria, C.I., Variables, "Global terrorism database, study of terrorism and responses to terrorism. http://www.start.umd.edu/gtd/downloads/Codebook.pdf, 2015, online: Accessed 22 Apr 2020

8. START, "Global terrorism database," http://www.start.umd.edu/gtd/, 2015, online: Accessed 22 Apr 2020
9. Felix, N., Hruschka, E., Hruschka, E.: Tweet sentiment analysis with classifier ensembles. Decis Support Syst. **66**, 07 (2014)
10. Umer, M., Ashraf, I., Mehmood, A., Ullah, D.S., Choi, G.S.: Predicting numeric ratings for google apps using text features and ensemble learning. ETRI J. **07**, (2020)
11. Korkmaz, M., GÜNEY, S., YİGİTER, Ş.: The importance of logistic regression implementations in the turkish livestock sector and logistic regression implementations/fields. Harran Tarım ve Gıda Bilimleri Dergisi **16**(2), 25–36 (2012)
12. Sadiq, S., Mehmood, A., Ullah, S., Ahmad, M., Choi, G.S., On, B.-W.: Aggression detection through deep neural model on twitter. Future Gener. Comput. Syst. (2020)

# Chapter 8
# 4T4R Performance Analysis in Live FDD LTE Network

**Jun Wang and Dejun Mu**

**Abstract** A clear analysis of 4T4R performance data in a commercial FDD LTE network is presented in this paper. The analysis shows several important performance indicators affecting 4T4R performance, such as RRC_connected user, PRB utilization, cell throughput, user throughput, and RI ratio, how these indicators are related to modulation and coding scheme and the gain 4T4R bring compared with 2T2R. For this purpose, the involved indicators are extracted from network performance counters. Results show that 4T4R bring significant gain for live FDD LTE network.

## 8.1 Introduction

Cellular network will be significant role in Internet of thing (IoT) era and 5G start to focus on IoT on the initial phase. Ultra-reliable, low latency, and massive connections will be 5G's advantage for IoT. With the mature of 5G technique, operators start to develop 5G network but 5G network still in its infancy due to limitation of high cost of 5G spectrum auction and 5G equipment, penetration of 5G UE, and commercial pattern. Undoubtedly, 4G network bears most of data service. 4G traffic will explosively increase in the next three years and reach more than 75% of all mobile traffic based on Global Data report. Operators have to continue to invest in the expansion of 4G network in high-traffic area and hotspot area and the expansion methods are adding new band, building new site, spectrum refarming and new technique, such as multi-sector, massive MIMO.

The purchase of new spectrum faces expensive spectrum renting costs and new equipment purchase, which will be increase operators' investment. New sites need to rent new land and land privatization is the problem that operators have to face. As with spectrum refarming, it still cannot satisfy capacity requirement in high-traffic and hotspot area. New technologies are still introduced to improve spectrum efficiency and existing deployment are 4T4R, split sector, and massive MIMO.

J. Wang (✉) · D. Mu
Northwestern Polytech. University, Xi'an 710000, China
e-mail: 648051693@qq.com

 
V. Chang et al. (eds.), *Modern Industrial IoT, Big Data and Supply Chain*,
Smart Innovation, Systems and Technologies 218,
https://doi.org/10.1007/978-981-33-6141-6_8

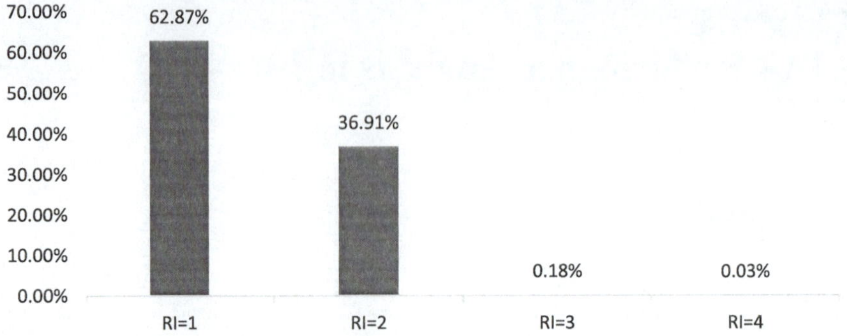


**Fig. 8.1** Ratio of different RI from 9 4T4R sites

Split sector is limited to deployment due to Mode 3 interference and massive MIMO equipment is very expensive and only are applied to high-value area. 4T4R will be a cost-effective method to improve spectrum efficiency for operator.

In commercial FDD LTE network, eNB mostly has two transmitters and two receivers (2T2R) and transmission mode always is configured as TM2, TM3, or TM4 [1]. 4 × 4 MIMO is triggered to greatly improve user's rate and experience with 4T4R eNB and 4R UE when the channel meet the correlation of four streams. Closed-loop TM4 is recommended for 4T4R eNB [1]. TM4 adopts wideband CRS to evaluate channel quality while TM9 adopt CSI-RS [1]. The signaling overhead of TM4 is reduced by 5% compared to TM9. TM4 can guarantee higher capacity when TM9 UE penetration is very low in commercial network. Currently, self-adaptive TM4 is activated in FDD LTE network.

Two scenarios will be considered based on UE's receivers when 4T4R are activated at eNB:

- 4 × 2 MIMO scenarios for 2R UE. Transmission mode includes space frequency blocking coding (SFBC), rank indicator (RI) = 1 closed-loop spatial multiplexing, RI = 2 closed-loop spatial multiplexing.
- 4 × 4 MIMO scenarios for 4R UE. Transmission mode includes SFBC, RI = 1 closed-loop spatial multiplexing, RI = 2 closed-loop spatial multiplexing, RI = 3 closed-loop spatial multiplexing, and RI = 4 closed-loop spatial multiplexing.

The method is based on combination of RI value and spectrum efficiency. Take 4 × 2 MIMO scenario as an example. If UE reports RI = 2, RI = 2 closed-loop spatial multiplexing will be determined. If UE reports RI = 1, eNB will chose SFBC or RI = 1 closed-loop spatial multiplexing based on spectrum efficiency.

As showed in Fig. 8.1 from 9 4T4R eNBs, RI = 3 and RI = 4 only occupy 0.21%, the penetration of 4R UE is still very low and 2R UE dominates in commercial network. 4 × 2 MIMO scenario dominates in 4T4R eNB. The max RI value of 4T4R eNB is the same as 2T2R eNB while 2R UE connects to eNB. In this paper, we focus on how much gain 4T4R eNB can bring for DL cell throughput compared with 2T2R eNB in current FDD LTE network (Fig. 8.1).

## 8.2 Principle and Definition

Link adaptation is a core process in FDD LTE network and its aim to select optimum modulation and coding scheme (MCS) [2] for users. Most existing link adaptation schemes consists of inner loop adaptation (ILLA) selector and outer loop link adaptation (OLLA) corrector [3]. ILLA selects suitable MCS based on channel quality indicator (CQI) UE reports. On the other hand, OLLA further adjusts MCS based on BLER to correct inaccuracy reported CQI [4–6] and effectively spectrum efficiency. Modified MCS can more accurately reflect channel quality and is used to evaluate cell throughput. In this paper, we carry on a detailed analysis on DL cell throughput on different DL average MCS values.

The number of RRC_Connected user needs to be considered in the calculation of DL average MCS, and it is big enough to effectively guarantee the accuracy of DL average value between different eNB. Therefore, the sampling data shall be meet the requirement that the number of RRC_Connected user is more than 60.

Cell traffic is cell throughput if cell physical resource block (PRB) utilization reaches 100%. The probability of 100% PRB utilization is extremely low, but we assume that cell traffic is cell throughput when PRB utilization is more than 99%.

In sum, DL average MCS, RRC_Connected user, PRB utilization, DL cell throughput, DL mean user traffic rate, RI = 2 ratio are key indicators in this paper:

- DL Average MCS
  DL Average MCS is defined as.

$$\text{DLAvgMCS} = \frac{\sum_{0}^{28} i * \text{NumMCS}i}{\sum_{0}^{28} \text{NumMCS}i} \tag{8.1}$$

  where NumMCS$i$ is scheduling number of MCS = $i$. in all considering time transmission intervals (TTI).
- Mean RRC_Connected user
  The numbers of users connect through RRC is incremented by on when eNB receives an RRC connection completion message from a UE. The number of users connected through RRC is decremented by one when eNB sends an RRC connection Release message to a UE. The number of users connected through RRC is incremented by one when a UE performs a handover from another cell to the local cell or perform reestablishment in the local cell. The number of users connected through RRC is decremented by one when UE performs a handover from the local cell to another cell or performs reestablishment in another cell. Sampling period is two seconds.
- Cell Throughput
  DL cell throughput is defined as

$$\text{CellTHP} = \frac{\text{TotalPDCPVol}}{\text{SchTime}} \tag{8.2}$$

where TotalPDCPVol is the total PDCP SDU data volume transferred in a cell with considering all TTIs [7]. SchTime is a counter. It counts the duration from the time data appears in the down link service buffer to the time the service buffer comes empty in a measurement period of all eNBs and sampling period is 1 ms.

- Average user throughput

  Average user throughput is defined as

$$\text{AvgUTHP} = \frac{\text{TotalPDCPU}}{\text{USchTime}} \tag{8.3}$$

where TotalPDCPU is the total downlink PDCP SDU dead load, not including packets lost at the Uu interface, tail packets and data completed through one data scheduling and counter UschTime counts the downlink service time during which eNB buffer becomes empty from be not empty, not including the data transmission time for tail packets and data competed through one data scheduling.

## 8.3 Results Analysis

2T2R sites and 9 4T4R sites are chosen from the same high-traffic area. Both of these sites are contiguous. 30 days' data is extracted from Band 3 network with 15 min gratuity. 406 samples of 2T2R and 765 samples of 4T4R meet our requirements that the number of RRC_Connected user is more than 60 and DL PRB utilization is more than 99%. The results analysis is as followed:

### *8.3.1 Sample Distribution of DL Average MCS*

As showed Fig. 8.2, the DL average MCS of 2T2R sites is better than that of 4T4R.

### *8.3.2 Mean RRC_Connected User*

As showed Fig. 8.3, the mean RRC_connected user of 4T4R sites is slightly more than that of 2T2R sites when DL average MCS 9, 10, and 12, and there are big differences between 4T4R and 2T2R when DL average MCS is 11 and above 12.

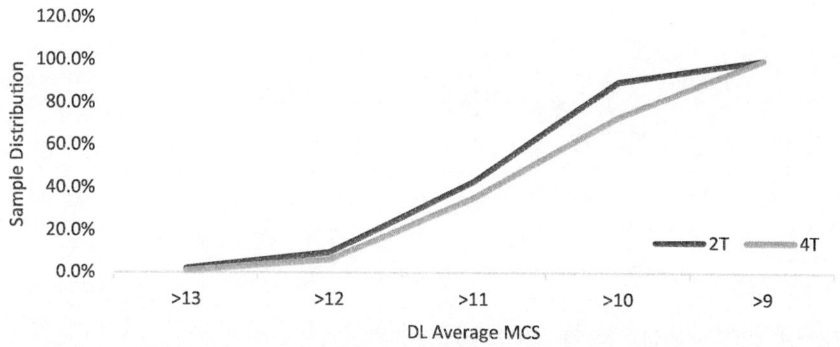


**Fig. 8.2** Sample distribution of different DL average MCS

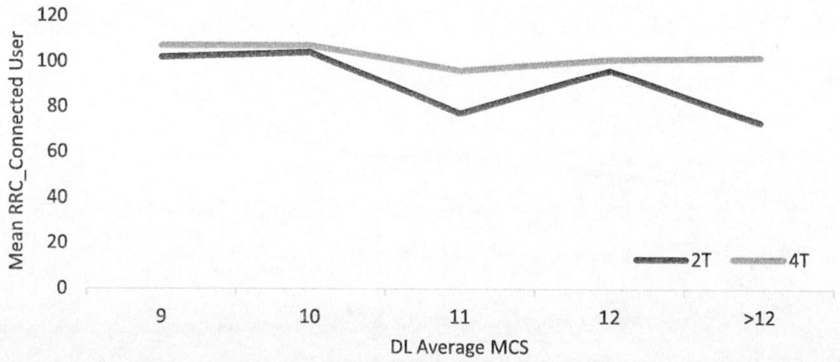


**Fig. 8.3** Mean RRC_Connected user of different DL average MCS

### *8.3.3 Cell Throughput*

As showed in Fig. 8.4, the cell throughput of 4T4R sites increases by 11.1, 17.1, 23.73, and 15.8% separately when DL Average MCS is 9, 10, 11, and 12. Overall, 4T4R cell throughput increases by 14.1%.

### *8.3.4 Average User Traffic Rate*

The average user traffic rate of 4T4R sites increases by 16.3, 34.9, 41.9, and 57.3% separately when DL Average MCS is 9, 10, 11, and 12. Overall, 4T4R cell throughput increases by 26.7% (Fig. 8.5).

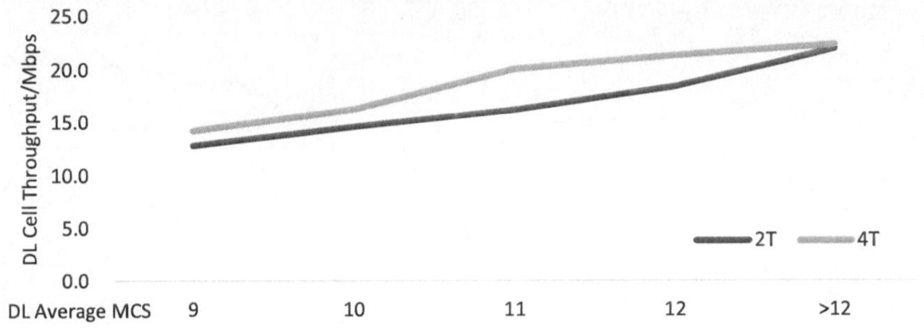


**Fig. 8.4** Cell Throughput of different DL average MCS

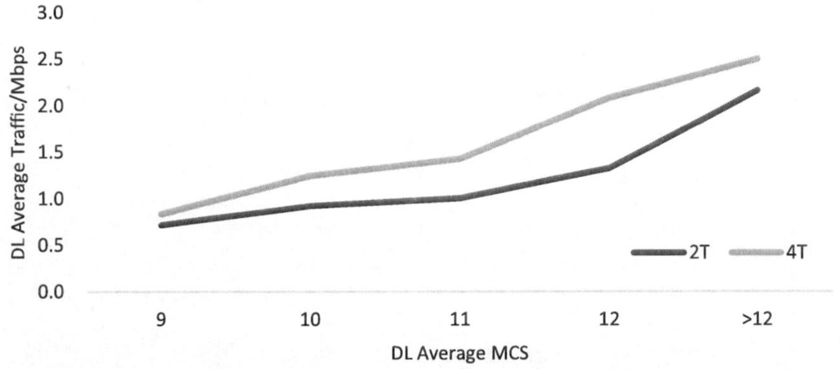


**Fig. 8.5** Average user traffic rate of different DL average MCS

### *8.3.5 RI = 2 Ratio*

As showed Fig. 8.6, undoubtedly, 4T4R greatly increase the ratio of RI = 2.

Although DL average MCS of 4T4R is slightly worse than that of 2T2R, 4T4R still bring 14.1% gain in cell throughput and 26.7% gain in user traffic rate compared with 2T2R. Based on Fig. 8.5, the significant gain come from the increasement of RI = 2 ratio, in other words, 4 transmitters significantly increase the probability of RI = 2.

## 8.4 Conclusion and Perspective

4T4R can bring 10–24% gain in cell throughput and 16–57% gain in user traffic rate when the penetration of 4R UE is still low. 4T4R significantly improve network capacity and user experience. 4T4R can be deployed in hotspot area and high-traffic

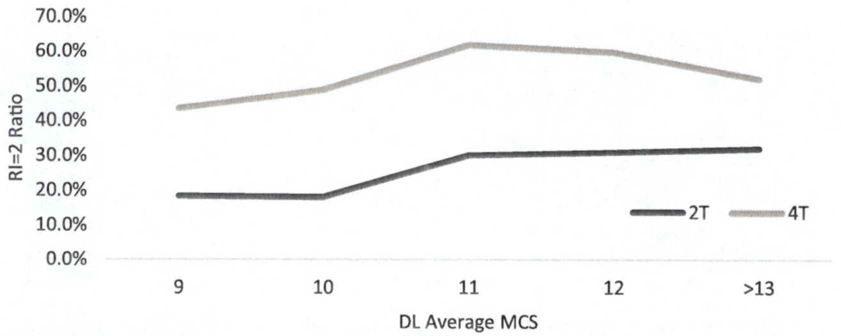


**Fig. 8.6** RI = 2 ratio of different DL average MCS

area to increase capacity and in high-value area to guarantee VIP user experience. With the development of 4R UE, the gain 4T4R bring will be more and more significant.

## Reference

1. 3GPP TS 36.212: Evolved Universal Terrestrial Radio Access (EUTRA); Multiplexing and channel coding (Release 15), 3GPP, https://portal.3gpp.org/desktopmodules/Specifications/SpecificationDetails.aspx?specificationId=2426# (2019)
2. 3GPP TS 36.211: Evolved Universal Terrestrial Radio Access (EUTRA); Physical channels and modulation (Release 15). 3GPP, https://portal.3gpp.org/desktopmodules/Specifications/SpecificationDetails.aspx?specificationId=2424 (2019)
3. Pedersen, K., et al.: Frequency domain scheduling for OFDMA with Limited and Noisy Channel Feedback. In: Proceedings of 66th IEEE Transactions Vehicular Technology Conference, pp. 1792–1796. IEEE, https://ieeexplore.ieee.org/document/4350027 (2007)
4. Morales-Jimenez, D., Sanchez, J.J., Gomez, G., Aguayo-Torres, M.C., Entrambasaguas, J.: Imperfect adaptation in next generation OFDMA cellular systems. J. Internet Eng. **3**(1), 202–209 (2009). CiteSeerx. https://citeseerx.ist.psu.edu/viewdoc/summary?doi=10.1.1.706.8626 (2009)
5. Ni, M., Xu, X., Mathar, R.: A channel feedback model with robust SINR prediction for LTE systems. In: Proceedings 7th EuCAP, pp. 1866–1870. ResearchGate. https://www.researchgate.net/publication/261313450_A_channel_feedback_model_with_robust_SINR_prediction_for_LTE_systems (2013)
6. Aho, K., Alanen, O., Kaikkonen, J.: CQI reporting imperfections and their consequences in LTE networks. In: Proceedings 10th International Conference Netw., pp. 241–245. IARIA. www.thinkmind.org/download.php?articleid=icn_2011_12_10_10192 (2011)
7. Buenestado, V., Ruiz-Aviles, J.M., Toril, M., Luna-Ramirez, S., Mendo, A.: Analysis of throughput performance statistics for benchmarking LTE networks. IEEE Commun. Lett. **18**(9), 1607–1610. IEEE. https://ieeexplore.ieee.org/document/6857393 (2014)

# Chapter 9
# An Optimal Equilibrium Solution Approach for Bi-level Nonlinear Programming Problem Using Lingo Software

**Qian Wang, Mingfa Zheng, and Bingjie Li**

**Abstract** In order to solve a bi-level nonlinear programming problem (BL-NLP) satisfying the certain convexity conditions, a new method is proposed in this paper. This method uses the concepts called optimal equilibrium solution and optimal equilibrium value for getting the Pareto optimal solution of BL-NLP. It has two main features. First, within lower-level decision-makers (LDM) [or within higher-level decision-makers (HDM)], for any fixed variable of HDM (or for any fixed variable of LDM), the optimal equilibrium solution is a Pareto optimal solution within LDM (or within HDM), and the optimal equilibrium value is a minimal uniform concession of all the decision-makers within LDM (or HDM) relative to their individual optimal objectives. Thus, the optimal equilibrium solution is an impartial solution within LDM (or HDM).

## 9.1 Introduction

We often regard the bi-level nonlinear programming problem (BL-NLP) as a static version of Stackelberg strategy which adopts the leader-follower game method. In the game, under the premise of giving the rational response of the follower, i.e., LDM, Stackelberg strategy is adopted by the leader, i.e., HDM. This question describes the hierarchy that naturally occurs in the real world such as firm budget, network facility location, electric power pricing, environmental regulation, credit allocation, optimal structure design, and so on (see Refs. [1–4]). The problem of bi-level programming is difficult to solve because the lower-level scheme has implicit dependence on the higher-level scheme. Therefore, many algorithms are limited to subclass problems with specific properties such as linear (see Refs. [5, 6]), quadratic, or convex objectives (or constraint functions) (see Refs. [7, 8]). In addition, the researchers put forward many heuristics or approximation approaches to solve the BL-NLP under different assumptions such as branch-and-bound technique (see Refs. [9, 10]),

Q. Wang (✉) · M. Zheng · B. Li
Department of Basic Science, Air Force Engineering University, Xi'an 710051, China
e-mail: wangqian5444@163.com



V. Chang et al. (eds.), *Modern Industrial IoT, Big Data and Supply Chain*, Smart Innovation, Systems and Technologies 218,
https://doi.org/10.1007/978-981-33-6141-6_9

decent methods (see Refs. [1, 2]), penalty methods (see Refs. [3, 4]), and trust-region method (see Refs. [8]). In our work, inspired by the above literatures, we propose a new technique by means of LINGO software package. As we all know, LINGO is a comprehensive tool for building and solving linear, nonlinear, and integer optimization models faster, easier, and more effective. A fully integrated package is provided in LINGO, which includes a powerful language for describing optimization models, a fully functional environment for building and editing problems, and a set of fast built-in solutions. In our study, LINGO is used to solve single-objective programming.

## 9.2 Solution Concept

BL-NLP problem is considered as follows:

$$\min_{x,y} g(x, y) = (g_1(x, y), g_2(x, y), \ldots, g_p(x, y)) \tag{9.1}$$

subject to

$$y \in \arg\min_{y'} h(x, y') = (h_1(x, y'), h_2(x, y'), \ldots, h_q(x, y')) \tag{9.2}$$

$$M = \left\{(x, y)\middle|m_j(x, y) \le 0;\ j = 1, 2, \ldots, N\right\} \tag{9.3}$$

where $x = (x_1, x_2, \ldots, x_n)$ is a vector variables indicating the choice of leader, and $y = (y_1, y_2, \ldots, y_m)$ is a vector variables represents the follower's choice. The function $f : R^n \times R^m \to R^p$ is the function of upper-level objective, and $h : R^n \times R^m \to R^q$ is the function of lower-level objective. The LDM has $p$ and $q$ objective functions, respectively. $g_i, h_i$, and $m_j$ are continuously differentiable functions. The name of $M$ is the constraint set.

The definitions of the solution for the BL-NLP can be given (see Refs. [5, 9])

**Definition 2.1** For any $x \in M_u = \{x|(x, y) \in M\}$ given by HDM, suppose that the variable $y \in M_L = \{y|(x, y) \in M\}$ is the LDM's Pareto optimal solution, then we call them $(x, y)$ as a feasible solution of the BL-NLP.

**Definition 2.2** If $(x^*, y^*)$ is a feasible solution of the BL-NLP's Pareto optimal solution, is $(x^*, y^*)$, no other feasible solution $(x, y) \in M$ exists, so that $g_i(x, y) \le g_i(x^*, y^*)$ for all $i \in \{1, 2, \ldots, P\}$ with $g_i(x, y) < g_i(x^*, y^*)$ for at least one index.

***Remark 2.1*** The equilibrium value denotes that all the objectives within LDM have a uniform ascend $s$ relative to their individual optimal objectives $\overline{h_i^*}$. In other words, all the decision-makers within LDM make a uniform concession value $s$. Obviously, if $\hat{y}$ is the $s$–equilibrium solution of the problem (9.4), then for any $s' > s$, $\hat{y}$ is also the $s'$–equilibrium solution of the problem (9.4).

***Remark 2.2*** The optimal equilibrium value $s^*$ is a minimal uniform concession of all the decision-makers within LDM relative to their individual optimal objectives $\overline{h}_i^*$. We are interested in the optimal equilibrium solution $y^*$. If there have the individual optimal solutions of the problems $(P_i)(i = 1, 2, \ldots, q)$ within LDM, respectively, then there exists the optimal equilibrium value $s^*$ and optimal equilibrium solution $y^*$ to the problem (9.4).

***Remark 2.3*** Theorem 3.1 explains that the method of the single-objective programming problem $(\overline{P})$ is equivalent to the optimal equilibrium solution to the problem (9.4).

***Definition 2.3*** Denote all the Pareto optimal solution to the problem (9.4) as the Pareto optimal set $(P^*)$. We define the Pareto front as follows (see [7]):

$$\mathrm{PF}^* = \left\{h(x, y) = (h_1(x, y), h_2(x, y), \ldots, h_q(x, y))\middle| y \in P^*\right\} \tag{9.4}$$

**Definition 2.4** Denote all the optimal equilibrium solutions as the optimal equilibrium solution set (OE*). We define the optimal equilibrium front as follows

$$\mathrm{OEF}^* = \left\{h(x, y) = (h_1(x, y), h_2(x, y), \ldots, h_q(x, y))\middle| y \in \mathrm{OE}^*\right\} \tag{9.5}$$

Theorem Suppose that $h_i(x, y), i = 1, 2, \ldots, q$ are convex functions, and the constraint set $H$ is compact convex set, then the optimal equilibrium solution set $(OE^*)$ is a connected convex set. Suppose that the functions $h_i(x, y), i = 1, 2, \ldots, q$ are strictly convex functions, and the constraint set $M$ is a compact convex set, then there exists a unique $s^*$-optimal equilibrium solution $y^*$.

***Proof*** Apparently, the set $M_L$ is also a compact convex set. Suppose that $y_1^*$ and $y_2^*$ are two $s^*$-optimal equilibrium solutions, then for any $(\bar{x}, \bar{y}) \in M$, we have $h_i(x, y_1^*) - s^* \le h_i(\bar{x}, \bar{y})$ and $h_i\left(x, y_2^*\right) - s^* \le h_i(\bar{x}, \bar{y}), i = 1, 2, \ldots, q$. For any $\rho \in (0, 1)$,

$\rho(x, y_1^*)+(1-\rho)(x, y_2^*) \in M$ and $\rho y_1^*+(1-\rho)y_2^* \in M_L$ hold due to the convexity of the set $M$, and $h_i(x, \rho y_1^* + (1 - \rho)y_2^* \le \theta h_i(x, \rho y_1^*) + (1 - \rho)h_i(x, \rho y_2^*)$ holds due to the convexities of the functions $h_i, i = 1, 2, \ldots, q$. Hence, we have

$$h_i(x, \rho y_1^* + (1 - \rho)y_2^* \le \rho(s^* + h_i(\bar{x}, \bar{y}) + (1 - \rho)(s^* + h_i(\bar{x}, \bar{y}) = s^* + h_i(\bar{x}, \bar{y})$$

Obviously, $\rho y_1^* + (1 - \rho)y_2^*$ is also a $s^*$-optimal equilibrium solution, which implies that the optimum equilibrium issue set (OE*) is a connected convex set.

From above, if $h_i(x, y), i = 1, 2, \ldots, q$ is strictly convex functions, we know that there exists a unique $s^*$-optimal equilibrium solution $y^*$.

## 9.3 Optimal Equilibrium Solution Method for BL-NLP

For any variables $y$ satisfying (12), we solve the problem

$$\min_{x \in M_U} g(x, y) = (g_1(x, y), g_2(x, y), \ldots, g_p(x, y)) \tag{9.6}$$

This problem is also a multiple objective optimization problem, whose optimal solution should be obtained. It is similar to Sect. 9.3 that the optimal equilibrium solution of the problem (9.6) should be obtained.

We consider the individual optimal solutions within HDM, namely the single-objective optimization problems $(P_{M,i})(i = 1, 2, \ldots, p)$

$$\min_{(\bar{x}, \bar{y}) \in M} f_i(\bar{x}, \bar{y}) \tag{9.7}$$

Let $\overline{g_i}^* > -\infty$ be the optimal objectives of the problems $(P_{M,i})(i = 1, 2, \ldots, p)$, respectively, we consider the single-objective programming problem $(\overline{P}_M)$ as follows

$$\begin{cases} \min \beta \\ \text{subject to} \\ g_i(x, y) - \beta \le \overline{g_i}^*, \ i = 0, 1, 2, \ldots, p \\ \beta \ge 0, \ x \in M_U \end{cases} \tag{9.8}$$

where $\beta, x$ are the variables of the problem $(\overline{P}_M)$. Obviously, by solving the problem $(\overline{P}_H)$, we can obtain the optimal equilibrium solution of the problem (9.6) for any variables satisfying (12). Hence, by replacing the lower-level problem with the Karush-Kuhn-Tucker conditions (12) (see Refs. [13–15]), we can get a Pareto optimal solution of the BL-NLP by solving the following problem …

$$\begin{cases} \min \beta \\ \text{subject to} \\ \beta > 0; \\ S > 0; \\ g_i(x, y) - \beta \le \overline{g_i}^*, \ i = 0, 1, 2, \ldots, p \\ 1 - \sum_{i=1}^{q} \lambda_i = 0; \\ \sum_{i=1}^{q} \lambda_i \nabla_y h_i(x, y) + \sum_{j=1}^{N} \mu_j \nabla_y m_j(x, y) = 0; \\ h_i(x, y) - s - \overline{h_1}^* \le 0; \ i = 0, 1, 2, \ldots, q \\ \lambda_i \ge 0; \ i = 0, 1, 2, \ldots, q \\ \lambda_i \left( h_i(x, y) - s - \overline{h_1}^* \right) = 0; \ i = 0, 1, 2, \ldots, q \\ m_i(x, y) \le 0; \ j = 1, 2, \ldots, N \\ \mu_j \ge 0; \ j = 1, 2, \ldots, N \\ \mu_j m_i(x, y) = 0 \ j = 1, 2, \ldots, N \end{cases} \tag{9.9}$$

***Example*** $h_1(x, y) = \frac{1}{2}y_1^2 + (y_2 - 1)^2 + xy_1 + 2xy_2 + (x - 4)^2$, $h_2(x, y) = x^2 + (y_1 - 3)^2 + (y_2 - 2)^2$, where $y = (y_1, y_2)$, for fixed variable $x = 1$ and $x = 2$, the following problem can be considered as

$$\min_{y_1, y_2 \in R_+} h(x, y) = (h_1(x, y), h_2(x, y))$$

Obviously, decision-maker 1 has the individual optimal solution $(x, y_1, y_2) = (4, 0, 0)$, and the decision-maker 2 has the individual optimal solution $(x, y_1, y_2) = (0, 3, 2)$. It is easy to get the individual optimal objectives $\overline{h_1}^* = 1, \overline{h_2}^* = 0$, respectively.

For fixed $x = 1$, by solving the problem $(\overline{P})$, we get the optimal equilibrium solution is $y_1^* = 0.4724$, $y_2^* = 0.4511$, and the optimal equilibrium value is $s^* = 9.7876$. It is easy to validate that $\omega \approx (0.7744, 0.2256)$ is its weight vector of corresponding problem (11). This weight vector is automatically formed in virtue of solving the optimal equilibrium solution. Similarly, for fixed $x = 2$, by solving the problem $(\overline{P})$, we get the optimal equilibrium solution is $y_1^* = 1.3541$, $y_2^* = 0.5140$, and the optimal equilibrium value is $s^* = 8.9172$. It is easy to validate that $\omega \approx (0.4952, 0.5048)$ is its weight vector of corresponding problem (11). This weight vector is automatically formed in virtue of solving the optimal equilibrium solution.

## 9.4 Conclusion

In this study, in order to solve BL-NLP problem, a new method, that is, the optimal equilibrium solution approach, was proposed. We also obtain that the Pareto optimal solution of the BL-NLP can be get by solving the single-objective programming problem $(\widetilde{P})$. This solution insures that impartiality for all the decision-makers within LDM (or HDM) have a uniform concession value—the optimal equilibrium value $s^*$ (or $\beta^*$) relative to their individual optimal objectives $\overline{h_i}^*, i = 0, 1, 2, \ldots, q$ (or $\overline{g_i}^*$, $i = 1, 2, \ldots, p$). Further, we take full advantage of the LINGO software package as a basic tool for solving all the single-objective programming problems. In this paper, the numerical examples are very preliminary. Further practical applications will be necessary.

**Acknowledgements** Author wishes to acknowledge assistance from Department of Air Force Engineering University Basic Sciences Air Force Engineering University Foundation (JK2020103).

## References


1. Bard, J.F.: Practical Bilevel Optimization Mathematical and Applications. Kluwer Academic Publisher (1998)
2. Migdalas, A.: Bilevel programming in traffic planning: models, methods and challenge. J. Glob. Optim. **7**, 381–405 (1995)
3. Vicente, L.N., Calamai, P.H.: Bilevel and multilevel programming: a bibliography review. J. Glob. optimi. **5**, 291–306 (1994)
4. Bard, J.F., Plummer, J., Sourie, J.C.: A bi-level programming approach to determining tax credits for biofuel production. Eur. J. Oper. Res. **120**, 30–46 (2000)
5. Hansen, P., Jaumard, B., Savard, G.: New branch and bound rules for linear bilevel programming. SIAM J. Sci. Stat. Comput. **13**, 1194–1217 (1992)
6. Wen, U., Hsu, S.: Linear bi-level programming problems-a review. J. Oper. Res. Soc. **42**, 125–133 (1991)
7. Bard, J.F.: Convex two-level optimization. Math. Program. **40**, 15–27 (1988)
8. Dempe, S., Brand, J.F.: Bundle trust-region algorithm for bilinear bilevel programming. J. Optim. Theory Appl. **110**, 256–288 (2001)
9. Edmunds, T.A., Bard, J.F.: Algorithms for nonlinear bilevel mathematical programs. Syst., Man Cybern., IEEE Trans. **21**, 83–89 (1991)
10. Thoai, N.V., Yamamoto, Y., Yoshise, A.: Global optimization method for solving mathematical programs with linear complementarity constraints. J. Optim. Theory Appl. **124**, 467–490 (2005)

# Chapter 10
# A Comparative Study on Transformer Versus Sequence to Sequence in Machine Translation

**Hao Jiang, Su Zhao, Da Fang, Chao Zhang, and Jianjin Duan**

**Abstract** Machine translation is the process of translating one natural language into another natural language. In the experiment, machine translation tasks were performed on the English to German data set and the English to Thai data set through the sequence-to-sequence model, the sequence-to-sequence model with attention mechanism and the transformer model. Through the analysis of the experimental data, it is concluded that the transformer model is not only better than the first two models in the performance of machine translation, but also the structural characteristics of the transformer model. When the data set is a relatively rare English to Thai data set, the transformer model is collected, and the impact of the result is less than the first two, which proves that the transformer model improves the quality of machine translation.

## 10.1 Introduction

Machine translation is an important research direction in the field of natural language processing. Machine translation, also known as automatic translation, is the process of using a computer to convert a source language into another natural language. It is a branch of computational linguistics and has important scientific research value.

H. Jiang · S. Zhao · D. Fang · C. Zhang · J. Duan (✉)
Yunnan Normal University, Kunming, China
e-mail: 529542202@qq.com

H. Jiang
e-mail: Jh072535@foxmail.com

S. Zhao
e-mail: zhaosusan@foxmail.com

D. Fang
e-mail: fangda0006@163.com

C. Zhang
e-mail: zhchynnu@foxmail.com

 
V. Chang et al. (eds.), *Modern Industrial IoT, Big Data and Supply Chain*, Smart Innovation, Systems and Technologies 218,
https://doi.org/10.1007/978-981-33-6141-6_10

The concept of machine translation was first proposed by the French scientist GB Artsouni in the early 1930s [1]; the first translation machine was invented by the Soviet scientist Troyansky [2], but limited by the level of science and technology, the translation machine was not really realized. The function of machine translation, with the emergence of the first modern electronic computer ENIAC in 1946, American scientist W. Weaver and British engineer AD Booth proposed the idea of automatic translation [3]. In 1954, Georgetown University in the USA cooperated with IBM and used the IBM-701 computer to complete the English–Russian machine translation experiment for the first time [4]. However, in 1996, the ALPAC committee published a report in November [5], which completely denied the feasibility of machine translation and brought machine translation research into a period of depression. However, as the exchanges between countries have become closer, the need for language communication has become more and more urgent, and the application of artificial intelligence in the field of natural language processing has revived machine translation research and has developed to this day. The original machine translation system can be divided into two types: rule-based [6] and corpus-based [7]. However, the rule-based machine translation system has always had the disadvantages of poor translation quality, the system is difficult to improve, and the design is cumbersome; the corpus-based machine translation system also has the problem of relying too much on the support of large-scale data sets. At the same time, with the development of deep learning, in 2013, machine translation based on artificial neural networks [8] surpassed the first two systems in translation performance, and the translation quality was significantly improved.

Judging from the history of the development of artificial neural networks, in 1943, W. S. McCulloch and W. Pitts established the neural network model for the first time [9]. In 1949, the perceptron was proposed [10]. However, after carefully analyzing the functions and limitations of the neural network system represented by the perceptron, M. Minsky et al. pointed out that the perceptron cannot solve the problem of higher-order predicates [11]. In 1982, J. J. Hopfield proposed the Hopfield neural network model and introduced the concept of "computational energy" [12]. In 1984, he proposed the Hopfield neural network model under continuous time [13]. In 1985, the Boltzmann model was born [14]. In 1986, scientists conducted cognitive microstructure research and proposed the theory of parallel distributed processing [15]. In 1986, Rumelhart, Hinton and Williams proposed the BP algorithm [16]. In 1988, Linsker proposed a new self-organization theory for perceptron networks [17] and formed the theory of maximum mutual information on the basis of Shanon's information theory [18]. In 1988, Broomhead and Lowe used radial basis functions to propose a hierarchical network design method [19]. In 1990, Jeffrey Elman used the fully connected recurrent neural network for the first time [20]. In the same year, Paul Werbos proposed the BP algorithm over time [21]. In 1991, Sepp Hochreiter discovered that the recurrent neural network has a long-term dependency problem. When learning a long sequence, the recursive neural network will have the phenomenon of gradient explosion and gradient disappearance, and it is impossible to grasp the nonlinear relationship of the long-term span [22]. In 1997, in response to this problem, the long and short-term memory network was born [23]. Entering the twenty-first century, Bengio proposed a sequence-to-sequence model in 2014 [24], through the basic recurrent neural network structure combined with long and short-term memory network algorithms to construct an encoder and decoder, and this model realizes translation in the field of machine translation quality improvement. In 2017, Google proposed an attention mechanism based on the sequence-to-sequence model [25], which improved the quality of translation. At the same time, Google, combining the sequence-to-sequence model and the attention mechanism, proposed a new model transformer based on the encoder and decoder structure and applying the attention mechanism to improve the quality of translation.

At present, recurrent neural networks and long short-term memory networks are still widely used in machine translation tasks, but the performance of transformers in machine translation tasks not only surpasses the former two but also achieves efficient parallelization.

In this paper, we compare the sequence-to-sequence model, the sequence-to-sequence model with attention mechanism and the transformer model on two data sets of English to German and English to Thai to evaluate the replacement performance [26]. The performance optimization range of the model and the sequence-to-sequence model with the attention mechanism added on the machine translation task and explored how the corpus of different language families affects the BLEU scores of the three models on the machine translation.

## 10.2 Basic Model

### *10.2.1 Sequence-to-Sequence Model*

Traditional recurrent neural networks have strict requirements on the length of the input sequence and output sequence in machine translation tasks, that is, the input and output sequences must be the same length. In order to solve this problem and flexibly perform machine translation tasks in multiple scenarios at the same time, the sequence-to-sequence model was born. The sequence-to-sequence model is an artificial neural network model based on the encoder and decoder. The two ends of the encoder and the decoder are constructed in the form of a convolutional neural network or recurrent neural network (Fig. 10.1).

In the sequence-to-sequence model, the output sequence $Y$ is determined according to the input sequence $X$. The encoder refers to the conversion of the input sequence into a fixed-length vector, and the decoder refers to the conversion of the fixed-length vector $C$ obtained from the input sequence into an output sequence. Because in the basic unit recurrent neural network of the sequence-to-sequence model, the hidden state $h_i$ of the coding layer at the current moment is determined by the hidden state of the coding layer at the previous moment and the input sequence at the current moment; the fixed-length semantic vector is hidden by each code in the coding stage The layer state is summarized; therefore, the output sequence at a

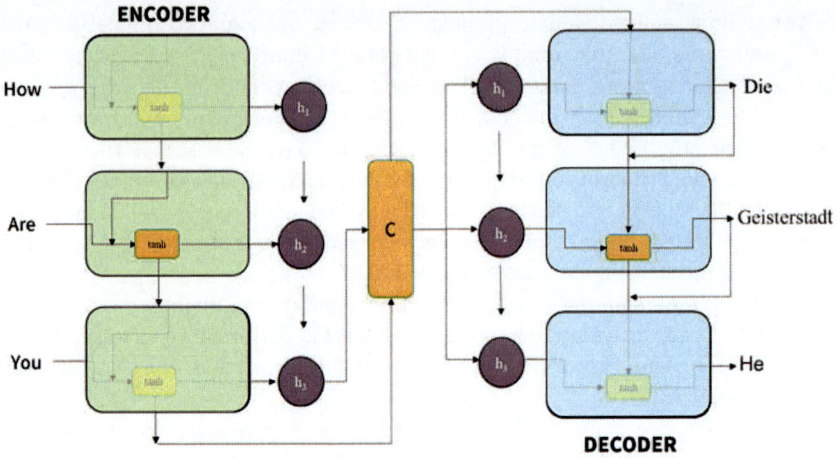


**Fig. 10.1** Sequence-to-sequence model based on RNN

certain moment in the decoding stage is obtained by the nonlinear activation function from the current decoding hidden layer state combined with the semantic vector output by the encoding layer and the output sequence of the decoding layer at the previous time. The above formula is as follows:

$$h_t = H(h_{t-1}, x_t) \tag{10.1}$$

$$C = \mathrm{Sum}(h_1, h_2, \ldots, h_i) \tag{10.2}$$

$$y_t = \mathrm{sigmoid}(y_{t-1}, C, h_t) \tag{10.3}$$

### *10.2.2 Attention Mechanism*

The basic unit of the sequential to sequential model is mainly composed of a recurrent neural network. Therefore, the sequence-to-sequence model has the advantages of a recurrent neural network. At the same time, the sequence-to-sequence model also solves the sequence-matching problem that the recurrent neural network is difficult to solve in the past (Fig. 10.2).

However, the sequential to sequential model relies too much on the final hidden layer information on the decoder side. At the same time, in the encoding process, the semantic vector may miss some important details because the input sequence is too long, which affects the quality of translation. Therefore, researchers hope that in the face of machine architecture, the sequential to sequential model can be like the

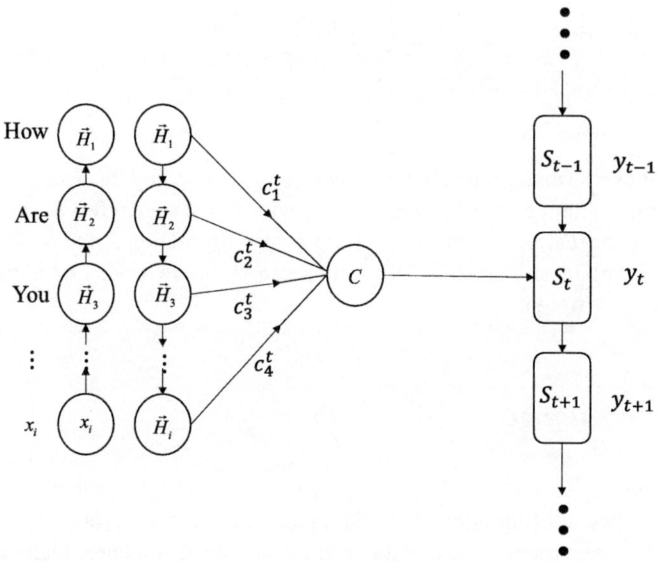


**Fig. 10.2** Basic attention model

human brain. When processing all information, it can select some local observation points, invest more attention, get more in the key areas and get more details of the information. In this paper, by adding an attention architecture to the basic sequence-to-sequence model, we observe how the attention mechanism architecture optimizes the property of the sequential to sequential model on machine translation tasks. Due to the characteristics of the attention mechanism, the system may not need to encode complete sentences into fixed-length semantic vectors. And the decoder can take use of different parts of the input sequence in every step of the output, let the model decide the position of intercepting the input sequence according to the input sequence and the generated sequence. In this model, the output sequence $y_t$ at any time is obtained by the sigmoid operation of the output sequence at the previous time $y_{t-1}$ combined with the semantic vector $C_t$ and the previous decoder's hidden state $S_{t-1}$. The semantic vectors at any time $C_t$ are obtained by the accumulation of the hidden state of the encoding layer $h_k$ and the time weight $W_{tk}$, where $e_{ij}$ represents the characteristic tag information of a certain time step, $f$ represents the nonlinear activation function, and $h_i$ represents a layer of code that hides information at any given moment. Above-mentioned formula is shown below:

$$y_t = \text{sigmoid}(y_{t-1}, C_t, S_{t-1}) \tag{10.4}$$

$$C_t = \sum_{k}^{T_i} W_{tk} h_k \tag{10.5}$$

$$W_{tk} = \frac{\exp(e_{tk})}{\sum_{j=1}^{T_i} \exp(e_{tj})} \tag{10.6}$$

$$e_{tk} = f(y_{t-1}, h_j) \tag{10.7}$$

In the above formula, refers to the translation generated by the decoder, refers to the translation output by the decoder at the previous time step, refers to the semantic vector of the current time step, refers to the label information at the previous time, refers to the nonlinear activation function operation, refers to the first hidden layer. Information is a weight.

### *10.2.3 Transformer*

Recurrent neural networks and convolutional neural networks have the problem of computing direction limitation. This limitation restricts the parallel ability of the model, and in the process of sequential calculation, there is a phenomenon of long-term dependence. Although the birth of the long short-term memory network solves the long-term dependence problem to a certain extent, in order to better avoid this risk, the transformer model came into being. The transformer essentially inherits the encoder-to-decoder structure of the sequence-to-sequence model, but uses six basic units of the self-attention mechanism model combined with the feed-forward neural network in the encoder-side module; in the decoder module, it uses six, and the self-attention mechanism combines the feed-forward neural network and encoder-to-decoder attention model and replaces the original recurrent neural network or convolutional neural network. For the first time, this structure completely abandons the convolution operation of the convolutional neural network and the recursive operation of the recurrent neural network and only uses the attention mechanism for feature extraction, achieving a highly parallelized network and hierarchical information capture function.

In the transformer model, due to its non-Turing complete characteristics, the transformer does not have the ability to capture the sequence by itself. In order to prevent translation quality problems caused by disordered sentence sequence, a position coding layer is added in the data preprocessing stage to encode the position. The latter information is summed with the information embedded in the position to obtain the relative position information (Fig. 10.3).

At the encoder side of the transformer, it will be composed of multiple unit layers, each unit layer is composed of two sub-layers, the two sub-layers are the multi-head attention mechanism and the fully connected feed-forward neural network, and the two sub-layers are connected by residuals.

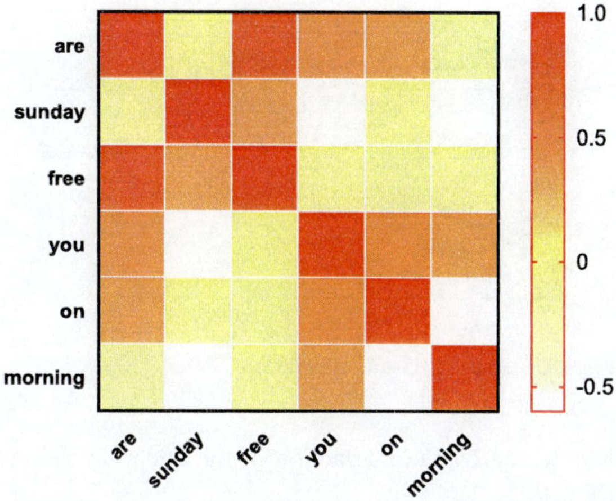


**Fig. 10.3** Example of a heat map of attention model

The mathematical essence of the traditional attention mechanism is a soft addressing operation. The information of the input sequence is used as the storage value. The element is composed of the meaning value and the address value. It is determined by calculating the similarity between the storage value and the address value of the element in the memory. How much information is extracted from the original stored value and the extracted stored value will be summed. Finally, the weighted sum of the meaning values is performed to obtain the attention value. Multi-head attention mechanism performs multiple different projections on the address value, storage value and meaning value. After a scaled-down dot product attention mechanism, the final result is spliced to obtain a linear mapping output (Fig. 10.4).

The feed-forward neural network has two dense layers in the fully connected form. The activation function used in the first layer is a linear rectification function [27], and the second layer is a linear activation function. When the output of the multi-head attention mechanism is $E$, the feed-forward neural network performs a reflow operation on the hidden layer after the operation and finally adds the residual network which performs the normal operation at the same time. The final text expression

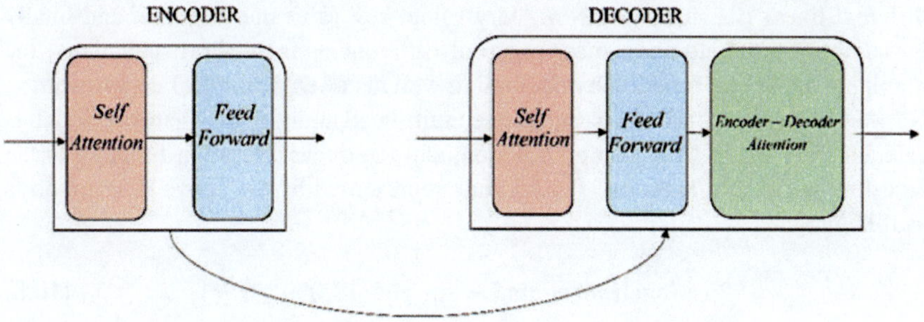


**Fig. 10.4** Basic transformer model

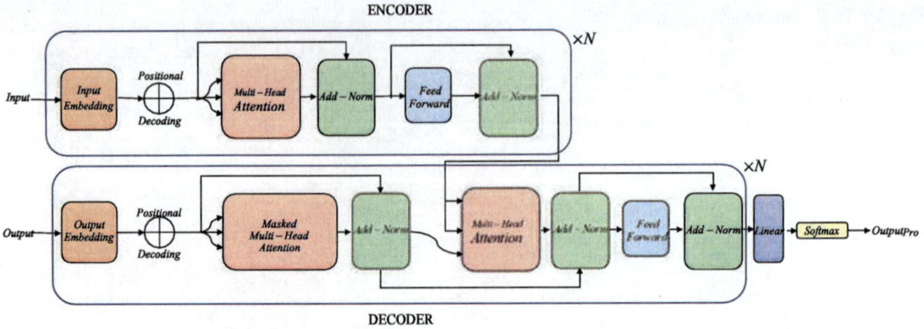


**Fig. 10.5** Specific transformer model

is obtained by the interaction of the attention mechanism and the nonlinear layer (Fig. 10.5).

The basic structure of the decoding layer of the transformer is the same as that of the coding layer, and only a sub-layer guided by the attention mechanism is added. And in the first multi-head attention layer of the decoder, the self-attention mechanism represents the relationship between the current translation and the translated sequence; the attention layer encoded to the decoder represents the relationship between the current translation and the encoded feature vector. At the same time, in order to ensure the predicted position, the system will rely on the output of the previous position to eliminate the interference of future information. It is necessary to add shielding protection in this multi-head attention layer.

The attention mechanism in the transformer is to pay attention to the interaction of each word in a sequence with all the words in the sequence and to determine that the relationship between words reflects the relationship between different words in this sentence to a certain extent sex and importance. Using this relationship, the corresponding translation vocabulary can be obtained.

Attention performs a nonlinear activation function by combining the key value $K$ with the query value $Q$ divided by the scaling factor $\sqrt{d_k}$ and multiplies the value $V$. PositionalEmbedding is obtained by sinusoidal operation of word position pos and word dimension, i.e., Sub_layer is obtained by residual connection of two sub-layers and layer standardization Layer_Norm. MT − Attention is obtained by a number of different linear transformations of query value, key value and $V$ value, and finally by combining the attention mechanism of different moment. $\text{Port}_i$ calculated the nonlinear activation function by combining weights of $Q$, $K$ and $V$. FeedForward − Network is derived from the output of the multi-head attention mechanism, which is calculated by the Relu activation function, and the linear activation function in the feed-forward neural network. The formula represented by the above description is as follows:

$$\text{PositionalEmbedding} = \sin(\text{pos}/10{,}000^{2i/d_{\text{model}}}) \tag{10.8}$$

$$\text{Sublayer} = \text{Layer\_Norm}(N + (\text{Sublayer}(N))) \tag{10.9}$$

$$MT - Attention = Concat(Port_1, Port_2, \ldots, Port_i)W^o \tag{10.10}$$

$$\text{Port}_i = \text{Softmax}\left(Qw_n^Q, Kw_n^K, Vw_n^V\right)O \tag{10.11}$$

$$\text{FeedForward} - \text{Network} = \max(\text{zero}, Ew_1 + g_1)w_2 + g_2 \tag{10.12}$$

$$\text{Attention} = \text{Softmax}\left(\frac{\text{Querie} \cdot \text{Key}^T}{\sqrt{d_k}}\right) \cdot \text{Value} \tag{10.13}$$

## 10.3 Experiment

### *10.3.1 Data Sets and Evaluation Performance*

In order to obtain the performance of the model objectively and accurately, in the process of model training, this article chose the official open-source corpus of English to German and English to Thai used in the top evaluation competition of machine translation in 2014, including English to German training data. The set contains 201,070 sentence pairs; the English to Thai training data set contains 190,166 sentence pairs. In the process of model testing, this article chose the official open-source corpus of English to German and English to Thai used by the International Spoken Translation Evaluation Contest in 2015. Among them, there are two English to German test data sets, the first one Contains 1347 sentence pairs, the second one contains 1449 sentence pairs; there are also two test data sets from English to Thai, the first one contains 1355 sentence pairs, and the second contains 1432 sentence pairs. Due to the source of the corpus, the corpus will not be cleaned up in this experiment but will be quoted directly.

This experiment uses a bilingual evaluation to replace index scores to translate model performance. The bilingual evaluation replacement index is a text estimation method, which reflects the quality difference between machine translation and professional human translation. Because of its close to human translation characteristics, it is now widely used to evaluate the quality of machine translation models.

The bilingual evaluation replacement index evaluation method is realized by combining the sentence obtained by machine translation and the reference sentence with $N$-gram and counting the number of word matches to calculate. Although the bilingual assessment replacement index evaluation method is not aimed at word order, because the data set used in this article is characterized by short sentences and uncomplicated word order, this risk is avoided to a certain extent. In the process of implementing the bilingual evaluation replacement index evaluation method, the penalty mechanism of too short will solve to a certain extent the problem of excessively high bilingual evaluation replacement index caused by too short sequence. The $N$-gram model approaches the actual accuracy rate by increasing the number of elements. The bilingual evaluation replacement index has the characteristics of language openness and closes to professional manual translation, which meets the needs of this article to obtain accurate performance indicators of the model, so this indicator will be used in the performance evaluation in the experiment.

BP refers to the penalty factor for short sentences, $c$ represents a translation produced by machine translation; $r$ stands for standard reference translation. $P_n$ represents the calculation of grammatical accuracy. $n-$ gram represents $n$-element mold processing method. Count represents the number of times an $n$-element mold appears in a machine translation. $w_n$ is to make the weights of $n$-element modules obey uniform probability distribution.BLEU represents the bilingual evaluation score of the final machine translation, Above formula is as follows:

$$\mathrm{BP} = \begin{cases} 1 & \text{if}, c > r \\ e^{1-r/c}, & \text{if}, c < r \end{cases} \tag{10.14}$$

$$P_n = \frac{\sum_{c \in \mathrm{Candiate}} \sum_{n-\mathrm{gram}} \mathrm{Count_{clip}}(n - \mathrm{gram})}{\sum_{c' \in \mathrm{Candiate}} \sum_{n-\mathrm{gram}' \in c'} \mathrm{Count}(n - \mathrm{gram})')} \tag{10.15}$$

$$\mathrm{BLEU} = \mathrm{BP} \cdot \exp\left(\sum_{n=1}^{N} w_n \log P_n\right) \tag{10.16}$$

## 10.3.2 Experimental Results and Analysis

In this experiment, the bilingual evaluation performance of the transformer model, sequence-to-sequence model and sequence-to-sequence model with additional attention mechanism on machine translation tasks is compared. Explore the performance differences of the three models on machine translation tasks when using data sets of different languages. The average bilingual evaluation scores of the three models from 20 time units on the two data sets are as follows. By comparing the average score of the model, the error can be eliminated to the maximum extent (Table 10.1).

According to the bilingual evaluation score table in the above figure, it is verified that the bilingual evaluation scores of the transformer on the two data sets are

**Table 10.1** BLEU score

| Data/model | Seq2seq | Seq2seq-attention | Transformer |
|---|---|---|---|
| En-Ge | 15.016 | 29.556 | 33.248 |
| En-Th | 13.455 | 27.998 | 32.334 |

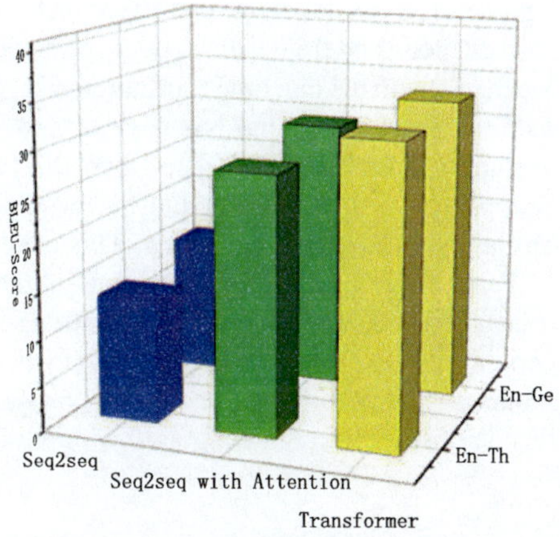


**Fig. 10.6** Three-dimensional BLEU scores of the three models

better than the sequence-to-sequence model and sequence-to-sequence-based attention mechanism model. Compared with the sequence-to-sequence model, the transformer has more than doubled the performance of the machine translation tasks in the two corpora; compared to the sequence-to-sequence model with the added attention mechanism, the transformer's bilingual evaluation score has also increased by four the above (Fig. 10.6).

At the same time, combine the above table with the three-dimensional bilingual assessment score chart below. This experiment found that the bilingual evaluation scores of the three models on the two corpora have the same difference, that is, for the English to German corpus of the Indo-European language family, the bilingual evaluation scores of the three models are slightly higher than the English to the East Asian language family. Thai language corpus. The specific difference is that in the sequence-to-sequence model, the bilingual evaluation score for English–German materials is 1.561 points higher than the bilingual evaluation score for English–Thai corpus; in the sequence-to-sequence model with the added attention mechanism, the score for English and German the bilingual evaluation score of the data is also 1.558 points higher than the bilingual evaluation score for the English–Thai language family; on the transformer model, the bilingual evaluation score for the English–German corpus is 1.014 points higher than the bilingual evaluation score for the English–Thai corpus.

## 10.4 Conclusions

This experiment is based on the use of sequence-to-sequence model, sequence-to-sequence model with added attention mechanism and transformer model to complete machine translation tasks. By comparing the average bilingual evaluation indicators of the above three models on the English to German data set and the English to Thai data set, the results show that the transformer model is better than the sequence-to-sequence model and the sequence-to-sequence model with the added attention mechanism on the machine translation task. At the same time, when the three models face the English to Thai language corpus of East Asian languages, the bilingual evaluation indicators are lower than the English to German corpus of the Indo-European language family, and the average difference between the three models on the two corpora is integrated. This experiment concludes that the difference in language family in the corpus will affect the performance of the model on machine translation tasks. The bilingual evaluation score in the machine translation task has a certain degree of negative impact. Due to the structural advantages of the transformer model, the impact of the uncommon corpus on its bilingual evaluation has a certain degree of improvement compared with the former two models.

## References

1. Mohamed Zakir, H., Shafeen Nagoor, M.: A brief study of challenges in machine translation. Int. J. Comput. Sci. Issues (2017)
2. Nida, E.: Theory of translation (2006)
3. Locke, W.N., Booth, A.D.: Machine translation. J. IEE **2**(2), 109–116 (1956)
4. Johnson, F.C.: An IBM-701 computer analysis of a reservoir by means of the Muskat equation. Metal Finish. **95**(3), 94–95 (1958)
5. ALPAC: Language and Machines: Computer in Translation and Linguistic (1966)
6. Koehn, Philipp: A parallel corpus for statistical machine translation. Proc. Third Workshop Statal Mach. Transl. **1**, 3–4 (2005)
7. Mathieu, G.: Toward Corpus-Based Machine Translation of Standard Arabic (2002)
8. Burges, C.J.C.: A tutorial on support vector machines for pattern recognition. Data Min. Knowl. Discov. (1998)
9. Mcculloch, W.S., Pitts, W.: A logical calculus of the ideas immanent in nervous activity. Biol. Math. Biophys. (1943)
10. Rumelhart, D.E., Hinton, G.E., Williams, R.J.: Learning Internal representation by back-propagation errors. Nature **323**, 533–536 (1986)
11. Minsky, M.L.: Computation: finite and infinite machines. Am. Math. Mon. **75**(4) (1968)
12. Hopfield, J.J.: Neural networks and physical systems with emergent collective computational abilities. Proc. Natl. Acad. Sci. (1982)
13. Hopfield J. Neural computation of decisions in optimization problems. Biol. Cybern., 1985, 52
14. Ackley, D.H.: A learning algorithm for Boltzmann machines. Cognit. Sci. 9 (1985)
15. Sengijpta, S.K.: Fundamentals of statistical signal processing: estimation theory. Control Eng. Pract. **37**(4), 465–466 (1994)
16. Rumelhart, D.E., Hinton, G.E., Williams, R.J.: Learning representations by back-propagating errors. Nature **323**(6088), 533–536 (1986)
17. Linsker, R.: Self-organization in a perceptual network. Computer (1988)
18. Akaike, H.: Information theory and an extension of the maximum likelihood principle. In: 2nd International Symposium on Information Theory. Akademiai Kaido, 1973
19. Broomhead, D.S., Lowe, D.: Multivariable Functional interpolation and adaptive networks. Complex Syst. **2**(3), 321–355 (1988)
20. Elman, J.L.: Finding structure in time. Cognit. Sci. **14**(2), 179–211 (1990)
21. Han, M., Xi, J., Xu, S., et al.: Prediction of chaotic time series based on the recurrent predictor neural network. IEEE Trans. Signal Process. **52**(12), 3409–3416 (2004)

22. Shewchuk, J.R.: An Introduction to the Conjugate Gradient Method Without the Agonizing Pain. Carnegie Mellon University, 1994
23. Hochreiter, S., Schmidhuber, J.: Long short-term memory. Neural Comput. (1997)
24. Sutskever, I., Vinyals, O., Le, Q.V.: Sequence to Sequence Learning with Neural Networks. NIPS. MIT Press (2014)
25. Vaswani, A., Shazeer, N., Parmar, N., et al.: Attention is all you need. arXiv (2017)
26. Papineni, K., Roukos, S., Ward, T.: BLEU: A Method for Automatic Evaluation of Machine Translation (2002)
27. He, K., Zhang, X., Ren, S., et al.: Delving Deep into Rectifiers: Surpassing Human-Level Performance on ImageNet Classification (2015)
28. Jaderberg, M., Simonyan, K., Zisserman, A., et al.: Spatial Transformer Networks (2015)
29. Lai, J.S., Maitra, A., Goodman, F.: Multilevel intelligent universal auto-transfomer (2007)
30. Chen, Q., Zhu, X., Ling, Z.H., et al.: Recurrent neural network-based sentence encoder with gated attention for natural language inference (2017)
31. Vig J.: A multiscale visualization of attention in the transformer model (2019)
32. Arkhangelskaia, E., Dutta, S.: Whatcha lookin' at? DeepLIFTing BERT's attention in question answering (2019)

# Chapter 11
# Sliding Mode Control of AGV Based on Extended State Observer

**Cuiping Pu, Jie Ren, and Gang Fang**

**Abstract** According to the error tracking model of automated-guided vehicle (AGV), considering the uncertain disturbances such as sideslip, ground roughness and friction, this paper designs sliding mode control extended state observer (SMC-ESO) control scheme. ESO describes all the uncertain dynamics in the system as additive total disturbances of the system and estimates the disturbances online and in real time. The SMC is designed to compensate and suppress these disturbances, thus reducing the uncertainty of the feasibility of AGV tracking and the influence of external random interference on the control performance in the tracking process and improving the rapidity and robustness of AGV tracking control. To verify the algorithm, SMC-ESO and active disturbance rejection control (ADRC) are compared by simulation, and the results show that SMC-ESO has better tracking ability, anti-interference ability and robustness.

## 11.1 Introduction

AGV refers to the installation of electromagnetic, laser, inertial, visual and other guiding devices on the transport trolley to guide and drive along the set path. Its main function is to transport and transfer goods. It has the advantages of high automation, simple and convenient scheduling management, safe and reliable performance and so on. It belongs to the category of wheeled mobile robots. AGV can realize efficient, economical and flexible unmanned production process because of its good automation, flexibility, system expansion and reliability. As a highly automated robot, it is widely used in flexible manufacturing, warehouse management, freight terminal and other scenarios. Navigation is the key to the autonomous movement of AGV, and tape navigation is widely used in practice because of its high reliability and flexibility [1–4].

C. Pu (✉) · J. Ren · G. Fang
Kunming University, Kunming 650214, China
e-mail: pucuiping@126.com



V. Chang et al. (eds.), *Modern Industrial IoT, Big Data and Supply Chain*,
Smart Innovation, Systems and Technologies 218,
https://doi.org/10.1007/978-981-33-6141-6_11

The magnetic guiding of AGV is to detect the magnetic field signal of the magnetic tape laid on the ground by the magnetic guiding sensor installed on AGV and correct the position deviation of AGV relative to the magnetic tape [5]. There have been many research results on the navigation control of tape-guided AGV. Qianyi [6] uses open-loop control to realize the tracking control of AGV, but its navigation reliability is poor.

Zhou [7] adopts PID to realize the closed-loop control of AGV tracking, but tracking chattering is easy to occur, and the dynamic response effect is poor. Hongjie [8] realized the correction effect of fuzzy PID controller on the path tracking of magnetic navigation AGV, but the fuzzy control rules need a lot of practical experience and data support, which will be affected by subjective factors, but cannot be completely summarized. In addition, there are predictive control, active disturbance rejection control, reinforcement learning and other intelligent control methods.

Researcher Han Jingqing reflected on the control law design method of "model theory", summarized the idea of "cybernetics" design method [9–11], and put forward ADRC. There are many parameters in nonlinear ADRC, so Zhiqiang [12–14] proposed linear ADRC(LADRC), which originated from the research and development of simplifying ADRC parameter adjustment and became another branch of ADRC's relatively nonlinear form. With the improvement of ADRC theory and parameter adjustment method, ADRC has been widely used in engineering practice. Shijie [15] designed a trajectory tracking controller based on LADRC for wheeled mobile robot AGV and realized the trajectory tracking of AGV by tracking the pose of the robot.

The essence of sliding mode variable structure control is a kind of nonlinear control. The advantage of sliding mode is that the motion on sliding mode is independent of uncertainty, and it has strong robustness to system parameter changes and external disturbances. Because of the uncertainty of system internal parameters and external interference, chattering in sliding mode variable structure control system will occur, which will induce the high-frequency dynamic response of the unmodeled part of the system, increase the risk of system instability and affect the control accuracy of the system.

In this paper, the uncertain dynamics of tracking control system is described as additive total disturbance of the system. An ESO is designed to estimate these disturbances in real time. A SMC is designed to compensate and suppress these disturbances. Combining the advantages of ESO and SMC, the uncertainty of AGV tracking feasibility and the influence of external random interference on control performance during tracking process are reduced. The rapidity of the system and the disturbance of internal and external uncertainties are improved, thus improving the robustness of AGV, thereby improving the control performance of AGV tracking.

## 11.2 Control System Design

### *11.2.1 System Transformation*

In this paper, two-wheel differential-wheeled AGV is considered. Two front wheels of differential AGV are used as driving wheels to realize differential steering, and two rear wheels are used as driven wheels to play a supporting role.

According to reference [1], the linearized system model can be obtained as follows:

$$\begin{bmatrix} \dot{e}_\theta \\ \dot{e}_d \end{bmatrix} = A \begin{bmatrix} e_\theta \\ e_d \end{bmatrix} + \text{Bu}, \tag{11.1}$$

where $A = \begin{bmatrix} 0 & 0 \\ v_c & 0 \end{bmatrix}$, $B = \begin{bmatrix} 1/D & 0 \end{bmatrix}^T$, $D$ is distance between two driving wheel centers of AGV. $v_c$ is AGV moving speed. $e_\theta$ is angle between AGV centerline and tape centerline. $e_d$ is vertical distance between AGV centroid and tape centerline Then

$$y = C[\, e_\theta \; e_d \,]^T, \tag{11.2}$$

where $C = [\,0\ 1\,]$.

By changing the state variables, the system can be transformed into an integral series system, and then the controller can be designed by using the active disturbance rejection control method. The input of the system is the speed difference between the left and right driving wheels of AGV. To control the vertical distance between the center of mass of AGV and the center line of tape, the system can be seen as a single input single output system (SISO). Make the following transformation

$$y = e_d \dot{y} =, v_c \sin e_\theta = \dot{e}_d, \tag{11.3}$$

when satisfied

$$y = x_1, \dot{y} = x_2, \ddot{y} = \dot{x}_2 = v_c/D * u \cos e_\theta, \tag{11.4}$$

where nonlinear terms can be controlled $u = D/v_c \cos e_\theta * v$, when $-\pi/2 < e_\theta < \pi/2$, Eq. 11.3 is meaningful. Use $[\, x_1 \; x_2 \,]$ to represent $[\, e_d \; e_\theta \,]$ as follows:

$$e_d = x_1, e_\theta = \arcsin(x_2/v_c). \tag{11.5}$$

In Eq. 11.4, there is a definition when $-v_c < x_2 < v_c$. The transformed state is

$$\dot{x}_1 = x_2, \dot{x}_2 = v_c/D * \cos(\arcsin(x_2/v_c)) * u. \quad (11.6)$$

### 11.2.2 ADRC Design

The AGV tracking system model is a second-order system. There are three states estimated by LESO. The estimated value $\hat{e}_{d1}$ of the vertical distance $e_d$ is denoted as $z_1$. The estimated value of vertical distance change rate $e_{d2}$ is $z_2$. The estimated value $\hat{f}$ of the total disturbance is denoted as $z_3$. The control structure block diagram of LADRC is shown in Fig. 11.1.

In the design of tracking differentiator (TD) as shown in Fig. 11.1, the transition process is arranged for reference input a and its differential signal is extracted. TD is

$$\dot{e}_{d1} = e_{d2}, \dot{e}_{d2} = -r(e_{d1} - e_d*) - 2re_{d2}, \quad (11.7)$$

where $r$ is the speed factor that determines the tracking speed.

LESO is designed as follows:

$$z_1 = z_2 - \alpha_1(z_1 - e_d), z_2 = z_3 - \alpha_2(z_1 - e_d) + b_0 u, z_3 = -\alpha_3(z_1 - e_d), \quad (11.8)$$

where $b_0$ is the compensation factor. The characteristic equation of LESO in formula 8 is

$$s^3 + \alpha_1 s^2 + \alpha_2 s + \alpha_3 = 0. \quad (11.9)$$

The ideal characteristic variance of formula 10 is $(s + \omega)^3 = 0$. Then, $\left[\alpha_1\ \alpha_2\ \alpha_3\right]^T = \left[3\omega\ 3\omega^2\ \omega^3\right]^T$ where $\omega$ is the bandwidth of LESO. According to the bandwidth requirement of the system or online setting, it is generally applicable to a wide range, so it is easier to adjust the appropriate value.

Linear state error feedback (LSEF) is designed as follows:

$$u_0 = k_1(e_{d1} - z_1) + k_2(e_{d2} - z_2), u = (u_0 - z_3)/b_0. \quad (11.10)$$

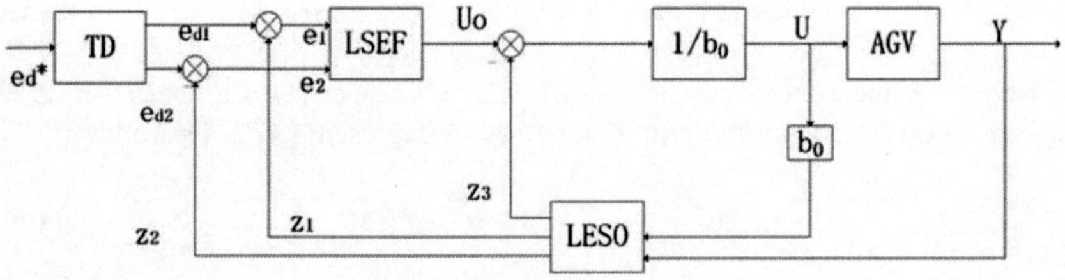


**Fig. 11.1** LADRC control structure block diagram

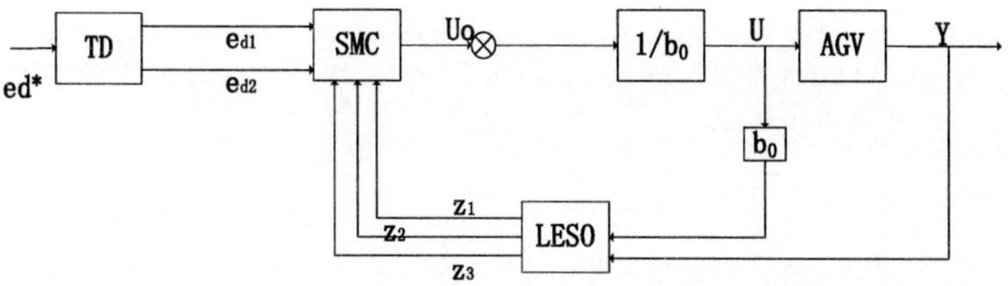


**Fig. 11.2** SMC-ESO control structure block diagram

### *11.2.3 SMC-ESO Design*

The high gain error feedback used in the extended state observer makes the observer's dynamics much higher than the system's dynamics. This is equivalent to the fast-changing subsystem in the system. It can ensure the fast convergence of observation error and high estimation accuracy and then provide available signals for feedback.

The block diagram of SMC-ESO control structure is shown in Fig. 11.2.

The sliding mode function designed for AGV error model is

$$s = ce - \dot{e}, e = e_{d1} - e_d*, \tag{11.11}$$

where $c > 0$.

The SMC based on ESO is designed as follows:

$$u = \frac{1}{b}(-\beta\hat{s} - \hat{\eta} - \hat{f}), \tag{11.12}$$

where

$$\hat{s} = c\hat{e} - \hat{\dot{e}}, \hat{e} = \hat{z}_1 - e_d*, \hat{\dot{e}} = \hat{z}_2 - \dot{e}_d*, \hat{\eta} = c\hat{\dot{e}} - \ddot{e}_d*, \hat{f} = \hat{z}_3, \tag{11.13}$$

Let Lyapunov function of sliding mode control be $V_c = \frac{1}{2}s^2$, and then

$$\begin{aligned} \dot{V}_c &= s\dot{s} = s(c\dot{e} + \ddot{e}) = s(c\dot{e} + b_0 u + f - \ddot{e}_d*) \\ &= s(c\dot{e} - \beta\hat{s} - \hat{\eta} - \hat{f} + f - \ddot{e}_d*) \\ &= s(\eta - \hat{\eta} + f - \hat{f} - \beta\hat{s}) = -\beta\hat{s}s + s(\tilde{\eta} + \tilde{f}) \\ &= -\beta s^2 + s(\tilde{\eta} + \tilde{f} + \beta\tilde{s}), \end{aligned} \tag{11.14}$$

where

$$\tilde{s} = s - \hat{s} = c\tilde{z}_1 + \tilde{z}_2, \tilde{\eta} = \eta - \hat{\eta} = c(\dot{z}_1 - \hat{z}_2) = c\tilde{z}_2, \hat{f} = f - \hat{f}, \tag{11.15}$$

Because $\tilde{f}$ is the estimated residual of interference, $\tilde{s}$ and $\tilde{\eta}$ are linear functions of state observation error. If $\beta$ is large enough, $\dot{V}_c \leq 0$ can be guaranteed, which shows that the system is stable.

## 11.3 Simulation Analysis

The tracking error model of tape-guided AGV is considered. Parameters are as follows: $v_c = 1$ m/s, $D = 0.4$ m, $e_\theta* = 0$, $e_d* = 0$, $e_\theta(t_0) = \pi/50$ rad, $e_d(t_0) = 0.05$ m. in which $v_c$ is the moving speed of AGV and $D$ is the distance between the centers of two wheels of AGV. The expected errors of AGV and tape centerline are $e_\theta*$ and $e_d*$. The initial values are $e_\theta(t_0)$ and $e_d(t_0)$. In order to verify the anti-interference ability of the controller, the disturbance term $\lambda(t)$ is added. Given $\lambda(t)$, a pulse signal with an intensity of 0.5 is added every 4 s to simulate the interference such as tire slip. Comparing the designed SMC-ESO with the traditional ADRC, their parameters are shown in Table 11.1.

The vertical distance deviation tracking diagram of the two algorithms is shown in Fig. 11.3.

It can be seen from Fig. 11.3 that SMC-ESO can track the route quickly, with small error and better anti-interference effect.

**Table 11.1** Parameters of two algorithms

| Name | Parameters |
|---|---|
| ADRC | $r = 7, b = 11.7, k_1 = 2.5, k_2 = 5, \omega = 60$ |
| SMC-ESO | $b = 12.2, c = 0.01, \beta = 5, \omega = 60$ |

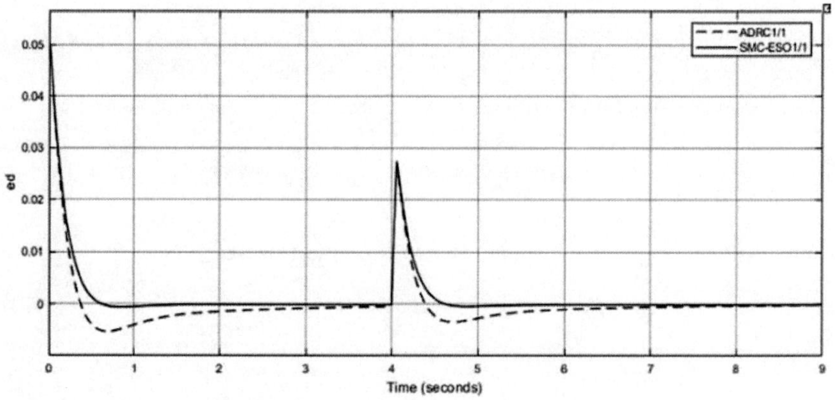


**Fig. 11.3** Comparison diagram of vertical distance deviation tracking

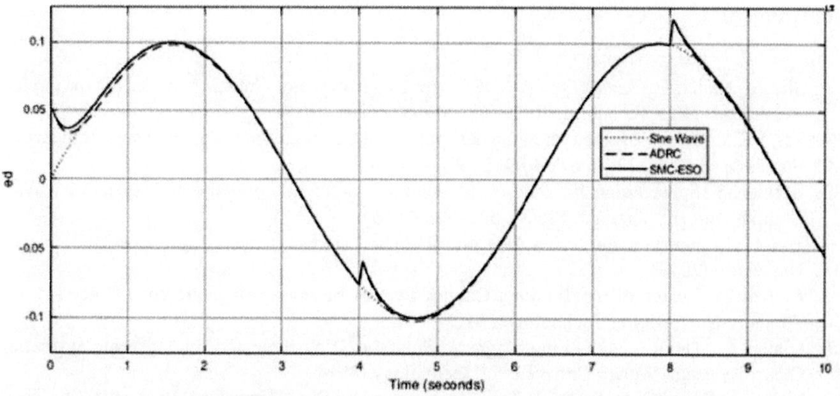


**Fig. 11.4** Comparison diagram of sinusoidal path tracking between two algorithms

The maximum deviation of ADRC is 0.53 cm, while that of SMC-ESO is 0.06 cm. It can be seen from Fig. 11.4 that SMC-ESO has better tracking ability and anti-interference ability.

## 11.4 Conclusion

According to the error tracking model of AGV, SMC-ESO control method is designed in this paper. In the design process, the uncertainties such as sideslip, ground roughness and friction are considered in the actual operation of AGV. The ESO is designed to describe all uncertain dynamics in the system as additive total interference of the system and estimate the interference online and in real time. Active disturbance rejection controller and sliding mode controller are designed to compensate and suppress these disturbances. The internal and external uncertainty of AGV tracking and the influence of external random interference on control performance during tracking are reduced. The rapidity and robustness of AGV tracking control are improved. The simulation results of SMC-ESO and ADRC show that SMC-ESO has better tracking ability, anti-interference ability and robustness.

**Acknowledgements** The authors acknowledge the joint special project of Yunnan Province (202001BA070001-055), Yunnan Education Department Project (2019J0560).

## References


1. Shu, Y., Haihai, Z., Xiang, Q.: An ADRC approach to trajectory tracking control of tape guided AGV. Chin. High Technol. Lett. **28**(06), 547–552 (2018)
2. Vis, I.F.A.: Survey of research in the design and control of automated guided vehicle systems. Eur. J. Oper. Res. **170**(3), 677–709 (2006)
3. Berman, S., Schechtman, E., Edan, Y.: Evaluation of automatic guided vehicle system s. Robot. Comput.-Integr. Manufact. **125**(3), 522–528 (2009)
4. Wei, L.: Research on the Design and Control Method of a Magnetic Navigation AGV. Xiangtan University (2019)
5. Haihai, Z.: Design of the Tracking Control System Based on Magnetic Tape Guided AGV. Zhejiang University of Technology (2017)
6. Qianyi, C.: The Research of the Magnetic Guided AGV (Automated Guided Vehicle) Central Control System. Wuhan University of Technology (2014)
7. Zhou, H.: Research on the AGV Technology Based on PLC Control and Its Application. East China University of Technology (2013)
8. Hongjie, T.: Design of Magnetic Guidance Differential AGV and Fuzzy PID Path Correction. Shandong University (2018)
9. Jinqing, H.: ADRC and its applications. Cont. Decis.-Mak. **13**(1), 19–23 (1998)
10. Jinqing, H.: ADRC Ttechnique—A Control Technique that Estimates and Compensates for Uncertainties. National Defense Industry Press (2008)
11. Jinqing, H.: An extended state observer for a class of uncertain objects. Cont. Decis. **10**(1), 85–88 (1995)
12. Zhiqiang, G.: Scaling and bandwidth-parameterization based controller tuning. Am. Contr. Conf. **6**, 4989–4996 (2003)
13. Zhiqiang, G.: Active disturbance rejection control: A paradigm shift in feedback control system design. In: American Control Conference, pp 2399–2405 (2006)
14. Junxiong, Y.: Research on Trajectory Tracking Control of Mobile Robot Based on Disturbance Observer. Tianjin Polytechnic University **44**(04), 95–100 (2017)
15. Shijie, Z., Qing, G.: LADRC-based AGV trajectory tracking controller design and parameter tuning. J. Beijing Univ. Chem. Technol. ( Nat. Sci.) **44**(04),95–100 (2017)
16. Xiangping, Z., Xingjiao, L., Jiang, L.: Research on path tracking control strategy of AGV based on magnetic guidance. Modul. Mach. Tool Autom. Manuf. Techn.**12**, 86–89+93 (2018)
17. Teatro, T.A.V., Eklund, J.M., Milman, R.: Nonlinear model predictive control for omnidirectional robot motion planning and tracking with avoidance of moving obstacles. J. Electr. Comput. Eng. Can. **37**(3), 151–156 (2014)
18. Shitao, Z.: Dissertation Submitted to Hangzhou Dianzi University for the Degree of Master. Hangzhou University of Electronic Science and Technology (2018)
19. Ren, Y., Shitao, Z.: Deep reinforcement learning based path following control of magnetic navigation AGV. J. Hangzhou Dianzi Univ. (Nat. Sci.) **39**(02), 28–34 (2019)
20. Kayacan, E., Kayacan, E., Ramon, H., et al.: Towards agrobots: trajectory control of an autonomous tractor using type-2 fuzzy logic controllers. IEEE/ASME Trans. Mechatron. **20**(1), 287–298 (2015)

# Chapter 12
# Stock Price Prediction Based on Machine Learning Algorithms

**Hanchen Wang**

**Abstract** Predicting the change of stock market is difficult, because the volatility of the price is acute and random. This predictive behavior includes physical factors, psychological factors, fixed or random behavior factors. These factors make share prices violate, which is not easy to forecast accurately. But, machine learning technologies have the energy to dig out the characteristics and traces that we undiscovered in the past. This paper based on the historical data of Microsoft's stock price, the period is from March 27, 2013, to March 27, 2018, combined with machine learning algorithm to predict the trend of the Microsoft stock price. The paper starts from the linear regression algorithm and then turned to the high-level model, such as auto-ARIMA, prophet and LSTM. We ultimately conclude LSTM model is best when predicting stock price of Microsoft, but because company shares will be affected by many uncertain factors, this paper predicted results that only can be used as a reference for investors' decision.

## 12.1 Introduction

Financial market is the most important part in the country's overall economic system. Among them, the capital market that includes stock market and bond market is vital. In recent years, due to the relevant economic policies of the country, most of the stock information and its historical data can be found in the network. However, the stock market is huge and has a lot of stocks. We can only find one stock of the market and predict its price. The data set is used as the research object, and the model is helpful for us to predict the tendency.

Forecasting the stock price in the future becomes one of the most attractive fields in capital market research. With the rapid development of technology, machine learning models have been gradually applied to study this problem. In recent years, machine learning algorithms such as LSTM model have been widely used in the prediction of

H. Wang (✉)
School of Finance, Central University of Finance and Economics, Beijing 102206, China
e-mail: 15163282198@163.com

 
V. Chang et al. (eds.), *Modern Industrial IoT, Big Data and Supply Chain*,
Smart Innovation, Systems and Technologies 218,
https://doi.org/10.1007/978-981-33-6141-6_12

financial time series data, and because of their excellent performance in prediction problems, LSTM model has been successfully applied to research the trend of stock price.

## 12.2 Related Work

There are many algorithms to predict stock market value. In previous study, back-propagation, LSTM method, SVM, CNN, merged hybrid model and systematic analysis are widely used to predict stock price. An overview of researches on comparing algorithms suggests the applicable use of technical indications for stock price prediction over the past 10 years.

Parimala [1] use back-propagation, long-short-term memory and support vector machine to predict stock price. They conclude that LSTM performs better in comparison to back-propagation and SVM.

Sim [2] use S&P 500 index, based on convolutional neural network, concluded the deep learning method CNN is useful for stock price forecast.

Chun [3] combined empirical analysis and qualitative analysis and build a hybrid model that can forecast stock trend more precisely.

Chonga et al. [4] used systematic analysis to forecast stock market. They find that neural networks can attain numerous information from the AR model and increase the precise of stock prediction.

Yan and Ouyang [5] proposed a combined time-series prediction model, which combined the wavelet analysis with LSTM. They conclude that the LSTM performs best when predicting financial time series data, and it shows gigantic influence on the price and trend prediction of the time series data.

In general, the most part of previous research is based on one or several algorithms such as ARIMA, SVM, and it is rare for researchers to use four or more methods to predict and compare. In addition, most previous research use index as their research object, but taking a specific company as the research object is little. Hence, this paper uses the Microsoft data and compares the prediction performance of four methods.

## 12.3 Algorithms

Stock prediction in my paper is based on five main algorithms, on analysis of which best algorithm for the prediction of stock market can be evaluated.

- Multiple linear regression
- Automatic ARIMA model
- Prophet
- LSTM model

### 12.3.1 Multiple Linear Regression

Linear regression is the most basic algorithm that can be implemented for prediction. Multiple linear regression model reflects the relationship between a dependent variable and multiple independent variables. The multiple linear regression model can be shown as:

$$Y = \theta_1 X_1 + \theta_2 X_2 + \cdots + \theta_n X_n$$

$X_1, X_2 \ldots X_n$ are the independent variables, and the coefficients $\theta 1, \theta 2, \ldots \theta n$ is the weight. In this study, the time column was used to show features, such as day, month, year, etc.

### 12.3.2 Automatic ARIMA Model

ARIMA model is a statistic method for time series forecast. It uses the past scores to predict future scores; *p*, *d*, and *q* are three coefficients in the model:

1. *p* (the previous value used to predict the next value)
2. *q* (the previous predicted error)
3. *d* (order of difference).

Parameter optimization of ARIMA needs to experiment many times. Therefore, automatic ARIMA is used in this paper to select the best combination with the minimum error (*p*, *q*, *d*).

### 12.3.3 Prophet

Prophet is designed by Facebook; it needs no data preprocessing and is easy to establish. The input of the prophet is a data box with date and destination (ds and *y*). The prophet tries to catch seasonal trend in previous time and performs well when the number of data is large.

### 12.3.4 LSTM Model

LSTM model is a very impactful to use the trend of time series data. They perform perfect because the LSTM abandon ineffective information and hold crucial knowledge.

The LSTM model has three doors:

**Table 12.1** Import result of Microsoft data

| Date | Open | High | Low | Close | Volume |
|---|---|---|---|---|---|
| 2017/10/02 | 74.71 | 75.01 | 74.295 | 74.61 | 15,210,338 |
| 2017/10/03 | 74.67 | 74.88 | 74.195 | 74.26 | 11,935,853 |
| 2017/10/04 | 74 | 74.72 | 73.71 | 74.69 | 13,287,346 |
| 2017/10/05 | 75.22 | 76.12 | 74.967 | 75.97 | 20,656,238 |
| 2017/10/06 | 75.67 | 76.03 | 75.54 | 76 | 13,692,791 |
| 2017/10/09 | 75.97 | 76.55 | 75.86 | 76.29 | 11,364,275 |

1. Input gate: It adds information to the model;
2. Oblivion gate: It removes useless information;
3. Output gate: It selects the output information.

## 12.4 Prediction Result

### *12.4.1 Data Processing*

The paper mainly talks about technical analysis. The author selected the data of the Microsoft of the NASDAQ market from Quandl data set. The period is from March 27, 2013, to March 27, 2018. The author predicted the stock price of the Microsoft based on Python, and the results of data import are shown in Table 12.1.

Table 12.1 shows that there are a lot of variables in the data set—date, opening price, high price, low price, closing price, total trade quantity. Because of the market is closed on weekends and holidays, there are missing of the 1/10/2018, 7/10/2018 and 8/10/2018 from the table. October 1 is National Day, and October 7 and October 8 are weekends. As the calculation of profit and loss is usually determined by the closing price of the stock on that day, this paper selects the closing price as the target variable and plots its change from March 27, 2013, to March 27, 2018, which is shown in Fig. 12.1.

In the next section, the author explores these variables in depth and uses different machine learning algorithms to predict the closing price of the Microsoft's stock.

### *12.4.2 Multiple Linear Regression Result*

Firstly, the paper sorts the data set in ascending order and then creates an independent data set to assure that the new features do not affect the original data. Furthermore, the paper assumed that the Monday and Friday have a deeper influence on the closing price of stocks than others. Therefore, we have established a mechanism to identify whether a given day is Monday or Friday.

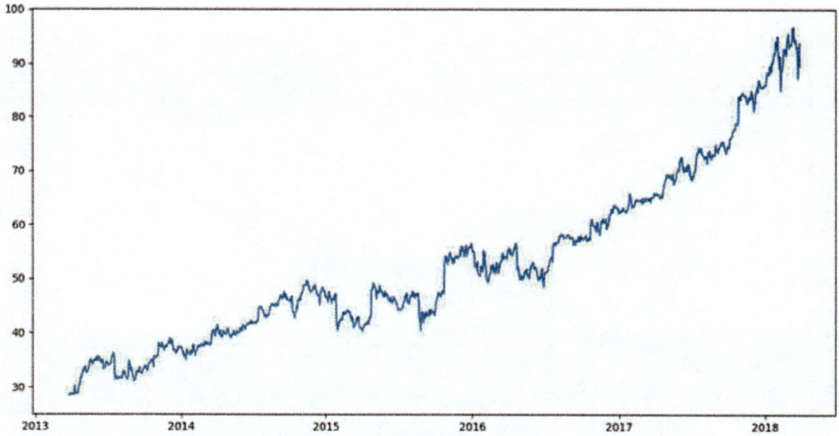


**Fig. 12.1** Closing price of Microsoft from March 2013 to March 2018

Finally, the data are divided into training set and validation set to assess the performance of the model. The RMSE value predicted by the model is 15.08, which is lower than that of the moving average model, which clearly indicates that the prediction results of the linear regression model are better. In this paper, the predicted value and the actual value are shown in Fig. 12.2. The 0 refers to the March 27, 2013, and it increases as the date increases.

As we can see from the chart above, Microsoft shares fell in January 2016 and January 2017. The model predicts the same for January 2018. Linear regression method is good for solving problems such as hypermarket sales. In that situation, the independent features are helpful for determining target values.

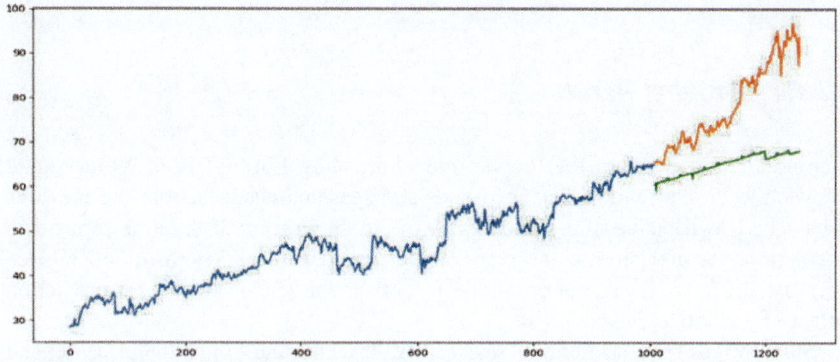


**Fig. 12.2** Prediction results of the linear regression model

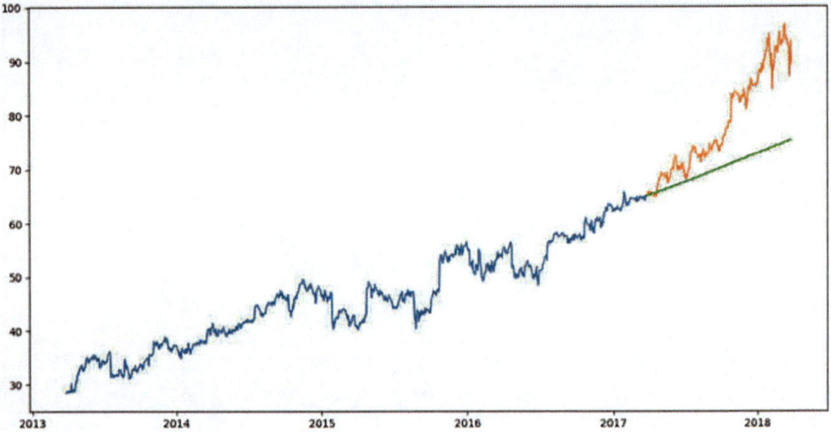


**Fig. 12.3** Prediction results of automatic ARIMA model

### 12.4.3 Automatic ARIMA Model Result

The predicted results are shown in Fig. 12.3. The RMSE value of this model is 10.41, which is smaller than the previous models, so the prediction effect is better than the previous models. But, the automatic ARIMA model uses the previous data to understand trends in the time series. The model captures the growth trends in the data by using the above tools. In this model, the predictions are much better than others, but the results are still far from the actual values.

As can be clearly seen from the figure, the model captures a trend in the sequence, but ignores the influence of seasons, so the predicted value differs greatly from the actual value.

### 12.4.4 Prophet Result

The prediction results of the Prophet model are shown in Fig. 12.4. As the figure shows, the Prophet model catches trends and seasonal changes from the previous data. The RMSE value of this model is 9.07. In the process of the time-series data prediction, the precision of the model is perfect, but in the Microsoft stock price forecast, the forecast value of this model is far from the actual value, so the prediction effect of this model is poor.

Stock prices have no trend or seasonality. The rise and fall of the price depend largely on current market conditions. Therefore, prediction methods such as MA, ARIMA and Prophet are not a good model to stock price. So, in the next section, this paper tries advanced machine learning algorithm, the LSTM network.

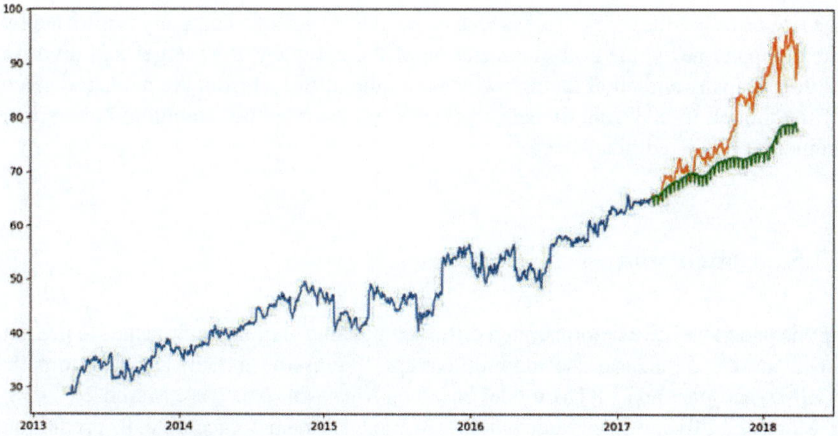


**Fig. 12.4** Prediction results of the prophet model

### *12.4.5 LSTM Model Result*

Now, this paper implements LSTM as a black box and examines its performance on specific data to achieve the purpose of forecasting Microsoft stock price. The prediction results are shown in Fig. 12.5. The RMSE value of this model is 1.77.

It can be seen from Fig. 12.5 that the forecast results of LSTM model are better than those of the above models and the forecast value of stock price is not far from the actual value. The LSTM model can change various parameters, such as changing the number of LSTM layers, increasing the dropout value or increasing the number of epochs in training. However, the prediction of LSTM cannot be used to determine

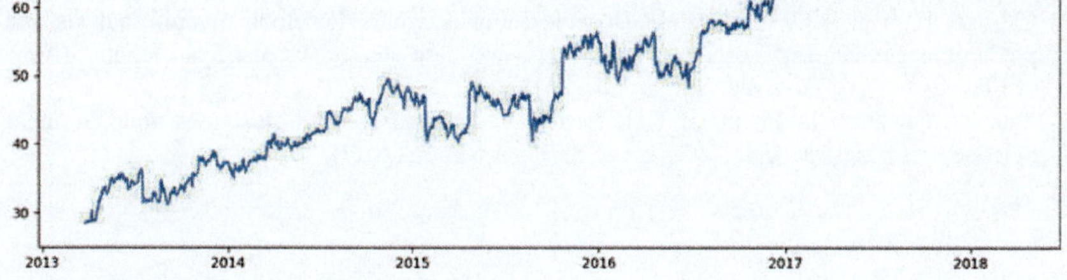


**Fig. 12.5** Prediction results of the LSTM model

the rise or fall of the stock price, because the stock price of a company is affected by the company news, the non-monetization of the company, the merger and division of the company and other factors, which are quite different from the predicted value of the model. In addition, the stock prices are influenced by intangible factors that cannot be predicted in advance.

## 12.5 Conclusion

In this paper, we have demonstrated different machine learning approaches to predict stock market trend. We use moving average, regression model, KNN, automatic ARIMA, prophet and LSTM model based on Microsoft data from March 27, 2013, to March 27, 2018; we conclude that LSTM model is the most accurate. Its prediction values are nearest to the real values.

For this result, we can conclude that if we include all the factors that affect performance of the stock and feed them to the network with data preprocessing and filtering, after training the network, we will be able to attain a model that predicts stock trend more precisely for the better idea of stock value so that firms may have increased profit ratio as compared to what is might be going currently at that time. This will also lead to more transparency regarding stock as it will be easier for firms to analyze loses and achieve great success.

**Future work**

In future, we plan to design a new algorithm that combines LSTM with other algorithms, which predict the stock market trend more accurately and efficiently.

## References

1. Arora, N., Parimala, M.: Financial analysis: stock market prediction using deep learning algorithms. In: International Conference on Sustainable Computing in Science, Technology and Management (SUSCOM-2019)
2. Sim, H.S., Kim, H.I., Ahn, J.J.: Is deep learning for image recognition applicable to stock market prediction? Hindawi Complexity, vol. 2019, Article ID 4324878, 10 pages
3. Chun, R.: Stock Price Prediction using Deep Learning. San Jose State University ScholarWorks
4. Chonga, E., Hanb, C., Park, F.C.: Deep learning networks for stock market analysis and prediction: methodology, data representations, and case studies. Expert Syst. Appl. · (April 2017)
5. Yan, H., Ouyang, H.: Financial Time Series Prediction Based on Deep Learning. Springer Science + Business Media, LLC, part of Springer Nature (2017)

# Part II
# Big Data Research and Multi-discipline Services

# Chapter 13
# Research on Precision Marketing of Big Data in Small- and Medium-Sized E-commerce Enterprises

**Yi Sun**

**Abstract** At present, with the rapid development of social information technology, data information presents an explosive growth. In the era of big data, the precision marketing of big data information has become the key to the competition of various e-commerce enterprises in the data age and began to pay attention to the research on precision marketing. By mining and analyzing the massive data, e-commerce enterprises can obtain more accurate user information and make personalized marketing strategies to achieve better marketing effect. This paper will analyze the precision marketing of small- and medium-sized e-commerce enterprises of big data and put forward corresponding marketing strategies.

The popularization of information technologies such as big data and artificial intelligence has brought great changes and influences to economic and social development and people's daily life. With the wide application of Internet and the popularization of electronic payment in commercial activities, people's consumption habits have changed greatly, and a large number of e-commerce enterprises rise in response to the proper time and conditions. In the operation process of e-commerce enterprises, a large number of economic business records have been produced, and the huge amount of data implies unlimited economic values. Through the analysis and mining of these data, it can help enterprises effectively improve market shares, reduce marketing costs and enhance profitability. As the commercial value of massive data is more and more obvious, the application is more and more common which indicates that human society has entered the era of big data. In the era of big data, based on the collection and accumulation of customer data and sales data, e-commerce enterprises have found data information to provide decision support for marketing activities. In the current stage of high popularity of the Internet, online shopping has become the main consumption behavior of people. E-commerce platform records a large number of consumer consumption data all the time. Through the collation, analysis and mining

Y. Sun (✉)
Liaoning Economic Vocational Technology Institute, Shenyang 110122, Liaoning, China
e-mail: mazda525@126.com

 
V. Chang et al. (eds.), *Modern Industrial IoT, Big Data and Supply Chain*,
Smart Innovation, Systems and Technologies 218,
https://doi.org/10.1007/978-981-33-6141-6_13

of these data, more effective consumption information can be extracted, and the characteristics of customers and potential customers can be identified accurately. Therefore, precision marketing strategy has been in the ascendant in the marketing activities of major e-commerce platforms and e-commerce enterprises. Big data not only provide practical basis for precision marketing of e-commerce enterprises, but also provide theoretical basis.

## 13.1 Theoretical Research on Precision Marketing

The concept of "precision marketing" is also the product of the development of the times. It is mainly the inheritance and further development of traditional marketing methods. In the current era of rapid development of big data, higher requirements are put forward for marketing methods. E-commerce enterprises need more accurate and measurable marketing communication methods to pay attention to the realization of marketing communication plan and to increase investment in direct sales communication. Some scholars' positioning of precision marketing is to establish marketing information dissemination and feedback system based on the accurate positioning of personalized customers with the help of modern information technology, so as to help enterprises realize the rapid expansion of marketing territory at lower cost. There are also views that precision marketing is based on the psychological and behavioral characteristics of consumers. Through the combination of quantitative research and qualitative analysis, it summarizes the customer characteristics of different consumers in the target market, so as to achieve effective communication with customers and potential customers. The implementation basis of precision marketing theory is big data. The further development of cloud technology, together with the corresponding software tools to realize the storage, management and analysis of data information, has effectively promoted the popularization of big data technology application. Big data not only has a large amount of information, but also has a large type, description range and commercial values. For example, the personal information and consumption records of customers stored by mobile communication companies belong to the category of big data. These data can accurately help communication companies formulate product packages that are more in line with customer needs. Making full use of modern information technology to process, analyze and predict customer consumption big data can help enterprises to push product information and launch specific products to specific users, effectively stimulate consumers' consumption desire and maximize the feasibility of product sales. Precision marketing based on database requires us to establish a normative and complete consumer database and use data mining technology to explore consumers' shopping habits and preferences, Then, according to the analysis results, design the most appropriate products, formulate accurate marketing plan, implement corresponding business activities and then realize precision marketing. In the Internet age, precision marketing is mainly to understand the consumer psychology and behavior characteristics of Internet users through the Internet, understand the potential needs and preferences of Internet users through their online records, ese big data technology for precise positioning and achieve the goal of precision marketing.

Marketing status of small- and medium-sized e-commerce enterprises in the era of big data.

With the in-depth development of network economy, more and more consumer's consumption concept tends to be rational from blindness. Big data marketing strategy is an important basis for modern enterprises to engage in network operation. For e-commerce enterprises, it is necessary to continuously establish personalized consumption service mode to meet the needs of different customers. For some small- and medium-sized e-commerce enterprises, they fully realize the important value of big data analysis in e-commerce marketing, but they do not have the strategic strength of big data marketing in terms of application technology, talents and funds. The data resources mastered by these e-commerce enterprises are scattered and static, and it is difficult to provide accurate and personalized product services for consumers. Therefore, it is necessary to tentatively adjust the strategy of precision marketing. In the process of big data marketing, small- and -sized e-commerce enterprises need to obtain sufficient data information. Whether it is user data and product data in internal data or enterprise data and social network information in external data, they need to establish relevant data model through big data analysis to do a good job in precision marketing.

## 13.2 Analysis on the Advantages of E-commerce Precision Marketing Under the Three Big Data Environment

### *13.2.1 Improvement of Advertising Accuracy*

The research of precision marketing strategy uses information technology, which can collect and process all kinds of information of consumers. After subdividing the market business environment, it classifies and locates the demand and shopping behavior characteristics of consumers. After accurately determining the target consumer groups, targeted product and service information for different target consumer groups are adopted for publicity and marketing, which can effectively improve the purchasing desire of consumers and improve the accuracy of electronic advertising.

### *13.2.2 Improve E-commerce Service Level*

Precision marketing business activities are based on understanding and grasping the needs of consumers. In order to improve their own economic benefits, e-commerce enterprises can meet the needs of consumers to the maximum extent according to their consumption needs and preferences. Choosing the suitable logistics company, shortening the delivery time and optimizing the circulation links can improve the service level of e-commerce.

### *13.2.3 Accurate Marketing Model to Effectively Improve the Transferring Value*

Consumer delivered value refers to the difference between the total value and the total cost of consumption. The total consumer cost refers to the money and time paid when consuming products or services. Consumers in the process of consumption, the relevant costs will be reduced to the minimum value; meanwhile, the relevant will also reach the maximum value. Therefore, they need to choose the way of "tranferring value." The precision marketing model adopts differentiated services, which can increase the adaptability of products and services and the satisfaction of consumers, and it also improves the total value of products and services. Precision marketing uses modern information technology such as big data to accurately deliver product and service information to target consumers and reduce the cost of marketing.

## 13.3 Collection and Analysis of Big Data of Small- and Medium-Sized E-commerce Enterprises

### *13.3.1 Collection of Business-Related Data and Information*

For small- and medium-sized enterprises, their own data information is limited. When analyzing big data, we need to first mine the existing product and customer information within the enterprise; at the same time, we also need to collect and analyze the external data of the enterprise. At present, many platforms provide data information for customers and individuals in big data. For example, if you input keywords on the search page of Baidu platform, you can get a lot of data and information, including the distribution of Internet users, and it can help us optimize our digital marketing program. In addition, Alibaba B2B market can upgrade the unified data product platform on the business side based on the data asset layer and get the demand and sales status in the industry from the results. Enterprises can also extract information from product forums and related social networks through web crawler technologies such as Python.

### 13.3.2 Analysis of Business Data Information

User basic information analysis. This paper analyzes the relationship between the user's basic information and the user's purchase in the order information. This kind of association analysis method can analyze customers' shopping habits and help enterprises to establish a most effective price promotion strategy. Cluster analysis is an effective tool for market segmentation. It can be used to study consumer behavior, looking for potential experimental market, conduct diversified analysis and processing, analyzing the characteristics of major consumer groups, help enterprises to establish the relationship model between the color and performance of products and the gender and age of users, complete the comprehensive analysis of customer's basic information.

Analysis of customer visiting behavior. The analysis of customer visiting behavior mainly focuses on the relationship between behavior and effect, analyzing information such as the pages visited by users and the duration of page stay, the types of search engines, the content of keywords. Through the analysis and modeling of the above data, customers' consumption habits can be more clearly defined. It help enterprises to find problems from the content of web pages and promotion channels. Those habits are conducive to make the enterprise marketing more accurate and effective, which improves the income of enterprises.

Analysis of customer social relationship. In the current social environment, more and more consumers obtain relevant products information through social media and generate potential consumer demand. Enterprises must constantly understand the impact of social media on the market, adjust the promotion strategy and product structure, and innovate in business model and expand customer resources. In the era of mobile Internet, increasing interactivity will enhance the overall effect of market strategy. The information interaction and social application based on user identity enable enterprises to grasp the information of potential customers' consumption demand, and at the same time, the flow of enterprise commodity sales platform increases significantly, which lays a solid foundation for business model innovation and profitability improvement. In the social network, users with mutual relationship have similar interests and hobbies and can recommend products to each other. Moreover, using the social relations of users for marketing can reduce the target cost customers. The close relationship network group of users will become the main target customers of the enterprise, and the social heterogeneity between them is strong, which can form a large social communication network, so that the data information obtained can be quickly spread and help enterprises establish strong user relationship.

## 13.4 Implementation Mode of Precision Marketing Based on Big Data for Small- and Medium-Sized E-commerce Enterprises Based on Big Data

In the current era of rapid development of e-commerce information, the third-party e-commerce platform is becoming more and more mature. The cost of e-commerce construction is low, and it is easy to promote. Many small- and medium-sized enterprises are applying the third-party platform to carry out e-commerce, laying a good foundation for the construction of big data of enterprises. More and more third-party e-commerce platforms provide big data marketing analysis tools to provide users with consumer characteristic information such as age, gender, purchase time, and some also provide data including industry data analysis and related product sales dynamics. Enterprises can integrate the data on different platforms and develop big data marketing analysis tools suitable for their own marketing needs through self-development or cooperative development. According to the actual situation of the enterprise, if the storage and analysis of big data must be run on a third-party platform, the data analysis results can be obtained through the mobile terminal, so as to carry out big data marketing at a lower cost.

Enterprises must segment customers according to their gender, age, consumption habits and other basic characteristics and then use big data analysis tools to divide the customers who have purchase behavior into a subdivided customer group. We extract all kinds of factors that affect the purchase behavior of users and classify similar customer groups through cluster analysis. Then, according to the appropriate standards, enterprises divide customers into customers, target customers, and potential target customers and formulate differentiated marketing strategies for different customer groups. Finally, targeted product development and promotion are carried out to achieve accurate marketing strategy.

According to a large number of customer transaction records, enterprises establish product models related to various types of customers and put the products into the target market for different target customer groups. As long as there is more user data, users' preferences and purchase habits can be analyzed, and by obtaining data on product transactions, we can further expand our target customer. The enterprise will select the corresponding customer group according to the basic characteristics of users and the model established between products. It is able to continuously explore the characteristics of potential customers and continuously upgrade products.

## 13.5 Conclusion

To sum up, with the popularization of the application of big data technology, it has been widely used in various industries, promoting the progress and development of all walks of life. In the big data environment, enterprises use the corresponding technology platform to obtain information with commercial value, help enterprises innovate and implement precision marketing business model, and bring greater development opportunities for enterprises. The development of digital marketing and the collection and accumulation of all kinds of data and information have laid a good foundation for the marketing activities of enterprises and promoted the development of enterprises.

## Bibliography


1. Zhang, Q., Shi, F.: Application of big data precision marketing based on supply chain network information in e-commerce field. J. Logistics Technol. **09** (2017)
2. Zhong, M., Cao, Y.: Research on marketing strategy of small and medium sized e-commerce enterprises in the era of big data. J. China Market **49** (2015)
3. Liu, S.: Analysis on marketing management strategy of iron and steel enterprises under e-commerce environment. J. Market. Field. **46** (2019)
4. Jiang, L.: Taxation. Research on the application of customer relationship management in enterprise marketing management. J. Taxation. **34** (2019)
5. Xie, H.: Research on marketing management of e-commerce enterprises based on data mining. J. Brand Res. **11** (2019)

# Chapter 14
# Big Data Platform for Oil and Gas Production Based on Apache Spark

**Peng Qing, Yi Li, Shuqin Luo, and Zhuoqun Xu**

**Abstract** For promoting the construction of intelligent oil field and applying big data technology to the daily production activities of oil field, this paper proposes a data mining platform based on Apache Spark multi-functional oil and gas production big data. This platform combines real-time and historical data processing-related framework and machine learning framework, designs massive oil and gas production data processing for different oil fields, supports real-time and offline processing data mining and other functions, supports the prediction of some data using related machine learning algorithms, supports the cleaning of large amounts of data, is used to train various deep learning models and will provide corresponding interfaces and permissions to relevant oil fields. Docker container is used to build the training environment of deep learning and machine learning, and the Kubernetes framework is used to complete the scheduling function. Struts, Spring and Hibernate relatively classic background processing framework is adopted at the Web framework level of the platform, different functions are decomposed into independent functional modules to reduce coupling, and agile development is used to improve the extended performance of the whole big data platform. This platform allows our data analysts to focus on data analysis and model training, without spending a lot of time and effort to deal with data problems, while the related calculation results are returned to other oil companies in an interface way or provide relevant operation authority to the field operation and maintenance personnel to access the data platform and data analysis.

## 14.1 Background Information

In recent years, with the rapid development of artificial intelligence [1, 2] and big data [3] technology, all walks of life have undergone rapid technological upgrading in data information technology processing, especially in traditional industries. In the field of oil, the extensive use of business systems within major oil enterprises puts forward

P. Qing · Y. Li (✉) · S. Luo · Z. Xu
Tsinghua University, Shenzhen 518000, GD, China
e-mail: liyi@sz.tsinghua.edu.cn



V. Chang et al. (eds.), *Modern Industrial IoT, Big Data and Supply Chain*,
Smart Innovation, Systems and Technologies 218,
https://doi.org/10.1007/978-981-33-6141-6_14

higher requirements for data processing, such as improving the efficiency of single well operation, rational planning of oil production capacity, reducing transportation costs, improving production efficiency, enterprise oil field data management and providing high-level decision support. At the same time, with the digitization of oil fields and the construction of intelligent oil fields [4], all kinds of Internet of things equipment are also used in oil production activities. These Internet of things equipment constantly produce large amounts of data in production activities, and these data show explosive growth. Such as in oil field multiphase [5, 6] flow measurement, only one multiphase flow meter a day of ECT [7], microwave [8], Venturi [9] and other data nearly 1 G. These multiphase flow-related oil and gas content data directly or indirectly reflects the output of a certain oil well and oil quality, and because of the very large amount of these data, the traditional data analysis method can no longer meet the current demand, so the multiphase flow of oil and gas big data in data mining [10] needs higher requirements. At the same time, there are also great problems in how to apply a large amount of data to deep learning. Because these data are collected directly from multi-phase flow meters, there are problems such as missing, error, redundancy and so on, which cannot be directly used in the training of deep learning [11] models; therefore, the data needs to be preprocessed with the help of big data platform, and then the processed data is trained and predicted by the advanced learning model.

Oil field enterprises have a large number of real-time oil and gas production data, and the application of mass production data is indeed a certain difficulty, especially for different enterprise data privacy cannot achieve data interaction, so the data is handed over to third-party companies and used the third-party big data platform for data analysis, the fused data is applied to the training of each model, and then the results of data analysis and the training model are returned to different enterprises to improve the reliability of the model.

Many oil companies, such as Shell, Chevron, Saudi Aramco and Schlumberger, have built their own big data platforms [12, 13] within their companies within their companies, but these big data platforms set up different business modules according to different needs, use data for data analysis, give on-site personnel decision support, reduce the possibility of manual intervention through automated decision making, and reduce the workload of data analysts. In the oil industry, there is no corresponding data analysis platform for the subdivision direction of multiphase flow. Therefore, the focus of this paper is on oil and gas static and dynamic multi-dimensional production big data based on multi-sensor technology means to support the intelligent management of oil and gas production and propose a set of solutions to these problems, establish a system for production of big data analysis and deploy it in the actual production environment.

## 14.2 Main Technical Analysis of Big Data

This platform mainly takes the Spark [14] as the computing core framework to realize the platform construction, Spark is the fast and general computing framework, compared with the traditional Hadoop [15], one of its main features is that the computation is based on memory, reduce the IO on disk, thus make the computing speed show fast growth, and provide the API, based on Python, Java, Scala and SQL, so it can be used in conjunction with other big data frameworks. And Spark core computing engine has fast and general features, and it can combine a variety of different application scenarios to specifically design different components, such as Spark SQL and machine learning [16] frameworks, these components can be understood as Spark encapsulated modules for common task scenarios, components can be called to each other, and various functional requirements can be realized through the combined use of these components. The components are shown in Fig. 14.1.

Spark core implements the functions of task scheduling, memory management, error recovery, interaction with storage systems and also contains the API definition of the Resilient Distributed Datasets (RDD); it mainly acts as a system administrator. Throughout the Spark, Spark SQL is a template in the Spark suite that converts the computing tasks of data into RDD computing in SQL form, similar to the Hive that converts the computing tasks of data into seamless integration of MapReduce, and Spark core in the form of data, and can configure the logic implementation when writing the entire RDD application, unified data access, Spark SQL provides standardized SQL queries, Spark succession to Hive, Spark SQL inheritance and operation of Hive syntax are realized through embedded Hive or connected to externally deployed cases, standardized connection mode, Spark SQL can support JDBC and ODBC access by starting the thrift server and use itself as a BI server; Spark Streaming mainly carries on the flow computation to the real-time data; MLlib is a parallel high-speed machine learning library built on the Spark and especially designed for big data processing, which is characterized by the use of more advanced iterative, memory storage analysis and calculation, so that the calculation and processing speed of data are much higher than that of ordinary data processing engine, used in the platform to achieve such as oil well similarity analysis, oil well anomaly classification; GraphX

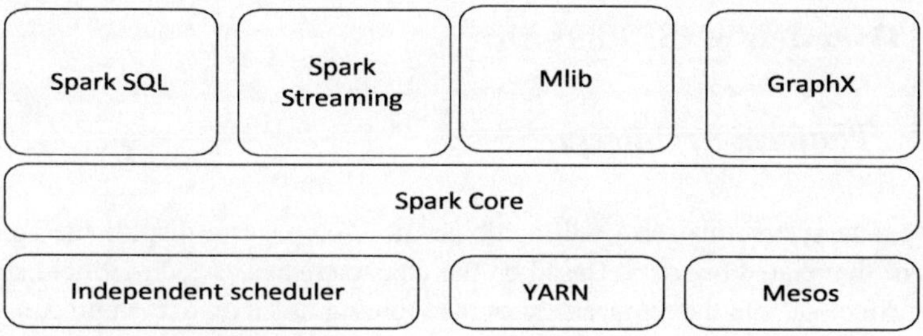


**Fig. 14.1** Spark component diagrams

is used to manipulate graphs, and graph computing can be carried out in parallel, supporting various operations of graphs.

Based on actual oil field production and data processing, the big data platform of oil and gas production is mainly considered from several aspects: storage, calculation, program scheduling and data presentation, but the big data platform will focus on storage and computing. Since the current amount of oil and gas data is very large, it is necessary to adopt the corresponding storage strategy, the data sources can be divided into two main categories: historical data and real-time data. Historical data, that is, contains historical data and historical data of each day, the historical data is stored on local disk, and the historical data of each day is transmitted through the Ali cloud and put into the Hbase through SQL script, where HBase needs to rely on the HDFS, historical data in the Hadoop is imported into the data of the pilot data platform for the first time, because the large amount of data requires disk space and backup mechanism, the real-time data is generated in real time, and the amount of data is less at this time. For some data that needs real-time processing, this is entered into the big data platform through the corresponding interface for processing, here HBase can dynamically add columns based on actual requirements, and the column is empty without storing data, saving storage space, automatically cutting data, making data storage automatic horizontal scalability, can provide high concurrent read and write operation support and other advantages, so used to store data. On the big data computing level, the current popular big data computing framework is mainly the MapReduce and Spark. In the Hadoop, because the speed of the whole computing framework is affected by the need to store the intermediate results in the disk regularly, it will be gradually abandoned in the future. Beyond this computing power Hadoop, it is essential more of a distributed data infrastructure: It distributes different data sets to a common computer cluster containing multiple nodes for storage according to distribution rules, meaning that no more hardware is needed in the actual production environment, Hadoop, it indexes and tracks the data to an unprecedented level of efficiency in big data processing and analysis; Spark is a tool dedicated to processing big data for those distributed storage, it does not store distributed data, but Spark will combine the underlying distributed storage to calculate, so in this platform will also use Hadoop HDFS as distributed storage part and YARN to do resource management.

## 14.3 Overview of the Platform

### *14.3.1 Platform Architecture*

The big data system platform will combine the Web page to display the analysis results of the related big data. Based on the characteristics of each oil field storing data on Ali cloud and the requirement of data mining based on data mining, and the whole platform can be used to train various AI models. This platform according to the software engineering design process advanced data mining and platform expansion of the overall requirements analysis, and for different business requirements to design different business modules, using agile development to achieve the functions of each module, the overall system architecture diagram is shown in Fig. 14.2.

1. Authority management: support to the employees of different departments within the company as well as other operators to establish the corresponding authority, for different roles also need to develop different authority.
2. Data management: The platform can preprocess the data by scripting language, or it can process files by code written automatically by the platform; it supports data processing from different sources, such as interface data, Ali cloud data, local data, real-time data, etc., and it supports various data forms, such as CSV, Excel, TXT text, json and Xml format data, database data, HDFS, Hive, Hbase and so on.

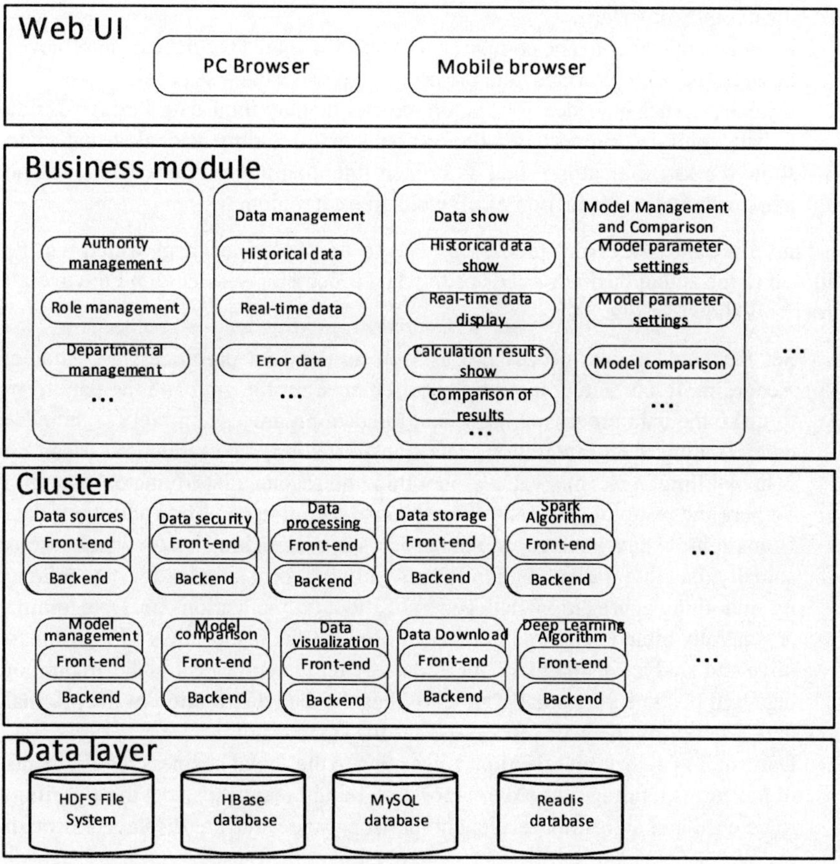


Fig. 14.2 Schematic diagram of the platform

3. Platform algorithm implementation: including regression, classification, cluster analysis, time series model and other machine learning algorithms, in addition to the platform can be trained with TensorFlow and presented results on the platform and various parameters of the same model are compared or compared with different AI models.
4. Data life cycle management: The platform supports data mining, model training, including data cleaning, data standardization, data visualization, etc., for all hydrocarbon data. Support to download and filter data that meets the requirements through the platform.
5. Real-time processing and offline processing: The platform according to different oil and gas production data sources and formats, for a large number of data offline processing and oil and gas data through the transmission interface real-time processing.
6. Friendly interaction: The platform design can drag, pull and other ways to adjust the complex interface.
7. Model management: The comparison of specific model results can be achieved through the Web interface, and the optimal model can be selected.
8. Friendly visual interface: The interface can display total data through Baidu Echarts plug-in, support two-dimensional or three-dimensional visual chart show the key indicators, such as broken line diagram, graph, radar map and map to show the production of oil wells in each region.

Platform data processing processes need to be phased, each of which can be tailored to the actual business needs of the data processing services, and the overall process is shown in Fig. 14.3,

1. Set the goal of data processing as well as the data preprocessing scheme: According to our actual business requirements, set the goal and the data form to make the data processing flow, and the data preprocessing process includes data cleaning, data integration, data transformation, data reduction. Clean up data by filling in missing values, smoothing noise data, identifying or removing outliers and resolving inconsistencies; data integration combines and stores data from multiple data sources, and the process of establishing a data warehouse is actually data integration; data is converted into a form suitable for data mining by smoothing aggregation, data generalization, normalization, etc. Data mining is generally aimed at a large number of production data. It takes a long time to mine and analyze a small amount of data. After data processing, the amount of data will become very small, but it still has the data distribution of the original data. Therefore, the later processing has the same results.
2. Data mining algorithm selection: According to the actual business requirements of the corresponding oil field or laboratory requirements to select the algorithm to be used and algorithm results comparison, the results are displayed through the Web.
3. Training of models and data mining: selecting the optimal and the most effective parameters according to the results of different models and using the platform

Fig. 14.3 Data processing flowchart

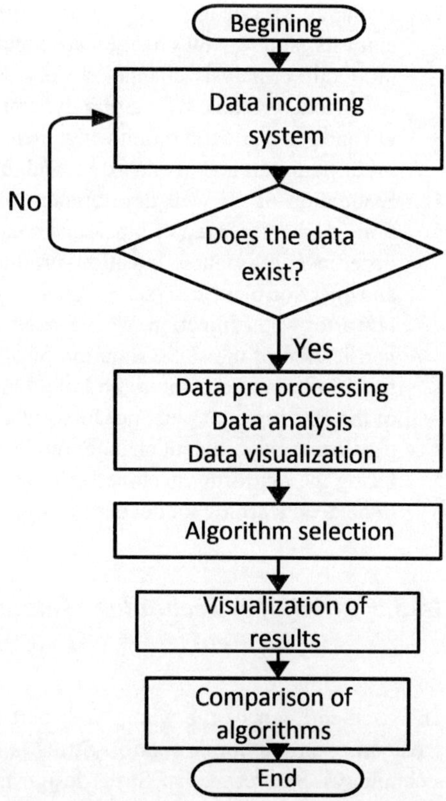


machine learning algorithm library and deep learning library to complete the corresponding data mining requirements.

4. Data visualization: With the help of the platform Web, front-end pages can display the results of data mining or intermediate results, which can be based on the data statistical analysis visualization results to select the appropriate preprocessing algorithm.

## *14.3.2 Main Functions of the Platform*

Various data analyses and projections can be made using the platform, such as.

1. According to different data sources, oil and gas production data management can be divided into historical data and real-time data processing; data types can be divided into ECT, Venturi and microwave data processing; data granularity can be divided into detail data and summary data processing.

2. Dynamic production analysis, including production analysis, measure effect analysis, single well change cause analysis, single well oil and gas dynamic production analysis change cause, oil and gas dynamic production analysis level (total level 5), well group dynamic analysis change cause.
3. Oil and gas well fault monitoring, including sucker rod break, piston resistance, oil or pump oil leakage, wax or sand, insufficient supply.
4. Evaluation of oil well development plan, determination of reasonable injection–production ratio of block and small layer, reasonable production pressure difference, reasonable injection–production strength, provision of distribution and injection plan; analysis and grasp of the oil well effect, see water and flooding law after water injection, work system to ensure stable production of oil well; verification of the static situation of oil field (oil layer together with the situation, fault location, nature and blocking effect), evaluation of the adaptability of the development plan, put forward adjustment plan; forecast the production, pressure, water content change trend and oil well flooding law.
5. Using the platform, machine learning library can be predicted and can be said to be deep learning model deployed to this platform and then in the prediction.

### *14.3.3 Platform Technology Implementation and Platform Deployment*

The back-end part of the whole Web part uses Struts, Spring and Hibernate data processing and authority control. Struts is a good framework for MVC, the main technology is Servlet and Jsp, Struts design makes it very easy for developers to understand, let us write the program hierarchical, playing a central control role throughout the MVC, Spring provides a unified approach to business and can help developers develop good habits of interface programming rather than class programming, thus realizing decoupling principles, Hibernate is mainly used to persist data, which is an object-oriented database operation method, which encapsulates the JDBC, so that users can put their best efforts on business logic as much as possible, making interaction with relational databases very lightweight. The SSH architecture is as follows (Fig. 14.4).

Web front page uses TML, CSS, JavaScript and so on as the main frame of the page, in the whole front-end HTML determines the structure and content of the page, CSS sets the presentation style of the page, JavaScript controls the behavior of the page, besides, we need to use Baidu Echarts, jQuery and other third-party plug-ins to realize the special function of the page and beautify the page. After the completion of the front end, it can be used to display part of the data, data through the back-end implementation, part of the page implementation as shown in Figs. 14.5 and 14.6.

On the whole big data platform level, Apache Spark is chosen as the computing core, and Spark can process historical data and real-time data. The oil and gas production data of the whole platform is divided into dynamic data and static data according to the principle of aging, the dynamic data and static data need different processing

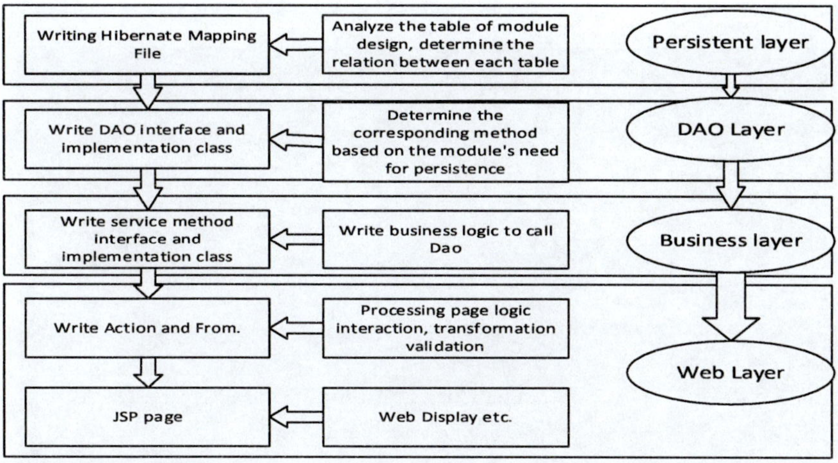


**Fig. 14.4** SSH architecture

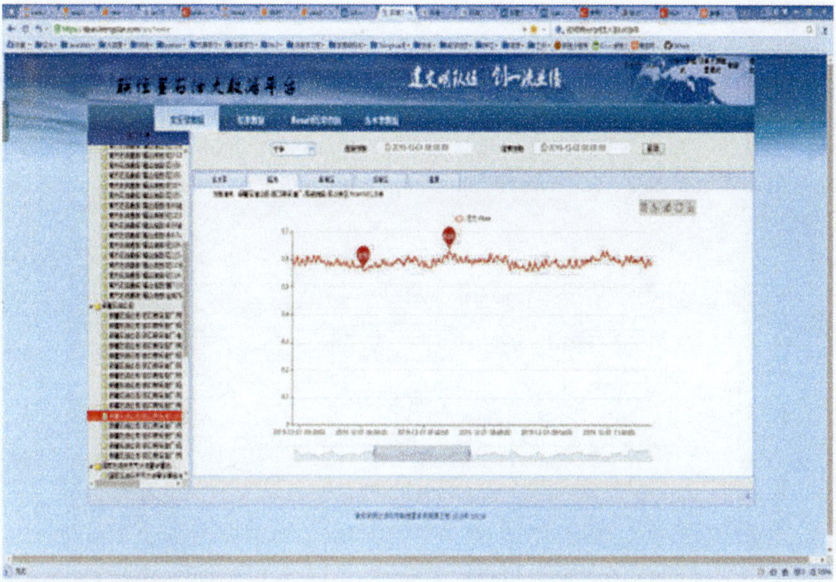

**Fig. 14.5** Historical data view

methods, in the platform for a large number of historical data can be stored in the HBase, each day of historical data through Ali cloud temporary storage, then in the script language to download and store data in the HBase, real-time data can be transmitted through the interface and calculated using the Spark computing framework, the intermediate results are stored in the Redis memory database, the results of the

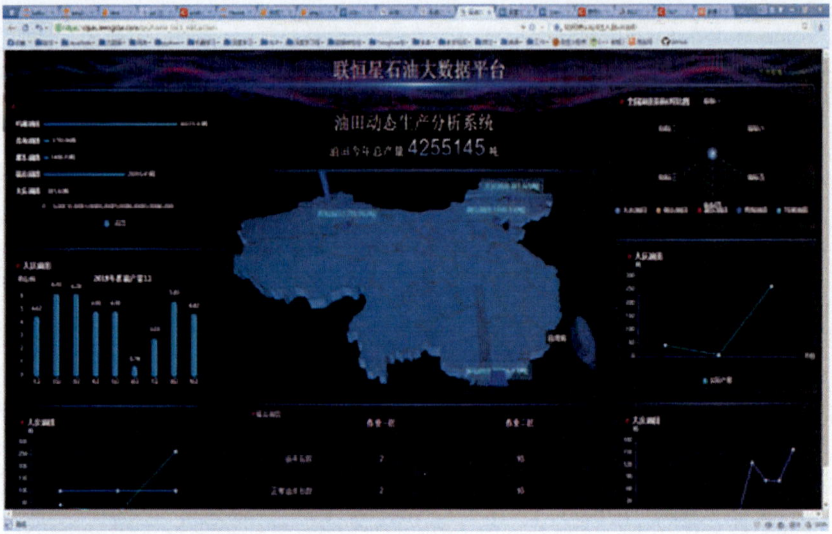


**Fig. 14.6** Oil and gas production data display

calculation are stored in the MySQL database and then displayed on the page through the Web, using hive it can analyze big data, because the whole platform has multiple job, it needs Azkaban job scheduling, and the whole data platform provides interfaces for external calls. The overall big data platform system architecture diagram is shown in Fig. 14.7.

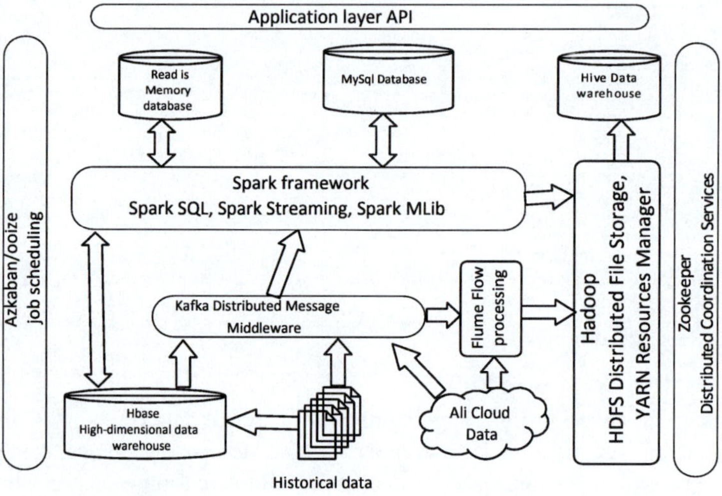


**Fig. 14.7** Big data framework

## 14.4 Experiment

After the whole platform is built, part of the display interface can refer to Figs. 14.5 and 14.6, the experimental part uses Spark MLlib to achieve clustering operation, because most of the raw data have no labels, and the data can be classified and labeled in practical application, and we can use clustering to label each label-free data. In order to get better results, we can use some attributes of Venturi data to cluster, select the average value of different wells in the same time period and select five attributes of Venturi data: pressure, front differential pressure, back differential pressure, temperature and moisture content. Firstly, the original data was normalized and normalized, then the K-means was used to cluster the oil wells into five classes, and the actual comparison was five grades from high to low, as shown in Fig. 14.8.

In addition, neural network regression can be used for long-term prediction when the oil content of a certain moment can be predicted using big data platform, and mobile average prediction can be used for short-term prediction as shown in Fig. 14.9.

In addition, the platform can deploy other deep learning models to the platform for data analysis, oil and gas production data management, oil and gas dynamic production analysis, oil and gas well fault monitoring, development scheme evaluation and so on. Through the customized data mining function provided by the platform, oil and gas mining enterprise professionals can make use of their professional knowledge to carry out data mining and data processing conveniently, without having to spend their energy studying the complex technology implementation related to big data processing.

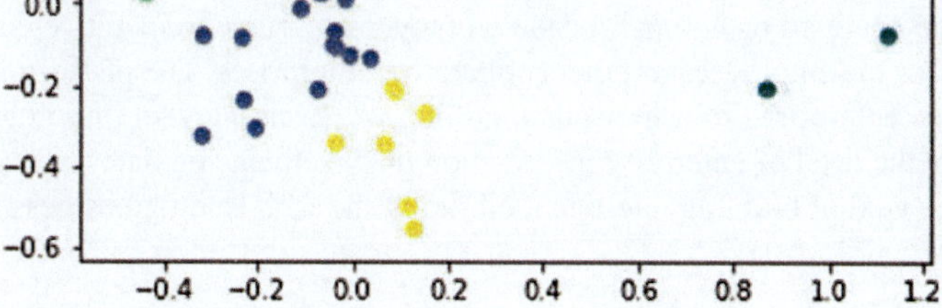


**Fig. 14.8** Clustering of oil output by clustering

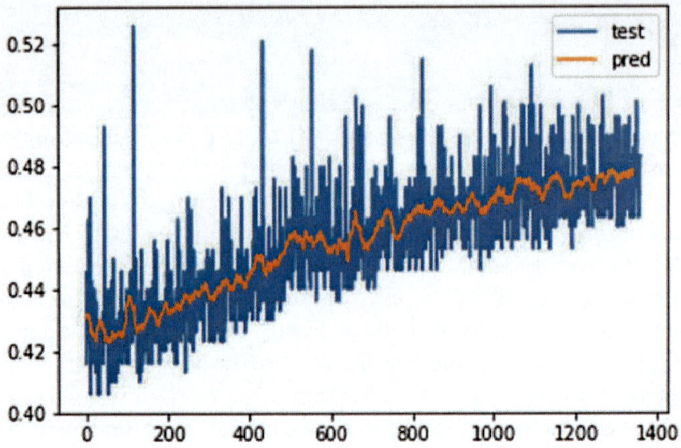


**Fig. 14.9** Use moving averages to predict output values for the next few minutes

## 14.5 Conclusion

This paper proposes, designs and implements an oil and gas big data platform based on Apache Spark as the core of calculation. This platform provides a friendly data processing and data mining service for data analysts and IT operators from data acquisition, data preprocessing, data mining, data visualization and later in-depth learning model online training and model results comparison module. Data management and download module can make the enterprise internal personnel will not spend too much time in the data preprocessing stage, the system will be more intelligent to remove the wrong data, more convenient to find data without spending too much time in data processing, data display module can make the data analyst pay more attention to the quality of the data itself, discover the regularity of the data, present part of the calculation results on the Web page, and support real-time display of data, and the permission management module can make the personnel of different departments within the enterprise have access to different modules and whether the data can be operated or not and for the employees of other cooperative enterprises resources for the employees of other cooperative enterprises. The platform has been deployed to enterprises to achieve data mining work and play an important role in it, improve the level of enterprise information management and data processing and achieve the goal of building intelligent oil fields and data leading business.

## References


1. Ligeza, A.: Artificial Intelligence: a modern approach. Appl. Mech. Mater. **263**(2), 2829–2833 (2009)
2. Weiss, G.: Multiagent Systems: A Modern Approach to Distributed Artificial Intelligence. MIT Press (2000)
3. Baaziz, A.: Analyse bibliométrique de la publication: how to use big data technologies to optimize operations in upstream petroleum industry. Working Papers (2020)
4. Zhi, L.I., Qiang, C.: Design and applications of digital oilfield. Geo-inf. Sci. (2002)
5. Abou-Kassem, J.H., Islam, M.R., Ali, S.M.F.: Introduction to modeling multiphase flow in petroleum reservoirs. Pet. Res. Simul. (Second Edition), 397–456 (2020)
6. Sabzevari, F.M., Winter, R.S.C., Oloumi, D., et al.: A Microwave sensing and imaging method for multiphase flow metering of crude oil pipes. IEEE J. Select. Top. Appl. Earth Observations Remote Sens. **13**(99), 1286–1297 (2020)
7. Chaturvedi, P.K.: Microwave measurement: instruments and techniques (2018)
8. Zhao, C., Wu, G., Li, Y.: Measurement of water content of oil-water two-phase flows using dual-frequency microwave method in combination with deep neural network. Measurement (2018)
9. Shuangxi, L., Renlong, S., Caihong, X., et al.: Sensitivity analysis of oil and gas two-phase venturi flowmeter.China Pet. Mach. (2018)
10. Gholba, N.D., Babu, A., Shanmugapriya, S., et al.: Application of various open source visualization tools for effective mining of information from geospatial petroleum data **XLII-5**, 167–174 (2018)
11. Schmidhuber, J.: Deep Learning in Neural Networks. Elsevier Science Ltd. (2015)
12. MapR Technologies, Inc. Mtell and MapR Deploy Big Data Platform for oil and gas to manage real-time and historical sensor data. Inf. Technol. Newsweekly (2015)
13. Zhao, R., Shi, J., Zhang, X., et al.: Research and application of the big data analysis platform of oil and gas production. In: International Petroleum Technology Conference (2019)
14. Zaharia, M., Chowdhury, M., Franklin, M.J., Stoica, S.S. Spark: cluster computing with working sets. Technical Report No. UCB/EECS-2010-53 (2010)
15. AAO, BJD, BRDS. 'Big data', Hadoop and cloud computing in genomics. J. Biomed. Inf. **46**(5), 774–781 (2013)
16. Mitchell, T.M.: Machine Learning. McGraw-Hill (2003)

# Chapter 15
# Single-Well Yield Prediction Based on LSTM and MA Combination Model

**Peng Qing, Haifeng Zhang, Yi Li, Shuqin Luo, and Zhuoqun Xu**

**Abstract** On the basis of the characteristics of oil single-well production and oil well abnormal events, it is proposed to use the combination of long short-term memory network (LSTM) and moving average (MA) to predict oil single-well production, to use time series model with LSTM long-term memory ability to predict oil single-well production and to use MA to predict output after abnormal events fluctuation, and MA can therefore be used to predict output after abnormal events, and then to use LSTM to predict subsequent output after actual output is stable. The experimental oil and gas production data shows that the combination model of LSTM and MA is compared with the BP neural network and the single LSTM. The prediction results of combination model of LSTM and MA are significantly better than individual LSTM and BP neural networks for the situation after abnormal events in oil wells. The combined model can obviously reduce the yield prediction error within a period of time after the occurrence of abnormal events, make it more real close to the actual yield value and improve the prediction accuracy, which is of great significance in practical application.

## 15.1 Background Information

Oil is a very important raw material in the daily industrial production activities, and the amount of oil production is of great significance to guide the production of chemical enterprises and the business activities planned by enterprises. At the same time, accurate prediction of oil production can make enterprises make reasonable production tasks and avoid blind decision-making. At present, the main methods to predict the single factor time series variables are regression method, grey system prediction and fuzzy system prediction. The traditional model and prediction methods, such as autoregressive model (AR), moving average [1] (MA) and autoregressive integrated moving average, can be used to predict time series when the linear relationship is

P. Qing · H. Zhang · Y. Li (✉) · S. Luo · Z. Xu
Tsinghua University, Shenzhen 518000, GD, China
e-mail: liyi@sz.tsinghua.edu.cn



V. Chang et al. (eds.), *Modern Industrial IoT, Big Data and Supply Chain*,
Smart Innovation, Systems and Technologies 218,
https://doi.org/10.1007/978-981-33-6141-6_15

satisfied, or unsatisfied linear relationships, and BP [2–4] neural networks can be used to predict. In recent years, there are other prediction methods which can reflect the intrinsic structure of the data to some extent, which play a certain role in solving the problem of nonlinear and non-normal time series, but it is hard to implement code. In addition, the single-well yield prediction method does not take into account the influence of abnormal events on the yield prediction, especially in the period after the abnormal events occured in the well. Along with the development of deep learning application in oil field, the correlation algorithm of deep learning has also been used to predict the time series variables and achieved good results, especially in the single-well yield prediction, when the output of the time series is relatively stable. However, there will be a large difference between the predicted and actual oil production values when the abnormal event occurs. This paper judges the abnormal events according to the characteristics of the original Venturi data measured by the multiphase flowmeter and combines the yield training LSTM model of time series to predict the overall yield trend, after the abnormal event, the MA is used to predict the yield value for a period of time in order to obtain a higher accurate value.

## 15.2 Theoretical Basis of Related Algorithm

### *15.2.1 LSTM Circulatory Neural Networks*

LSTM [5–8] first appeared in 1997 is a special case of a cyclic neural network and LSTM developed on the basis of RNN [9], but RNN is a chain structure with repetitive neural networks and long-term dependence, which in turn leads to the problem of gradient explosion and gradient descent, resulting in weak memory of excessive time. The problem caused by continuous multiplication of matrices in RNN network can be solved by using the LSTM memory principle. The model is improved for the problem of hidden layer, and the network memory function is extended so that the model can obtain more lasting information and slow down the rate of information decay.

The core of the LSTM is a memory unit, relative to the RNN added forget gate, input gate and output gate, the function of various gates is to reduce the output to a minimum of 0, a maximum of 1, the signal is discarded when the output is 0, and signal is completely retained when the output is 1, and the number between 0 and 1 indicates that part of the model is retained. In the LSTM network structure, there are three main doors, the first is the forgot gate, and it can reduce information that the model does not think is needed. The second gate is the input gate that allows the selected input information to enter the model. The third is the output gate that allows the information to be the output. LSTM structural units are as follows.

In Fig. 15.1, i is the input gate, the $f$ is the forget gate, the o is the output gate, the $c$ is the state of the whole memory cell, $x$ represents the vector of the input signal, and $h$ represents hidden layer vectors. As shown in (15.1)–(15.6), LSTM six formulas

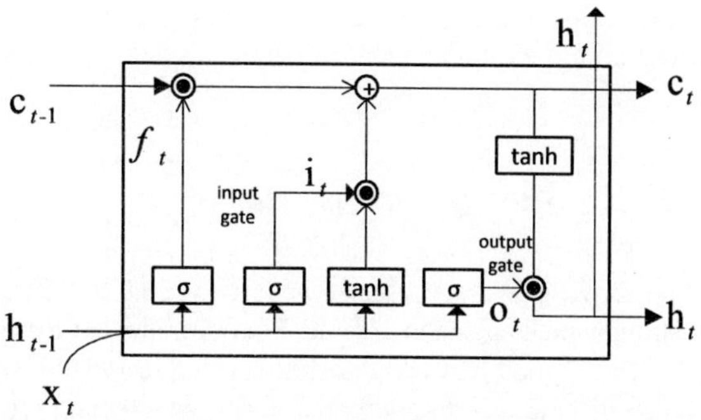


**Fig. 15.1** LSTM memory unit structure

for the three gates and the implied and cellular states are as follows:

$$I_{t=\delta(W_{xi}x_t+W_{hi}h_{t-1}+b_i)} \quad (15.1)$$

$$f_{t=\delta(W_{xf}x_t+W_{hf}h_{t-1}+b_f)} \quad (15.2)$$

$$o_{t=\delta(W_{xi}x_t+W_{ho}h_{t-1}+b_o)} \quad (15.3)$$

$$c_t = f_t * c_{t-1} + i_t * \tanh(w_{xc}x_t + w_{hc}h_{t-1} + b_c) \quad (15.4)$$

$$h_t = o_t * \tanh(c_t) \quad (15.5)$$

$$\widetilde{C}_t = \tanh\left(w_c \cdot \left[h_{t-1}, x_t\right] + b_C\right) \quad (15.6)$$

In all the formulas, where w represents the weight matrix, the $b$ is biased, and the $\delta$ is a sigmoid function, $\tanh(x)$ as hyperbolic tangent function, $\delta$ and $\tanh(x)$ as activation function, and there are four sets of parameters in the whole memory cell that need to be solved during the training.

LSTM model uses a BPTT algorithm based on back-propagation algorithm to update all weights during training. LSTM whole model training can be divided into four steps: (1) The output value of each layer LSTM is calculated by forward calculation method; (2) the error term of each LSTM unit is calculated by reverse calculation according to the result, mainly including the direction of LSTM depth and LSTM time series by transverse and longitudinal direction; (3) the gradient is calculated according to the loss rate; (4) the corresponding algorithm is then used to update all the weight parameter w and bias parameter b. During the selection of gradient optimization algorithm, optimization acidification includes SGD, AdaGrad, RMSProp and so on. This paper selects adaptive moment estimation algorithm (Adam), and Adam is a gradient-based stochastic optimization method among all optimization algorithms, which combines the advantages of AdaGrad and RMSProp algorithms, that is, momentum and environment.

### 15.2.2 Moving Average Model

The moving average algorithm is a typical statistical model that can be used to estimate the local mean of yield. For one hand, compared with other dynamic test data processing methods, the algorithm is simple and the computation is small, especially it can be calculated in recursive form, which can save storage unit, fast and easy to process non-stationary data in real time. For the other hand, there is some subjectivity and arbitrariness in the moving average method, and the parameter $N$ of the moving average algorithm can be reasonably selected by experience. The updated formula for variable $y$ recorded as $y_t$, $y_t$ at the $t$ time is as follows:

$$y_t = \frac{1}{N} \sum_{i=t-N}^{t-1} y_i \tag{15.7}$$

The coefficient for each $y_i$ is $1/N$ in the preceding formula (15.7). But the disadvantage of moving average is that it is only suitable for short-term calculation, because MA calculates the average value, the predicted value is based on the past period of time, so the calculated average value cannot be too long. When the calculation time is too long, all the predicted values will come close to a fixed value, so the moving average is not suitable for long-term yield prediction.

### 15.2.3 Model Combinations

Since the LSTM has the advantage of long-term memory for processing yield prediction, the MA algorithm can make it fluctuate less after dealing with abnormal events. In this paper, two algorithms are combined to predict the yield:

The LSTM prediction model can be obtained by training the time series data into the LSTM model according to the characteristics of the time series, but the LSTM prediction model alone does not deal with the abnormal events or the production changes caused by the operation of the well by the injection process, that is, yield prediction in an ideal state without considering external abnormal events, so the results after abnormal events are predicted using MA, and then the predicted results of the MA model are used to replace the predicted values of LSTM in the corresponding time periods. Finally, the revised single-well predicted output is obtained.

Using MA prediction instead of LSTM prediction model combination process, that is, selecting the most basic eight columns from the original Venturi data based on three-phase flowmeter measurements, which can reflect oil well production or react to abnormal events, data attributes are pressure, front differential pressure, back differential pressure, temperature, moisture content, mass instantaneous flow (oil), volume instantaneous flow (gas), mass instantaneous flow (water), the data is processed into the sum of every hour and then the Venturi data is used to determine whether there is an abnormal event; this paper uses more than 20% of output change to judge abnormal events combined with 8 columns of Venturi raw data to judge abnormal events if 8 columns of Venturi raw data change greatly, then judge the type of abnormal events and the proportion of output change of this event relative to normal conditions according to the current data K and then use the MA model to the prediction. Since there are two ways to calculate the production ratio K after an abnormal event occurs in an oil well, the K value can be calculated directly using the contrast value of the output change before and after the abnormal event or using the original data of Venturi and other models to learn the ratio K. Here the first method is simpler, and the second is more accurate; then the K is multiplied by formula (15.7) to obtain the output value after the prediction. The combination of LSTM model and MA model applied in practice is shown in Fig. 15.3.

The advantage of the combined model is to use the LSTM to predict the overall output of a single well when the output volatility is small, to continue the prediction after the abnormal event because the actual output will fluctuate, not suitable for the LSTM to predict the period after the abnormal event, so the MA is used to make the output region stable for a period after the abnormal event and then to use the LSTM to predict when the actual output after the abnormal event tends to be stable.

## 15.3 Model Description and Experiment

The LSTM and MA models used in this paper can be divided into four processes: data processing, feature selection, model training and combination prediction. To verify the performance of the LSTM and MA combined model under abnormal conditions, the yield results were compared with those predicted by BP neural network regression and individual LSTM models, which shows that the combined model has better prediction yield results.

### *15.3.1 Data Processing and Feature Selection*

The experimental data is from real data of an oil field in the Southern China, which contains the daily cumulative data of production of a certain oil well and the corresponding date of Venturi data and the output data of the original oil field per minute. The data of Venturi data measured by the multiphase flowmeter is 10 frequency samplings per second [10, 11]. After the output data processing, the data is two columns, respectively, time and output corresponding to that time. For the LSTM model, the stability of the data needs to be checked, and the output data will change greatly after the abnormal event, but the stability before the change is not changed much, so the data stability cannot be processed before sending the data into the model.

### 15.3.2 LSTM Model Training

An overall framework of the LSTM prediction model constructed in this paper is shown in Fig. 15.2, and the whole structure contains several parts. For the input layer, the output per hour after processing is used as the input, the hidden layer part uses the LSTM cells represented in Fig. 15.2 to build circulating neural network, and the output layer is to predict the output value of the next hour.

Based on the results of the oil well output per hour, a total of 229 days of data are included, some of which are missing or wrong and finally processed into 5103 data, each data corresponding to an hour of output, and then processed into LSTM

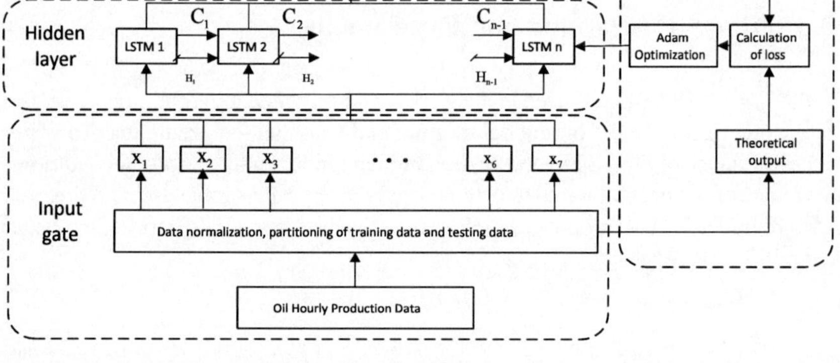


Fig. 15.2 LSTM forecast structure

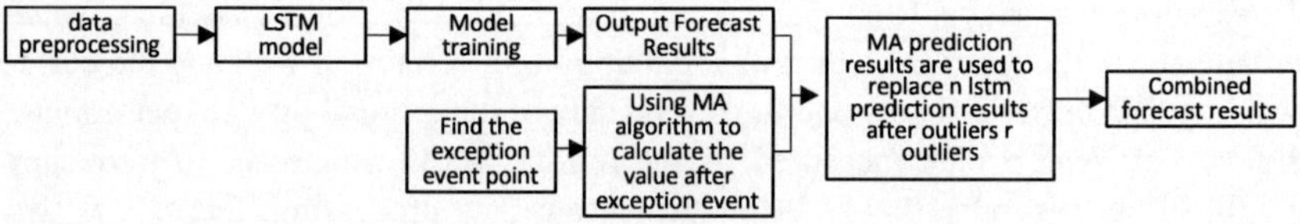


Fig. 15.3 LSTM and MA model combination forecasting flowchart

input vector. The data is input into the LSTM model, LSTM network uses the function that comes with in the Keras function library, the training index of the whole network model is the predicted root mean square error (RMSE) error, the network model structure of each layer is a single layer of LSTM layer, the loss function uses the RMSE function, the optimizer selects the Adam algorithm, the model batch processing length is 100. LSTM, the model training, is completed, and the next hour output can be predicted. Based on the model error, the model error is evaluated quantitatively by RMSE of the training set and the test set. Formula (15.8) is as follows:

$$RMSE = \sqrt{\frac{1}{n}\sum_{i=1}^{n}(p_i - y_i)^2} \tag{15.8}$$

In formula (15.8), $p_t$ represents oil and gas production forecast,$y_t$ is the measured value, and the smaller the RMSE value, the better the prediction effect of the whole model on the oil target output. The sustainability model is used to predict the entire oil production over the next few hours. In the sustainability model, t-n observations to the t-1 moment as observations t at the next moment as the prediction result of the persistence model are the performance baseline of the data set, which can be used as the evaluation index to judge the effect of the LSTM model. The model training process divides data sets into training sets and test sets, takes the first 70% of the data from 2019-04-25 to 2020-01-13 as the training set, and the latter 30% data as the test set, $F_o = \{f_1, f_2, \ldots, f_n\}$for hourly time series of oil production, and then the partitioned training set and test set can be expressed as $F_{tr} = \{f_1, f_2, \ldots, f_m\}$ and $F_{te} = \{f_{m+1}, f_{m+2}, \ldots, f_n\}$, meet the constraints $m <$ nand $m, n \in N$. Element $f_t$ in the training set is then standardized, by using the classical z-score standardization formula (mean 0, standard deviation 1), a standardized training set can be expressed as $F'_{tr} = \{f'_1, f'_2, \ldots f'_m\}$, the $f'_t = \left(f_t - \sum_{t=1}^{n} \frac{f_t}{n}\right) \Big/ \sqrt{\sum_{t=1}^{n}\left(f_t - \sum_{t=1}^{n} f_t/n\right)^2 \Big/ n}$ $1 \le t \le m, t \in N$ and the processed data is then fed into the entire model for training.

### 15.3.3 *LSTM and MA Model Combinations Forecast Yield Values*

The MA model does not require training, in predicting yield, the past $N$ hours yield at the current time point is used, here the weight for each past hour is ($1/N$), where the $N$ is the number of yield values for the past $N$ hours, and then the next hour yield is calculated by combining the average of the last $N$ hours yield. Since the characteristics of moving average are not suitable for continuous prediction of output over a long period of time, it is very important to select the appropriate $N$ and the number of l hours in the future. When the prediction period is too long, the predicted results will approach the same value with little significance, and the best $L$ according to experience is not more than 10, so the $N$ and $L$ choices in this paper are 7.

### 15.3.4 MA Model Forecast Yield

The LSTM forecast of output can be used to predict the output value for each hour in the future according to the output trend, but the output will also change when the injection and production process is adjusted or the oil well is abnormal. It is generally considered that the output change is more than 20% of the possible abnormal events. To determine the time point of occurrence of an anomaly based on the Venturi data measured using a multiphase flowmeter, then calculate the ratio value of the output data before and after the time point of the abnormal event, the ratio value K directly using the output of one hour after the anomaly and one hour before the abnormal time then multiply the output value of the past N hours with the ratio value and then use the MA to predict the output value of the next few hours. The combination of the LSTM and MA models is shown in Fig. 15.3.

## 15.4 Analysis of Experimental Results

### 15.4.1 BP Neural Network Prediction Results

In Fig. 15.4, BP neural network regression selects four hidden layers, each layer contains a network of 16 neural units, and finally the fully connected layer is a single neuron, for a more realistic trend of fitting yield, and the data is used processed raw data. After training 200,000 times, the overall results are shown in Fig. 15.4. It can be seen from Fig. 15.4 that the fitting ability of BP neural network is limited, and the fitting prediction above the details is not very good, which can only reflect the prediction trend of the whole data.

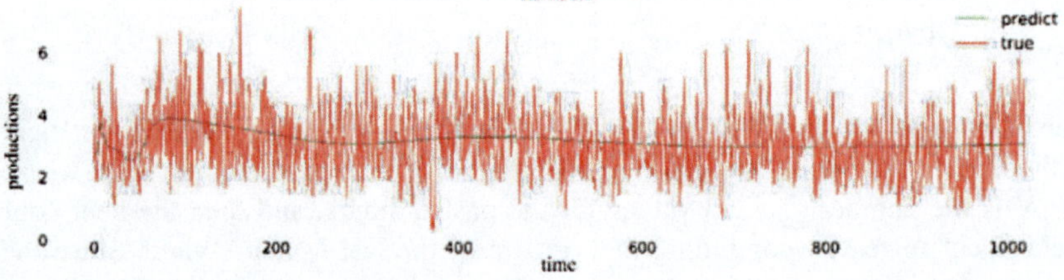


**Fig. 15.4** BP neural network regression simulation

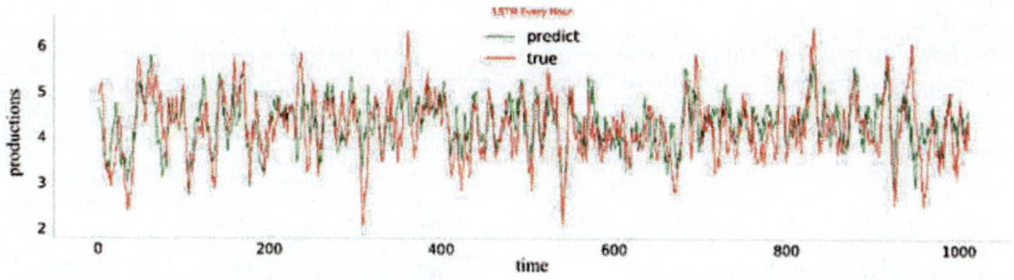


**Fig. 15.5** LSTM model projections separately

### *15.4.2 LSTM-Independent Forecast Results*

LSTM persistence model predicts the number of steps set to 7, the number of hidden layer units 150 and the number of neurons 4. It can be seen from Fig. 15.5 that the green curve is the real data curve, the red curve is the predicted value, and the yield value can be predicted well for most of them. Because the LSTM yield forecast is based on the previous period of production forecast, the output will fluctuate with the change of time, the larger the prediction error will become later, and the abnormal event cannot be predicted in the long term, so the long-term yield forecast will deviate from the true value.

### *15.4.3 LSTM and MA Combination Projections*

After using the original data and the corresponding yield data for anomaly point discrimination in the prediction, or can use the yield change more than 20% to judge the abnormal time point change from Fig. 15.6, we can see that the yield after using moving average to correct the abnormal event can better fit the actual yield curve, higher accuracy relative to BP neural network regression, compared with individual LSTM, since only two abnormal events occurred in the whole prediction range, the overall prediction accuracy is not improved, and the prediction accuracy of the

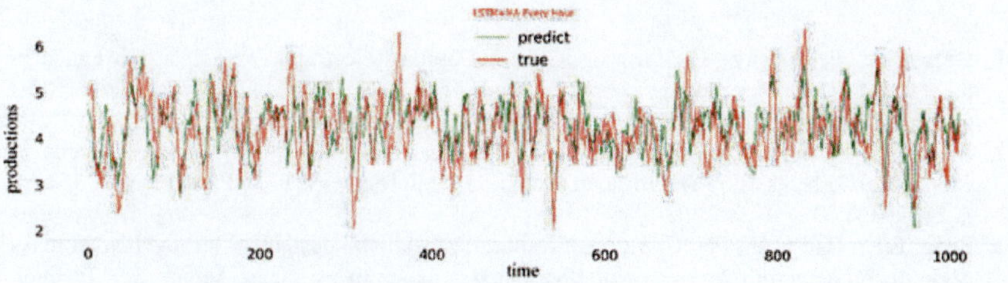


**Fig. 15.6** Results of LSTM and MA models combined

**Table 15.1** Accuracy of three algorithm models predicting after abnormal time

| Name of algorithm | Accuracy rate (%) |
|---|---|
| BP neural network | 58.23 |
| LSTM | 65.20 |
| LSTM + MA | 85.56 |

LSTM and MA combination model is 20.36% higher than that of the LSTM model (Table 15.1).

## 15.5 Conclusion

Through the principle of single-well oil field production and the characteristics of data, the LSTM and MA combination models are used to predict single-well oil production and compare the predicted output value with the actual output value. LSTM and MA combination models have obvious advantages over BP neural network regression and individual prediction models, LSTM can predict the trend of output per hour as a whole, and the output will change when the abnormal event occurs. Because the output change fluctuates greatly after the abnormal event occurs, the ratio of output before and after the abnormal event change can be calculated using multiphase flow data and then combined with the yield and MA model N hours before the current time to predict the output. Since LSTM prediction will combine yield data over the past period, the prediction value depends on the historical yield value, and if only the MA prediction is used, there will be an average value; when the prediction time is longer, the yield prediction is very close to a very flat curve, and the yield cannot be predicted separately; when the actual yield value tends to be stationary LSTM, then it is used for subsequent prediction. This paper uses LSTM and MA model in the field of reliability prediction and extends the application of deep learning techniques.

## References

1. Drevinskas, T., Telksnys, L., Maruska, A., et al.: Capillary electrophoresis sensitivity enhancement based on adaptive moving average method. Anal. Chem. (2018). https://doi.org/10.1021/acs.analchem.8b00664
2. Bastani, D., Hamshaw, M.E., Davardoost, F., et al.: Prediction of $CO_2$ loading capacity of chemical absorbents using a multi-layer perceptron neural network. Fluid Phase Equilib. **354**(5), 6–11 (2013)
3. Song, L.K., Bai, G.C., Fei, C.W.: Multi-failure probabilistic design for turbine bladed disks using neural network regression with distributed collaborative strategy. Aerosp. Sci. Technol. **92**(SEP.), 464–477 (2019)
4. Songlin, D., et al.: Predicted SAGD development effects by BP neural network for the high-watercut reservoir. Petroleum Geology & Oilfield Development in Daqing (2019)

5. Gers, F.A., Schmidhuber, J., Cummins, F.: Learning to forget: continual prediction with LSTM. Neur. Comput. **12**(10), 2451–2471 (2000)
6. Shi, X., Chen, Z., Wang, H., et al.: Convolutional LSTM network: a machine learning approach for precipitation nowcasting. Comput. Vision Pattern Recogn., 802–810 (2015)
7. Sugiyarto, A.W., Abadi, A.M.: Prediction of Indonesian palm oil production using long short-term memory recurrent neural network (LSTM-RNN). In: 2019 1st International Conference on Artificial Intelligence and Data Sciences (AiDAS). IEEE (2020)
8. Wu, Y.X., Wu, Q.B., Zhu, J.Q.: Improved EEMD-based crude oil price forecasting using LSTM networks. Phys. A: Stat. Mech. Its Appl. **516**, 114–124 (2019)
9. Shi, Z., Liang, H., Dinavahi, V.: Direct interval forecast of uncertain wind power based on recurrent neural networks. IEEE Trans. Sustain. Energy, 1–1 (2017)
10. Ramli, M.F., et al.: Multiphase flow measurement by electrical capacitance tomography and microwave cavity resonant sensor. In: 2019 IEEE International Instrumentation and Measurement Technology Conference (I2MTC). IEEE (2019)
11. Yun, Z.: Current status of development and application of multiphase flowmeters. China Petroleum Machinery, Beijing (1999)

# Chapter 16
# Comparative Study of Think Tank on Domestic and Overseas Based on WOS and CSSCI

**Su Nan**

**Abstract** Taking the WOS and CSSCI as data source, using meta-analysis and visual methods build knowledge maps, to carry out comparative study of think tank on domestic and overseas. The results show that the research focus of the global think tanks has remained relatively stable at present, Chinese think tank research is the lack of cooperation network, research of China's think tank was narrow and less, research method of China's think tank is outdated, and China's think tank research is the lack of study on microcosmic, concrete, and future issues. In short, China's think tank research should learn from Western experience and focus on improving the quality and innovation.

## 16.1 Introduction

There have been relevant literature on think tanks at home and abroad, but from the perspective of journal subject groups, geographic cooperation networks, and dual clustering topics, the results of comparative research on domestic and foreign think tanks are still very rare. Therefore, it is urgent to compare the similarities and differences in research focus, cooperation networks, core strengths, and hot topics of think tanks at home and abroad and explore the historical experience of think tank construction in developed countries, to make up for the lack of research and construction of think tanks in China, to provide support and reference for improving the level of think tank construction and enhancing think tank research capabilities, and to provide intellectual support for accelerating the construction of new type think tanks with Chinese characteristics in the twenty-first century and realizing the Chinese dream of the great rejuvenation of the Chinese nation.

S. Nan (✉)
School of International Studies, Sichuan University, Chengdu 610065, Sichuan, China

Collaborative Innovation Center for Security and Development of Western Frontier China, Sichuan University, Chengdu 610065, Sichuan, China



V. Chang et al. (eds.), *Modern Industrial IoT, Big Data and Supply Chain*,
Smart Innovation, Systems and Technologies 218,
https://doi.org/10.1007/978-981-33-6141-6_16

## 16.2 Data Sources and Research Methods

With Web of science core collection database as data sources, and with "think tank", "brain tank" as key words, the retrieval period is from 1998 to 2019. After accurate retrieval, 684 pieces of literatures were found. With "think tank", "brain tank" as the key words, the retrieval period is from 1998 to 2019. With CSSCI core collection database as data sources, 396 pieces of literatures were found. Based on these literatures, establish a local database for domestic and foreign think tank research and comparison. After carrying out data removal and standardization, import bibliometric tools for visual analysis.

The research adopts methods such as journal group analysis, spatial geographic analysis, co-citation analysis, and co-word analysis to explore the core of knowledge, cooperation networks, core strengths, and hot topics of domestic and foreign think tank research and use the literature meta-analysis and visualization tool CiteSpace4.4 to interpret the above research questions in the form of a network map [1, 2].

## 16.3 Knowledge Mapping of Think Tank Research at Home and Abroad

### *16.3.1 Focus Shift of Think Tank Research Based on Double-Picture Superposition Analysis*

Periodical double-graph overlay analysis is a new method to dig out important information such as the distribution direction, citation trajectory, and center of gravity drift of research in various disciplines [3]. The analysis of the double-image overlay analysis of journals of think tank research is shown in Fig. 16.1.

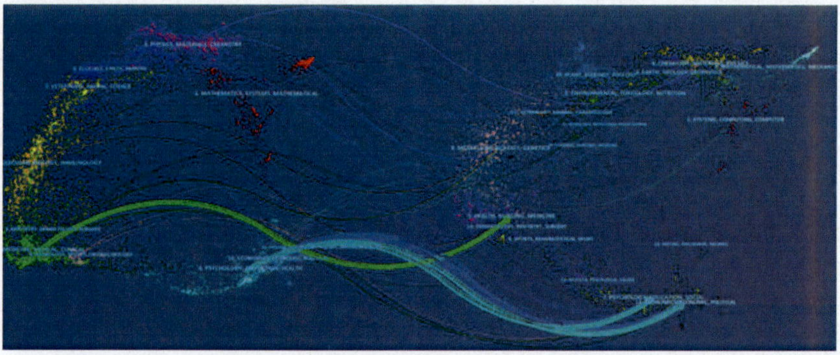

**Fig. 16.1** Superposition analysis of the journal of think tank research

The ellipse on the left shows that the research is mainly located in the journal category, economics–economic–political, psychology–Education–health, and medicine–medical–clinical. The ellipse on the right shows that the research is mainly located in the journal category economics–economic–political, psychology–education–social, and health–nursing–medicine. The line is the citation relationship between journals, and the citation direction is from the cited journal on the right to the citing journal on the left. Through Z-Score processing, three band-shaped significant relationships are obtained, showing the knowledge source of think tank research, that is the core citation journal groups are economics–economic–political, psychology–education–social, and health–nursing–medicine.

It can be seen that think tank research originated from economics, political science, psychology, education, sociology, medicine, and other fields, and in the subsequent development process, it continued to strengthen the research foundation in these fields. After comparison, it is found that the journal subject group of think tank research has changed very little, which indicates that the research field is still in its infancy stage with a short history, and there has not been any significant research diffusion or knowledge transfer, so it is difficult to have the discipline attribute. However, from the branch connection, we can see that the knowledge diffusion of think tank research has shown a multi-threaded development trend. In the future, global think tank research results may become an important literature source and theoretical support for other disciplines. According to the statistical analysis of literature source journals, think tank research is mainly distributed in public administration, policy science, international relations, economy and finance, sociology, psychology, higher education, medicine, cancer clinic, and so on.

### 16.3.2 Think Tank Research Cooperation Network Based on Google Earth

In order to understand the research cooperation of global think tanks, the documents recording spatial geographic information were imported into Google Earth after being processed by KML2.0 to draw the atlas of global think tank research cooperation network. Figure 16.2 shows the cooperation network of think tanks in

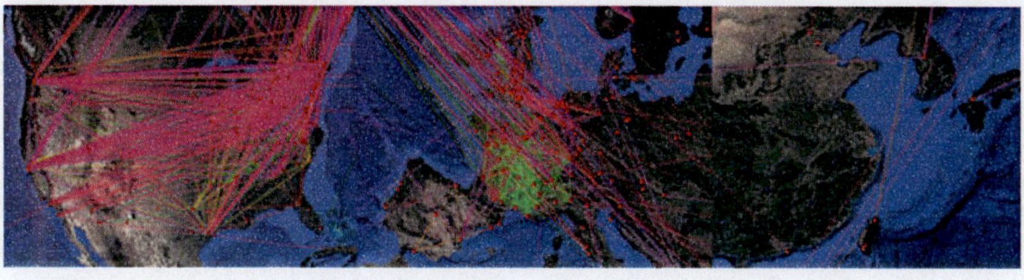

**Fig. 16.2** Think tank research collaboration network in the USA, Europe, and East Asia

the USA, Europe, East Asia, and other countries or regions. The nodes in the figure represent the geographical location of the researchers, and the connections represent the cooperative relationship between the researchers. The network formed by the two shows the density of cooperation between the researchers.

The map shows that the number of nodes, the number of connections, and the network density vary greatly between different countries and regions, indicating that the global distribution of think tank research is very uneven. Especially in the Lakes Basin and northeast areas of the USA unusually intense cooperation network, regional distribution with numerous well-known universities, research institutions such as the best carrier of research, a think tank, at the same time, the political center, economic center in New York, Washington, industrial center international cities, such as Chicago, also have a large number of professional think tank, Among them, the USA has the largest network density, the largest number of nodes, and the most complex connections. Especially the cooperation network in the Great Lakes Basin and the northeastern region of the USA is extremely dense. There are a large number of well-known universities and research institutions in the region that have become the best carriers for think tank research. At the same time, international metropolises such as the political center Washington, the economic center New York, and the industrial center Chicago also have a large number of professional think tank institutions. Washington alone has the Brookings Institution, the Center for Strategic and International Studies (CSIS), and Woodrow Wilson International Center for Scholars. The Wilson Center, the Carnegie Corporation of New York, the World Bank, etc., Seattle, San Francisco, and Los Angeles on the west coast of the USA are also gathering places for think tanks. For example, RAND is located in Santa Monica, Los Angeles. European think tanks are concentrated in France, Germany, the UK, Belgium, the Netherlands, Denmark, and other countries. From the perspective of cooperation network, cooperative research in the USA and Europe is very frequent, and cross-domain cooperative research between the USA and Europe is also very close. These studies promote the exchange of information, the spread of technology, and the collision of ideas, which is conducive to the production of innovative research results. In comparison, East Asia has not formed a trans-regional cooperative research network, and there are few internationally representative think tanks in each country. In particular, the international results of China's think tank research are basically in a blank state. The main manifestations are as follows: first, the lack of global competitive and influential local think tank institutions or representative results; second, the cooperation network of think tank research has not yet been established, and the cooperation mechanism has not yet been established. A model and cooperative ecology remain to be incubated.

### *16.3.3 Cluster Atlas of Think Tank Research Based on Co-word Analysis*

Co-word analysis is to count the frequency of a group of words appearing in the same document in pairs. Based on this co-occurrence frequency to measure the affinity relationship, it is often used to analyze the hot content of the subject, the distribution of the subject, and the structure of the subject. In this paper, co-word analysis is carried out on the basis of word frequency statistics. By using Sati2.0 for word frequency statistics, 437 effective keywords were extracted from WOS database literature, with a total frequency of 2,753, averaging 4.3 per article. 391 effective keywords were extracted from CSSCI database literature, with a total frequency 1,610, averaging 4.1 per article. Tables 16.1 and 16.2 show the high-frequency keywords of think tank research at home and abroad. Based on these keywords, we will carry out co-word analysis and draw the co-word cluster atlas of domestic and foreign think tank research, as shown in Figs. 16.3 and 16.4, respectively.

It can be found from Table 16.1 and Fig. 16.3 that the core labels are think tank, politics, policy, USA, governance, foreign policy, democracy, public policy, climate change, China, innovation, education, NGO, information, etc.

On the whole, the research content of foreign think tanks is rich and diversified, emphasizing pragmatism, involving medical and health, education, poverty management, economic development, non-profit organizations, foreign policies, climate change, public security, China issues and many other fields, these researches focus on theory research and empirical research, quantitative research, and qualitative research, they are much skillful in the integrated research of historical stories and modern stories, macro and micro-research, abstract, and specific problems, and the

**Table 16.1** List of high-frequency keywords studied by foreign think tanks (top 30)

| No. | Key words | Frequency | No. | Key Words | Frequency | No. | Key words | Frequency |
|---|---|---|---|---|---|---|---|---|
| 1 | Think tank | 100 | 11 | Democracy | 9 | 21 | Education | 5 |
| 2 | Politics | 21 | 12 | Neoliberalism | 9 | 22 | NGO | 5 |
| 3 | Policy | 25 | 13 | Knowledge | 8 | 23 | Decision-making | 5 |
| 4 | Science | 18 | 14 | Public policy | 7 | 24 | Community | 5 |
| 5 | United States | 14 | 15 | Discourse | 7 | 25 | Information | 5 |
| 6 | City | 11 | 16 | Climate change | 7 | 26 | Government | 5 |
| 7 | Management | 10 | 17 | China | 6 | 27 | Policy research | 5 |
| 8 | Governance | 10 | 18 | Quality of life | 6 | 28 | Expertise | 5 |
| 9 | Foreign policy | 10 | 19 | Innovation | 6 | 29 | Poverty | 4 |
| 10 | Advocacy | 9 | 20 | Ethics | 5 | 30 | International relation | 4 |

**Table 16.2** List of high-frequency keywords studied by domestic think tanks (top 30)

| No. | Key words | Frequency | No. | Key words | Frequency | No. | Key words | Frequency |
|---|---|---|---|---|---|---|---|---|
| 1 | Chinese characteristics | 138 | 11 | College think tank | 36 | 21 | Scientific decision making | 18 |
| 2 | New think tanks | 138 | 12 | G20 | 32 | 22 | Local universities | 18 |
| 3 | Think tank | 119 | 13 | Science technology think tank | 31 | 23 | Southeast Asia | 17 |
| 4 | University think tank | 92 | 14 | Educational think tank | 30 | 24 | Public decision-making | 17 |
| 5 | Think tank construction | 88 | 15 | China think tank | 27 | 25 | American think tank | 17 |
| 6 | Think tank with Chinese characteristics | 76 | 16 | Collaborative innovation | 25 | 26 | Government decision-making | 17 |
| 7 | New educational think tank | 72 | 17 | Public policy | 23 | 27 | Information research | 16 |
| 8 | America think tank | 57 | 18 | Library and information | 22 | 28 | Sports think tank | 16 |
| 9 | Decision-making consultation | 50 | 19 | Private think tank | 21 | 29 | Research university | 16 |
| 10 | Influence | 40 | 20 | Policy process | 20 | 30 | RAND | 15 |

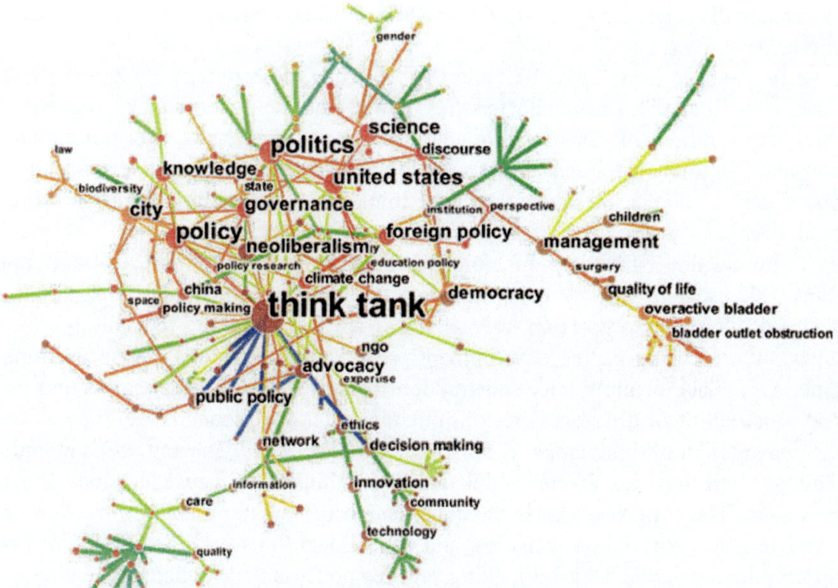


**Fig. 16.3** Co-word clustering map of foreign think tank research

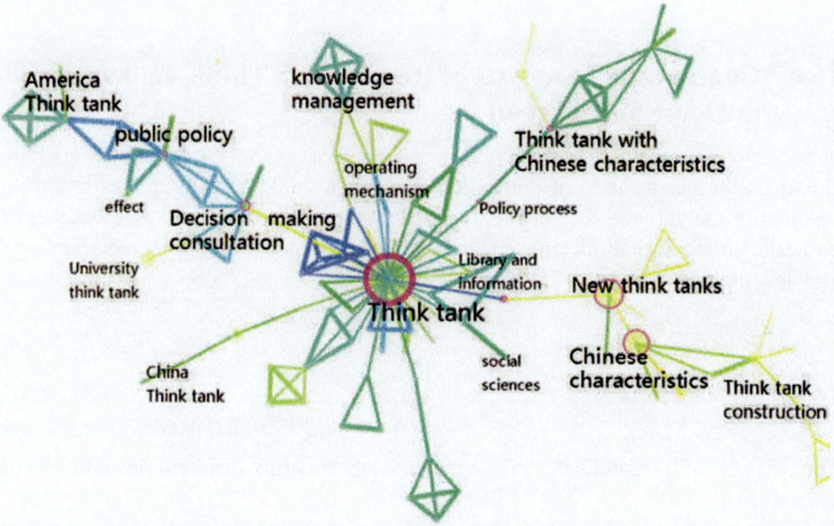


**Fig. 16.4** Co-word clustering map of domestic think tank research

systematically scientific system of modern think tanks researches are becoming more and more perfect.

It can be seen from Table 16.2 and Fig. 16.4 that the core tags of China's think tank research are Chinese characteristics, news think tanks, think tanks, university think tanks, think tank construction, new educational think tanks, decision-making consultation, influence, science and technology think tanks, collaborative innovation, public policy, books intelligence, private think tanks, scientific decision-making, public decision-making, American think tanks, intelligence, etc.

There are three directions for domestic think tank research: first, international think tank research. The key nodes include American educational think tanks, Southeast Asia, RAND Corporation, private think tanks, science and technology think tanks, etc., focusing on the development process of international advanced think tank, experience of think tank construction in European and American countries, and construction of different types of think tank, etc.; the second is the research on the construction of think tanks with Chinese characteristics. The key nodes include Chinese characteristics, Chinese think tanks, new think tanks, new educational think tanks, etc. The main research is the construction of new think tanks with Chinese characteristics, research universities, and educational think tanks, etc.; the third is think tanks and public decision-making, and key nodes include collaborative innovation, public policy, public decision-making, scientific decision-making, government decision-making, etc. The main research is the influence of think tanks on public policy.

## 16.4 Comparative Analysis of Research on Think Tanks at Home and Abroad

Based on the cluster map of domestic and foreign think tank researches based on co-citation analysis and co-word analysis, this paper makes a comparative analysis of domestic and foreign think tank research from three aspects: hot topic, methodology, and development trend.

### *16.4.1 Hot Topic*

The research of foreign think tanks are extensive, covering different disciplines and specialties.

In the spectrum, politics and policy appear the most frequently, indicating that political policies and government decisions attract the most attention. Similar nodes are specific policy fields, such as education services, public security, medical and health care, social management, international relations, economic policies, basic security, and scientific and technological development. The policy focus of think tanks in European and American countries focuses more on public decision-making, government governance, foreign policy, energy security, public medical care, transportation, aging protection, science and technology, family development, etc. For example, RAND Corporation issues annual policy analyses focusing on areas such as basic health care, the child assistance act, cybercrime prevention, poverty assistance, human health, workforce and population development, public disaster prevention and management, and public skills training. Some think tanks focus on terrorism and homeland security, economic policy, Internet finance, manufacturing development, and other topics. For example, Brookings Institution, the most influential think tank in the USA, has been long concerned about the development mechanism, prevention mechanism, and cooperative combat of terrorism. Topics, the national counter-terrorism investigation report and the Middle East national military security report formulated by it are often adopted by the US Pentagon and become the external brain support for formulating counter-terrorism policies.

In addition, foreign think tank research emphasizes innovation and foresight. This feature is mainly reflected in think tanks' attention to new technologies and products, such as satellite technology and space flight technology in the 1950s, Internet and supercomputer in the 1970s, and system science since the 1990s. Since the twenty-first century, the frontier topics of European and American think tanks have shifted to artificial intelligence, big data, cloud computing, automation, and other aspects to help their home country maintain a long-term lead in the high-tech field. For example, under the joint efforts of think tanks such as Brookings Institution and the Carnegie Peace Foundation, the US approved the "Big Data Research and Development Plan", three years ahead of China's big data strategy.

The think tank researches content in China is relatively concentrated in the theme of the international think tank research, the think tank construction with Chinese characteristics, the government public decision-making, and other field. In terms of international think tank research, it is actually focused on the construction and development of the think tank in Western developed countries. By introducing, learning and digesting advanced foreign practices improve the quality of Chinese think tank construction. In the research on the construction of think tanks with Chinese characteristics, think tanks are required to pay attention to China's issues and the Chinese field, respond to the urgent needs of the party and the state for think tank intellectual support, and give play to the enthusiasm and initiative of different subjects. In terms of the nature of institutions, role orientation, operational mechanism, talent pool, technological innovation, and performance appraisal, we will explore solutions that suit China's national conditions and build a new type of think tank with Chinese characteristics that is inclusive and diversified. In the field of government policies and public decision-making, it mainly focuses on research on decision-making issues such as social governance, basic public services, medical system reform, administrative approval reform, supply-side structural reform, targeted poverty alleviation, migrant workers, industrial structure upgrading, opening to the outside world, and the Belt and Road Initiative. From the map, it is found that my country's think tanks still lack effective penetration in research on similar micro-, specific, and practical issues such as worker health and welfare, juvenile delinquency risk and prevention, women's health issues, and domestic violence prevention.

The comparison found that my country's think tank research is still relatively backward in terms of topic selection perspective, research depth, and innovative thinking, and it is difficult to meet the needs of comprehensively deepening reforms and modernizing national governance capabilities [4]. At present, my country's think tank research urgently needs to learn from Western experience in terms of problem orientation, targeted thinking, system models, and operating mechanisms to broaden research horizons, focus on research issues, reform research mechanisms, enhance research capabilities, and update research tools.

### *16.4.2 Methodology*

The west generally regards theoretical research, empirical research, empirical research, and its method tools as an important prerequisite for supporting think tank research. Theoretical models help them to clarify their thinking and explore the essence of the problem; historical experience is to explore solutions through horizontal and vertical comparative analysis; empirical research is to verify alternative paths through repeated mathematical information [5]. For example, RAND has successively created research methods such as Delphi method, critical path method, and quasi-quantitative analysis method [6]. McKinsey has proposed methods such as quantitative strategy matrix method, value tree, structure–behavior–performance model, [7] etc. Since twenty-first century, Western think tanks have continuously innovated methods and technical tools, etc., and proposed steady-state network models, equilibrium matrix analysis, mixed qualitative simulation analysis, institution–domain-applicable analytic hierarchy model, etc., [8] more importantly, some developed think tanks are sensitive to the importance of Internet information tools in information collection and problem research, so they have used data mining and data extraction technology methods earlier to use the power of big data to help solve real problems [9]. For example, as an emerging top think tank, the Asian Policy Research Institute of South Korea uses big data and cloud computing technology to build a massive data information database and creates numerous high-quality think tank reports in short time, which are adopted by the Korean government and relevant authorities.

According to the statistics of "Global Think Tank Report 2018", the number of think tanks in China has reached 735, firmly ranking the second-largest think tank in the world, but China only accounts for 9 of the top 100 think tanks. This shows that although the number of think tanks in China is huge, but international competition strength and research quality are still relatively backward. The weak competitiveness of my country's think tanks has a greater relationship with insufficient methodological innovation. Taking the empirical method as an example, the literature review shows that my country's think tank research relies more on regression analysis, panel analysis, fuzzy analysis, gray relational analysis, etc. [10]. The research methods of Western think tanks have been updated and upgraded from the aspects of informatization, automation, and intelligence. For example, the Brookings Institution has constructed Pascal and Erc databases and used diachronic meta-analysis, full-chain analysis, and fuzzy systems on a large scale. In analysis, disequilibrium analysis, system simulation test, integrated analysis, and other methods, McKinsey & Company has adopted crisis management simulation analysis and qualitative analysis in risk assessment.

### *16.4.3 The Development Trend*

Driven by the information society, Western think tanks have shown the following trend characteristics: First, the research of European and American think tanks is developing in the direction of diversified fields and rich levels. Through different perspectives such as politics, economy, culture, law, technology, finance, and even military, discuss many topics from national strategic decision-making to basic people's livelihood and propose mature research results; second, Western think tanks combine practical value, problem orientation, and theoretical research. While investing in energy, climate, social education, security governance, and other issues, it emphasizes the digging of in-depth laws from historical experience and basic theories to achieve the win–win of basic research and cutting-edge forecasting; third, the operating mechanism of think tanks is more modern, structured, and flexible. Take the "revolving door" of think tanks in the USA and Australia as examples. Think tank researchers can flexibly switch between the roles of officials and scholars. The existence of the revolving door makes it easy for the government to obtain external brains with rich research experience, and it also gives think tank researchers more advantages such as practical experience and problem awareness.

However, China's think tank development is facing some obstacles. First of all, the think tank research field in our country is relatively narrow, most think tank is the official think tank or semi-official think tank. They are susceptible to the will of the government in the selection of topics. They pay more attention to national strategy, economic policy, foreign relations, public management, party and government building, and other topics. Basic people's livelihood issues and social reality issues are more focused on macro-policy research. The lack of micro-level perspectives and in-depth discussions on specific issues has led to a lack of effectiveness and operability in think tank research on social issues. Secondly, the operating mechanism of think tanks in China retains a strong color of planning. A large number of think tank personnel have the status of civil servants or public institutions, the flow of think tank personnel is not smooth, the introduction of private think tank talents is relatively difficult, and the lack of a market-oriented operation mechanism for think tank research limits the "contending of a hundred schools of thought and a hundred flowers blossoming". Third, China's think tank research still faces the contradiction of autonomy and independence. Think tanks have strong dependence on the party and government departments. The hot spots that think tanks pay attention to are often set by the government. They generally lack the ability or courage to question and criticize the government. The consciousness of "cold learning" has not been formed, and it is difficult to maintain credibility, autonomy, criticality, and independence. Finally, China's think tank research is often short-sighted, utilitarian, far from predictive and forward-looking, it is difficult to "see the horizon a little further away", and it is difficult to balance the relationship between "solving current problems and meeting future problems" And the relationship between "theory and practice".

In addition, international think tank research has already broken through individual case studies and spread to the analysis of think tank mechanisms and think tank systems [11, 12]. However, domestic think tank research has just begun to shift from qualitative research to quantitative research, from individual cases to systematic analysis, so the gap still needs to be narrowed [13].

## 16.5 Conclusion

Based on the sample analysis of WOS and CSSCI, it is found that on the cooperative network, the American and European think tank research cooperative networks are relatively mature, while China lacks representative scholars or think tank institutions with international influence, and the think tank research cooperative network is almost blank [14, 15].

In terms of hot spots, foreign think tank research is wide-ranging, innovative, and forward-looking, while domestic think tanks have narrow research fields, outdated topics, and difficult to innovate; in terms of methodology, foreign think tank research methodology has the characteristics of informatization, automation, and intelligence, while domestic think tank research methodology is relatively old and lacks innovation; [16, 17] in terms of development trends, Chinese think tanks lack credibility, autonomy, criticism, and independence, and lack in-depth consideration of micro-issues, specific issues, and future issues [18].

To sum up, there is a big gap between domestic and international. It is urgent to learn from the mature experience of Western think tanks, such as research focus, operation mode, research topic selection, cooperation mechanism, etc., and promote domestic think tanks becoming diversified, modern, and flexible. Innovative research concepts, methods, technologies, and tools strengthens the independence, autonomy, and criticality of think tanks and encourages think tanks to conduct in-depth research on specific issues and micro-fields related to the national economy and people's livelihood. Iay a solid foundation for the construction of new type Chinese think tanks.

## References


1. Chen, C.: CiteSpace II: detecting and visualizing emerging trends and transient patterns in scientific literature. J. Am. Soc. Inform. Sci. Technol. **57**(3), 359–377 (2006)
2. Chen, C.: CiteSpace II: detecting and visualizing emerging trends and transient patterns in scientific literature. J. Am. Soc. Inform. Sci. Technol. **3**, 359–377 (2006)
3. Chen, C., Leydesdorff, L.: Patterns of connections and movements in dual-map overlays: a new method of publication portfolio analysis. J. Am. Soc. Inf. Sci. **65**(2), 334–351 (2014)
4. Zhao, R., Guo, F., Qiu, Y.: The development of corporation in the United States and its enlightenment to China's think tank construction. J. Chongqing Univ. (Soc. Sci. Ed.) **02**, 125–131 (2016)
5. Ding, Y.: Exploring the research methods of think tanks. China Dev. Observ. **08**, 14–16 (2012)
6. Li, Y.: Think tank 2.0 for deliberative policy analysis. Policy Sci. **48**(1), 25–50 (2015)
7. DaRosa, D.A., Bell, R.H., Dunnington, G.L.: Residency program models, implications, and evaluation: results of a think tank consortium on resident work hours. Surgery **133**(1), 13–23 (2003)
8. Zhao, L., Zhong, Y., Shi, H.: On the application and characteristics of the quantitative analysis new tools in international authoritative think tanks. J. Intell. **05**, 83–87 (2014)
9. Laffin, M.: Central-local relations in an era of governance: towards a new research agenda. Local Gov. Stud. **35**, 21–37 (2009)
10. Xu, X., Chen, Q.: Research on the basic problems of Chinese think tanks. 学术论坛 **11**, 178–184 (2012)
11. Stone, D.: Global public policy, transnational policy communities, and their networks. Policy Stud. J. **36**(1), 19–38 (2008)
12. Li, J., Chen, C.: CiteSpace: Text Mining and Visualization in Scientific Literature, 2nd edn., p. 93. Capital University of Economics and Business Press, Beijing (2017)
13. Kremer, A., Malzkorn, W. and Strobel, F.: Ratings Revisited: Textual Analysis for Better Risk Management. McKinsey & Company (2013)
14. Hart, P., Vromen, A.: A new era for think tanks in public policy? International trends, Australian realities. Aust. J. Public Adm. **67**(2), 135–148 (2008)
15. Liu, N.: The historical evolution, basic characteristics and trend of think tanks. Chongqing Soc. Sci. **03**, 103–109 (2012)
16. Lindquist, E.A.: Think tanks or clubs-assessing the influence and roles of Canadian Policy Institutes. Can. Public Adm. **36**(4), 547–579 (1993)
17. Xu, G.: Comparison of think tanks in Europe and America and its enlightenment to the development of think tanks in China. Compar. Econ. Soc. Syst. **02**, 77–83 (2010)
18. Lindquist, E.A.: Think tanks or clubs? Assessing the influence and roles of Canadian Policy Institutes. Can. Public Adm. **36**(4), 47–579 (2008)
19. Ahmad, M.: US think tanks and the politics of expertise: role value and impact. Polit. Q. **79**(4), 529–555 (2008)
20. Lawson, G.: Think tanks and power in foreign policy: a comparative study of the role and influence of the Council on Foreign Relations and the Royal Institute of International Affairs, 1939–1945. Millennium-J. Int. Stud. **34**(3SI), 998–999 (2006)

# Chapter 17
# Research on Big Data Analysis and Evaluation Method of In Use Vehicle Technology

XiaoJuan Yang, FuJia Liu, and ShuQuan Xv

**Abstract** An evaluation method of the technical condition of vehicle is put forward. Through the calculation and analysis of the big data recorded in the history of automobile maintenance and repair, the fuzzy analytic hierarchy process is used to establish the evaluation index system of the technical condition of the vehicle, so as to comprehensively evaluate the service condition of the vehicle, the performance of the vehicle, and the vehicle system It is helpful to evaluate the technical status of vehicles and promote the upgrading and improvement of equipment technology.

## 17.1 Introduction

The evaluation of automobile technical condition is a comprehensive evaluation of reliability, safety, economy, and other aspects in the use of automobile. The evaluation of the current automobile technology mainly uses expert on-site inspection, which consumes manpower and material resources and has limitations [1–3]. Moreover, the comprehensive evaluation system of the whole process of the technical status in the life cycle of the automobile has not yet been designed with pertinence [4]. This paper studies the performance characteristics and failure rules of the automobile in the use process by collecting the automobile maintenance data and forms the evaluation method of the automobile technical condition based on the maintenance data.

## 17.2 Collecting Maintenance Data

Firstly, basic vehicle information and maintenance record information are collected. Basic vehicle information includes license plate number, VIN code, registration date, fuel consumption of 100 km at the time of purchase and current mileage. Maintenance

X. Yang (✉) · F. Liu · S. Xv
Research Institute of Highway, Ministry of Transport, Beijing, China
e-mail: xj.y@rioh.cn



V. Chang et al. (eds.), *Modern Industrial IoT, Big Data and Supply Chain*,
Smart Innovation, Systems and Technologies 218,
https://doi.org/10.1007/978-981-33-6141-6_17

**Table 17.1** Vehicle basic information sheet

| | | | | | |
|---|---|---|---|---|---|
| Basic vehicle information | Fuel consumption per 100 km | Fuel consumption at the time of purchase | 7.8 L | | |
| | | Fuel consumption of current mileage | 8.24 L | | |
| Maintenance record information | Mileage | 3054 km | 5873 km | … | 59,558 km |
| | Maintenance items | Replace the rearview mirror | Tank cap replacement | … | Replace front bumper |
| | Fault information | The rearview mirror is damaged | Tank cap cannot be opened | … | Front bumper damage |
| | Service parts | Rearview mirror | Tank cap | … | Front bumper |
| | Maintenance costs | 100 Yuan | 190.5 Yuan | … | 210 Yuan |

record information includes mileage, maintenance items, fault information, maintenance accessories, and maintenance costs. Based on the extraction of mileage, maintenance items, maintenance parts, and other information in the maintenance records, the maintenance and replacement of vehicle parts are analyzed, and the failure rate of relevant parts is calculated according to this (see Table 17.1).

## 17.3 Establish the Evaluation Index System of Automobile Technical Conditions

### *17.3.1 Car Usage*

The usage of automobile is represented by two indexes: mileage and years.

### *17.3.2 Vehicle Performance*

Safety is represented by the failure rate of active and passive safety components. Power performance is expressed by the failure rate of power system. Economy is expressed by basic economy and failure rate of fuel supply system. Environmental protection is expressed by the failure rate of vehicle noise and pollution emission. Reliability is represented by first failure mileage, average failure interval mileage, and average maintenance cost.

### 17.3.3 Technical Status of Automobile System Device

The technical conditions of automobile system devices include the technical conditions of transmission system, driving system, steering system, instrument lighting, air conditioning system, electronic equipment, body accessories, etc.

## 17.4 Weight Calculation of Evaluation Index System of Automobile Technical Condition

The importance of each index in the evaluation system is different. The weight of evaluation index is determined by fuzzy analytic hierarchy process (FAHP) [5] and then establishes the priority matrix $S = \left(s_{ij}\right)_{m\times m}$ (The value of M is the total number of evaluation indexes).

$$s_{ij} = \begin{cases} 0.5 & s(i) = s(j) \\ 1 & s(i) < s(j) \\ 1.5 & s(i) > s(j) \end{cases} \tag{17.1}$$

Among them, $s(i)$ and $s(j)$ represent the relative importance of index $i$ and $j$, respectively. Then the priority relation matrix $s$ is transformed into a fuzzy consistent matrix $R = \left(r_{ij}\right)_{m\times m}$, i.e., sum s first, which is recorded as:

$$r_i = \sum_{k=1}^{m} s_{ik}, \qquad i = 1, 2, \ldots m \tag{17.2}$$

Here *i*, *m* equals *i*, *m* in $S = \left(s_{ij}\right)_{m\times m}$. Then the transformation is carried out to obtain:

$$r_{ij} = \frac{r_i - r_j}{2m} + 0.5 \tag{17.3}$$

At last, the index weight is calculated. First, the sum $c_i$ of every element except itself in the fuzzy consistent matrix $R$ is calculated

$$c_i = \sum_{j=1}^{m} r_{ij} - 0.5, \quad i = 1, 2 \ldots m \tag{17.4}$$

Find the sum of elements without diagonal $\sum_{i=1}^{m} c_i$ again:

$$\sum_{i=1}^{m} c_i = \frac{m(m-1)}{2} \tag{17.5}$$

Finally, the weight $\omega_i$ of each index can be obtained by normalizing $c_i$:

$$\omega_i = \frac{c_i}{\sum_{i=1}^{m} c_i} = \frac{2c_i}{m(m-1)} \tag{17.6}$$

Then we get the weight of technical condition evaluation index(see Table 17.2).

**Table 17.2** Evaluation index weight of automobile technical condition

| First-level indicators | Second-level indicators | Third-level indicators |
|---|---|---|
| Car usage $P_1 = 0.2$ | Mileage $B_1 = 0.75$ | |
| | Service life $B_2 = 0.25$ | |
| Car performance $P_2 = 0.7$ | Security $U_1 = 0.25$ | Active safety component failure rate $U_{11} = 0.50$ |
| | | Passive safety component failure rate $U_{12} = 0.50$ |
| | Dynamic property $U_2 = 0.22$ | Failure rate of power system $U_{21} = 1$ |
| | Economy $U_3 = 0.15$ | Basic economy $U_{31} = 0.60$ |
| | | Failure rate of fuel supply system $U_{32} = 0.40$ |
| | Environmental protection $U_4 = 0.10$ | Noise pollution $U_{41} = 0.75$ |
| | | Pollutant discharge level $U_{42} = 0.25$ |
| | Reliability $U_5 = 0.28$ | Average maintenance interval mileage $U_{51} = 0.50$ |
| | | Mean distance between failures $U_{52} = 0.33$ |
| | | Average maintenance cost $U_{53} = 0.17$ |
| Technical status of automobile system device $P_3 = 0.1$ | Driveline $V_1 = 0.20$ | Failure rate of transmission system $V_{11} = 1$ |
| | Driving system $V_2 = 0.20$ | Failure rate of driving system $V_{21} = 1$ |
| | Steering system $V_3 = 0.20$ | Steering system failure rate $V_{31} = 1$ |
| | Instrument lighting $V_4 = 0.10$ | Failure rate of instrument lighting device $V_{41} = 1$ |
| | Air conditioning system $V_5 = 0.05$ | Failure rate of air conditioning system $V_{51} = 1$ |
| | Electronic equipment $V_6 = 0.20$ | Failure rate of electronic equipment $V_{61} = 1$ |
| | Body accessories $V_7 = 0.05$ | Failure rate of body accessories $V_{71} = 1$ |

Each index value is multiplied by the weight of each index to get the total score $P$. the specific formula is as follows:

$$
\begin{aligned}
P = {} & P_1 * [M(1) * B_1 + M(2) * B_2] + P_2 * \{U_1 * [M(3) * U_{11} + M(4) * U_{12}] \\
& + U_2 * U_{21} * M(5) + U_3 * [M(6) * U_{31} + M(7) * U_{32}] \\
& + U_4 * [M(8) * U_{41} + M(9) * U_{42}] \\
& + U_5 * [M(10) * U_{51} + M(11) * U_{52} + M(12) * U_{53}]\} \\
& + P_3 * [V_1 * M(13) * V_{11} + V_2 * M(14) * V_{21} \\
& + V_3 * M(15) * V_{31} + V_4 * M(16) * V_{41} \\
& + V_5 * M(17) * V_{51} + V_6 * M(18) * V_{61} + V_7 * M(19) * V_{71}]
\end{aligned}
\tag{17.7}
$$

When calculating the index values of $M(3)$–$M(5)$, $M(7)$–$M(9)$, $M(13)$–$M(19)$about the failure rate.

$$
M(6) = \frac{\text{The fuel consumption of the car in 100 km}}{\text{100 km fuel consumption of the car at the time of purchase}} \tag{17.8}
$$

It is necessary to carry out min max standardization and then carry out forward processing for $M(10)$, $M(11)$. Let min ($x$), max ($x$) be the minimum and maximum values of $x$, respectively, and map the original value $x$ to the value $y$ in the interval [0,1] through min–max standardization. The formula is as follows:

$$
y = [x - \min(x)] \big/ [\max(x) - \min(x)] \tag{17.9}
$$

Then $\max[M(1)] = 600000$; $\min[M(1)] = 0$; $\max[M(2)] = 15$; $\min[M(2)] = 0$; $\max[M(10)] = 10000$; $\min[M(10)] = 0$; $\max[M(11)] = 10000$; $\min[M(11)] = 1000$; $\max[M(12)] = 5000$; $\min[M(12)] = 0$.

The forward processing of $M(10)$, $M(11)$ is as follows

$$
M(10) = 1 - M(10),\ \ M(11) = 1 - M(11) \tag{17.10}
$$

$$
M(1) = \frac{M(1) - 0}{600000 - 0},\ M(2) = \frac{M(2) - 0}{15 - 0} \tag{17.11}
$$

$$
M(10) = 1 - \frac{M(10) - 0}{10000 - 0},\ M(11) = 1 - \frac{M(11) - 1000}{10000 - 1000},\ M(12) = \frac{M(12) - 0}{5000 - 0} \tag{17.12}
$$

By analyzing the registration date in the basic information of the vehicle to be tested, the service life of the vehicle $Y = 3$ is obtained. By analyzing the driving mileage in the maintenance record information, the total driving mileage of the vehicle is $L = 59558$ km, the first failure mileage is $\mathrm{l} = 3054$ km, the average failure interval mileage is MTBF $= 3809.2$ km, and the average maintenance cost is $\overline{D} =$

**Table 17.3** Scoring standard of failure rate

| Description of failure probability | Scoring of failure frequency |
|---|---|
| The probability of failure is very high, and the failure frequency is more than 20% | 1 |
| The probability of failure is general, and the failure frequency is between 10 and 20% | 0.8 |
| The occurrence of faults is accidental, and the failure rate is between 1 and 10% | 0.6 |
| The failure rate is between 0.1 and 1% | 0.4 |
| The probability of failure is almost zero, and the failure rate is less than 0.1% | 0.2 |

553.7 yuan. The fuel consumption of the car is 8.24L in 100 km. At the time of purchase, the fuel consumption of the vehicle is 7.8L at the speed of 100 km. The failure rate scoring standard of the evaluation system (see Table 17.3).

The corresponding indicators and the calculated values of various indicators of the technical condition evaluation of the vehicle (see Table 17.4).

Then

$$\begin{aligned} P = {} & 0.2 * [0.75 * (59558/600000) + 0.25 * (3/15)] \\ & + 0.7 * \{0.25 * (0.5 * 0.4 + 0.5 * 0.2) + 0.22 * 1 * 0.6 + 0.15 \\ & * (0.6 * 8.24/7.8 + 0.4 * 0.4) + 0.1 * (0.75 * 0.2 + 0.25 * 0.4) \\ & + 0.28 * [0.5 * (1 - 3054/10000) \\ & +0.33 * \left(1 - \frac{3809.2 - 1000}{10000 - 1000}\right) + 0.17 * \left(\frac{553.7 - 0}{5000 - 0}\right)]\} \\ & + 0.1 * (0.2 * 0.6 + 0.2 * 0.4 + 0.2 * 0.8 + 0.1 * 0.8 \\ & + 0.05 * 0.6 + 0.2 * 0.6 + 0.05 * 0.6) = 0.4487 \end{aligned} \tag{17.13}$$

## 17.5 Evaluation of Automobile Technical Status

The corresponding technical status level and description are obtained (see Table 17.5).

According to the description of the technical status level failure rate of the car, the average failure rate of each system assembly on the whole car is more than 10% and less than 20%, and the frequency of failure is too high Level.

**Table 17.4** Calculation value of evaluation index of automobile technical condition

| First-level indicators | Value | Second-level indicators | Value | Third-level indicators | Value |
|---|---|---|---|---|---|
| Car usage | 0.124 | Mileage | 0.099 | | |
| | | Service life | 0.2 | | |
| Car performance | 0.517 | Security | 0.3 | Active safety component failure | 0.4 |
| | | | | Passive safety component failure rate | 0.2 |
| | | Dynamic property | 0.6 | Failure rate of power system | 0.6 |
| | | Economy | 0.794 | Basic economy | 1.056 |
| | | | | Failure rate of fuel supply system | 0.4 |
| | | Environmental protection | 0.25 | Noise pollution | 0.2 |
| | | | | Pollutant discharge level | 0.4 |
| | | Reliability | 0.593 | Average maintenance interval mileage | 0.695 |
| | | | | Mean distance between failures | 0.688 |
| | | | | Average maintenance cost | 0.111 |
| Technical status of automobile system device | 0.62 | Driveline | 0.6 | Failure rate of transmission system | 0.6 |
| | | Driving system | 0.4 | Failure rate of driving system | 0.4 |
| | | Steering system | 0.8 | Steering system failure rate | 0.8 |
| | | Instrument lighting | 0.8 | Failure rate of instrument lighting device | 0.8 |
| | | Air conditioning system | 0.6 | Failure rate of air conditioning system | 0.6 |
| | | Electronic equipment | 0.6 | Failure rate of electronic equipment | 0.6 |
| | | Body accessories | 0.6 | Failure rate of body accessories | 0.6 |

**Table 17.5** Description of vehicle technical status

| Automobile technical status grade | Detailed description of technical status classification | Technical condition level failure rate description |
|---|---|---|
| Level 1 | The total evaluation score is less than or equal to 0.2 | During normal use, the average failure rate of each system assembly is less than 1%, and it is in good condition |
| Level 2 | The total evaluation score is greater than 0.2 but less than or equal to 0.4 | In the normal use process, the average failure rate of each system assembly is more than 1% but less than 10%, which can basically maintain the normal working state |
| Level 3 | The total evaluation score is greater than 0.4 but less than or equal to 0.6 | During normal use, the average failure rate of each system assembly is more than 10% but less than 20%, and the frequency of failure is too high |
| Level 4 | The total evaluation score is greater than 0.6 but less than or equal to 0.9 | The vehicle is basically unable to use normally, and the average failure rate of each system assembly is more than 20%, which is basically unable to work |
| Level 5 | The total evaluation score is greater than 0.9 | The vehicle is completely unable to drive and is in the state of being scrapped |

## 17.6 Conclusion

This paper focuses on solving the shortcomings of the existing technology and provides a method to evaluate the technical status of the in use vehicle. Using this method can accurately evaluate the reliability, safety, economy, and other comprehensive indicators of the vehicle in the use process and analyze the performance characteristics and failure rules of the vehicle, which is helpful to comprehensively evaluate the overall technical level of the vehicle, including the use status and performance of the vehicle, etc.

## References

1. Xin, Z., Zhijun, Y.: Research on expert system of automobile performance evaluation. Mod. Mach. **6**, 43–44 (2004)
2. Lijun, Z., Jiguang, L., Zhanhui, T.:Comprehensive performance evaluation method of car based on fuzzy comprehensive evaluation. Autom. Technol. **4**, 31–33 (2008)

3. Qin, S., Zhaoyang, C., Xin, L.: Comprehensive evaluation method of car design quality. Shanghai Auto **8**, 28–33 (2004)
4. Weimin, Y.: Reliability, Maintainability and Supportability. National Defense Industry Press, Beijing (1995)
5. Shubai, X.: Principle of AHP. Tianjin University Press, Tianjin (1988)

# Chapter 18
# Forecast of Hydroelectricity Generation in Guangxi Based on ARIMA Model

**ZhiXiang Lu, JiChang Chen, and ShiMing Ma**

**Abstract** Hydropower is a kind of clean energy with sustainable development. Hydropower generation has strong periodic regularity, so the prediction and control of power generation are particularly important. Real-time and accurate hydroelectric power generation forecasting is a key issue in the energy industry. In this paper, the time series analysis method of ARIMA ($p$, $d$, 0) model structure is used for data preprocessing, and the ARIMA (2, 1, 2) model was finally constructed by identification, model comparison, and testing. After comparison, it can be seen that compared with linear regression and exponential prediction models, the ARIMA model has good data fit and adequate extraction. The prediction method is simple and concise, and the short-term prediction is accurate, which can better adapt to the cyclicality of hydropower generation in all seasons. The model predicts the hydroelectric power generation of Guangxi in 2020 to be 5.3 billion kWh in May, 5.5 billion kWh in June, 41 kWh in July, 40 kWh in August, and 36 kWh in September. Based on this forecast, it can be seen that Guangxi's hydroelectric power generation will continue to develop in a sustainable trend, and its quantitative forecast results will provide Guangxi with the energy industry and provide useful references for production planning and industrial layout.

## 18.1 Introduction

Water energy is an inexhaustible, renewable, and clean energy source. The basic principle of hydroelectric power generation is to use the water level difference, with the water wheel generator to produce electricity. In other words, it is to use the water potential energy into the water wheel's mechanical energy, which is then used to drive the generator to generate electricity. Scientists use the natural conditions of the water level difference to effectively utilize hydrodynamic engineering and machine to achieve the highest level of generation of electricity for people to use cheap and non-polluting electricity.

Z. Lu · J. Chen (✉) · S. Ma
School of Information and Engineering, Nanning University, Nanning, Guangxi, China



V. Chang et al. (eds.), *Modern Industrial IoT, Big Data and Supply Chain*,
Smart Innovation, Systems and Technologies 218,
https://doi.org/10.1007/978-981-33-6141-6_18

The water resources in the Guangxi region are very rich as it is located in the sea corridor of China, with an average annual total of 1880, accounting for 7.25% of the total water resources [1]. The abundance of water resources has also promoted the formation of peak forest landscape to some extent. The abundance of water resources also facilitates the formation of peak forest landscape to a certain extent. In the water system structure of Guangxi region, the geographic environment is very important. The main composition is peak and karst cave. Water resources and stone geomorphology put forward higher requirements for the development of water conservancy and hydropower technology. Hydropower is clean and renewable energy. It is cheap, with no fuel, simple operation and maintenance. Integrated water conservancy project can simultaneously meet the needs of power generation, flood control, irrigation, and shipping.

Hydropower generation has a strong cyclical regularity, so the prediction and control of the generation capacity are particularly important. Seo et al. [2] proposed a real-time method for transforming the variational assimilation of hydrological and hydrometeorological data into operational hydrological forecasting. And the hourly real-time observed water flow and precipitation, as well as the potential evaporation, was used for climatological estimation. However, this method is more applicable to the European geographic environment and does not fully fit the hilly basin landscape of Guangxi's land. The time-segmented model proposed by Skamarock et al. [3] also failed to satisfy the Guangxi karst landform of the river characteristics. The method used in similar articles [4–6] is more suitable for hydroelectric power generation prediction in the plain area and fails to meet the requirements of the Guangxi where are rich in unique geographic features, such as hilly basins and estuaries.

Hydroelectric power generation is an important part of the energy industry. Through the scientific forecast of hydroelectric power generation, it will have a clear perception of the energy supply of Guangxi. The problem is how to build a reasonable model to predict Guangxi's hydroelectricity generation. While in a complex geographic and meteorological environment, how the government and enterprises plan the industrial layout and production of energy has always been a challenge for us. This paper uses the monthly data of hydropower generation in Guangxi Zhuang Autonomous Region from March 2017 to April 2020. The time series analysis procedures, such as data processing, model identification, and model optimization, are used to find the model with the best fit for the next five months. The application of the ARIMA model in the forecast of hydroelectricity generation in Guangxi has good forecasting effect, high degree of fit, and effective data utilization. The ARIMA model is used to statistically analyze the data of hydroelectric power generation in Guangxi, in order to provide reference and basis for the industrial layout and production layout of energy in Guangxi.

## 18.2 Data and Methods

### 18.2.1 Data

The data stem from the National Bureau of Statistics of China [7] and collected from information published on the Web site "Energy"—"Production of major energy products"—"Hydropower", January 2017—April 2020 current monthly value of hydropower generation in Guangxi (billions of kWh).

### 18.2.2 Analytical Method

The ARIMA predictive model is a time series predictive statistical analysis method that fits a time series which is smooth. ARIMA $(p, d, q)$ model, where d denotes the number of times the timing is different, p denotes the number of observations, and $q$ denotes the number of residuals. ARIMA $(p, d, q)$ model is an extension of ARIMA $(p, q)$ model. The ARIMA $(p, d, q)$ model can be represented as.

$$\left(1-\sum_{i=1}^{p}\phi_i L^i\right)(1-L)^d X_t=\left(1+\sum_{i=1}^{q}\theta_i L^i\right)\varepsilon_t$$

where $L$ is the lag operator,

A non-stationary time series, after eliminating its local level or trend, shows a kind of homogeneity. Known as some parts of the time series are similar to other parts. The non-stationary time series can be converted to a stationary time series after differential processing, and such a time series is said to be uniformly non-stationary, where the number of differences is the order of unity. It will be recorded as a difference operator, then there are

$$\nabla^2 y_t=\nabla(y_t-y_{t-1})=y_t-2y_{t-1}+y_{t-2}$$

For the delay operation,

$$y_{t-p}=B^p y_t, \forall p\geq 1$$

there are thus

$$\nabla^k=(1-B)^k$$

If we have a non-stationary time series $y_t$, of order d, then $\nabla^d y_t$ is a stationary time series, we can set it to be ARIMA $(p, q)$ model, i.e.,

$$\lambda(B)\left(\nabla^d y_t\right) = \theta(B)\varepsilon_t$$

where $\lambda(B) = 1-\lambda_1 B-\lambda_2 B^2-\cdots-\lambda_p B^p, \theta(B) = 1-\theta_1 B-\theta_2 B^2-\cdots-\theta_p B^p$ are the autoregressive coefficient polynomial and the sliding mean coefficient polynomial $\varepsilon_t$ is the zero-means white harmonic series. The model can be called the ARIMA (*p*, *d*, *q*) sliding summation model.

## 18.3 Results of Data Analysis

### *18.3.1 Data Preprocessing*

Some data have been missed from the original data, such as January 2017, February 2017, January 2018, February 2018, January 2019, February 2019, January 2020, and February 2020. For this purpose, the sample bar interpolation method is used to fulfill the missing data. The monthly data of hydropower generation in Guangxi were temporalized and the temporal graph was plotted Fig. 18.1. From Fig. 18.1, it can be seen that the incidence of hydropower generation in Guangxi has a random trend and does not fluctuate around a certain value. It is seen that hydropower generation has non-stationary characteristics. Guangxi hydropower generation mortality has a linear level trend.

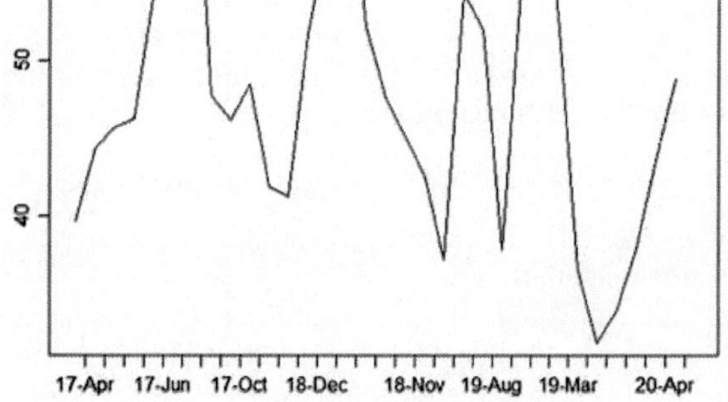


**Fig. 18.1** Time series of hydroelectricity generation in Guangxi province

## 18.3.2 Model Identification

The autocorrelation plots and partial autocorrelation plots of Guangxi hydropower generation are shown in Figs. 18.2 and 18.3, respectively.

From Fig. 18.2, it can be seen that the hydroelectricity incidence number autocorrelation has first-order trailing and the partial autocorrelation plot has second-order trailing. Therefore, the prediction model of hydroelectricity generation incidence number is ARIMA (2, 1, 2). In order to prevent the subjective selection of the model from causing errors in the prediction model, several alternative models were selected from different perspectives according to Fig. 5. The alternative identification models are provided as ARIMA (2, 1, 1), ARIMA (1, 1, 1), and ARIMA (1, 1, 2).

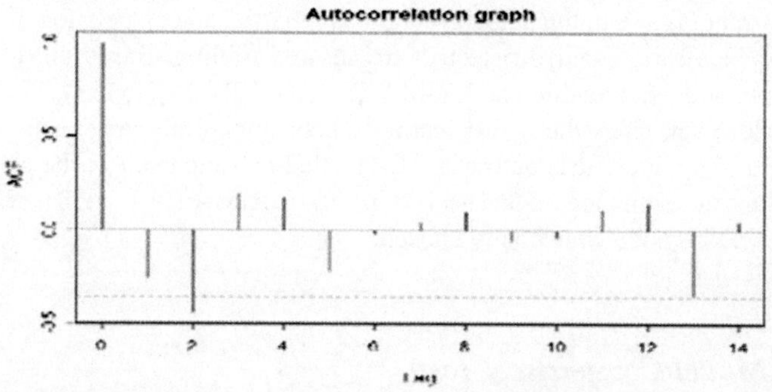


**Fig. 18.2** Autocorrelation chart

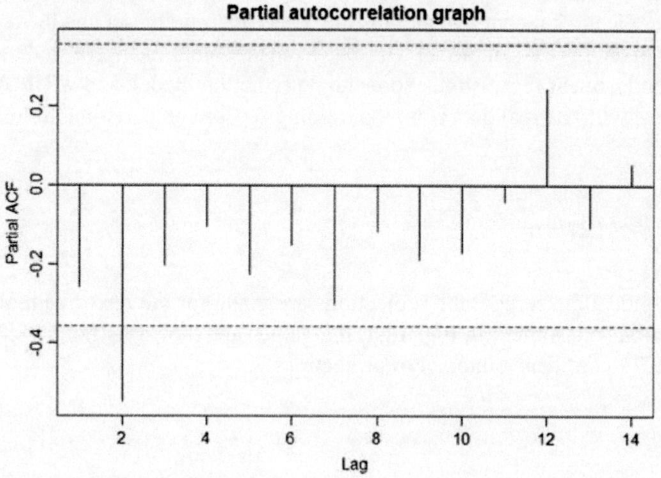


**Fig. 18.3** Bias correlation plot

**Table 18.1** Comparison of model

| Prediction method | | | | Prediction error | | |
|---|---|---|---|---|---|---|
| True | Linear | ETS | ARIMA | Linear | ETS | ARIMA |
| 31.8 | 48.2315 | 43.3127 | 36.64251 | 16.4315 | 11.5127 | 4.84251 |
| 33.9 | 48.2315 | 41.2576 | 37.55548 | 14.3315 | 7.3576 | 3.65548 |
| 37.9 | 48.2315 | 43.3127 | 44.19623 | 10.3315 | 5.4127 | 6.29623 |
| 43.6 | 48.2315 | 50.4439 | 41.30412 | 4.6315 | 6.8439 | −2.29588 |
| 48.8 | 48.2315 | 51.7489 | 52.06914 | −0.5685 | 2.9489 | 3.26914 |
| | | | | Sum of errors | | |
| | | | | 46.2945 | 34.0758 | 20.35924 |

As shown in Fig. 18.3, the autocorrelation and partial autocorrelation of the first-order difference series of hydroelectric are showed trailing. Based on the trailing features, the model is identified as ARIMA (2, 1, 1), ARIMA (1, 1, 1), and ARIMA (3, 1, 1). However, since the hydroelectricity bias autocorrelation function trailing results are not obvious, it is automatically graded and modeled at the same time. The recognition results are ARIMA (1, 0, 0) and ARIMA (0, 1, 1). Therefore, the ARIMA (2, 1, 2) model was finally chosen.

### 18.3.3 Model Comparison Test

In order to compare the advantages and disadvantages of the models, after selecting the ARIMA (2, 1, 2) model, the hydroelectricity generation capacity of the data was retained for the most recent five months of data as the training set and the remaining 48 months of data as the sample set. The data from the sample set were predicted using three methods, linear regression, exponential prediction model, and ARIMA model, respectively, with the real data from the training set for comparison (Table 18.1).

### 18.3.4 Prediction

Using the model developed, the projections are made for the next five months and the projections are plotted in Fig. 18.4, the values are shown in Table 18.2, giving 80% and 95% confidence interval of projections.

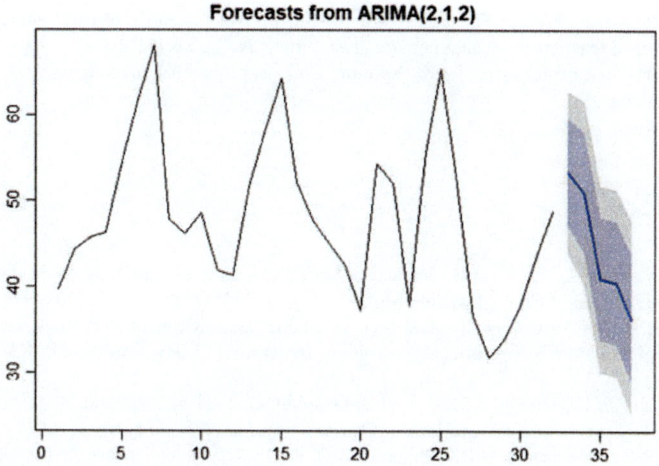


Fig. 18.4 Time series plot of predicted results

Table 18.2 Predicted values

| Data | Forecast | Lo 80 | Hi 80 | Lo 95 | Hi95 |
|---|---|---|---|---|---|
| 20-May | 53.34887 | 47.11298 | 59.58476 | 43.8119 | 62.88584 |
| 20-Jun | 50.97487 | 43.95573 | 57.99411 | 40.24003 | 61.70971 |
| 20-Jul | 40.96418 | 33.86555 | 48.06282 | 30.10776 | 51.82061 |
| 19-Aug | 40.45421 | 33.32982 | 47.57857 | 29.55841 | 51.34998 |
| 20-Sep | 36.33321 | 28.97911 | 43.68731 | 25.08608 | 47.58034 |

## 18.4 Conclusion

The above calculations and analysis show that the ARIMA model fits the data of hydroelectric power generation better than other models. It has sufficient extraction, simple prediction method, simple and effective model, and accurate short-term prediction results. In this paper, a new real-time adaptive prediction algorithm for short-time traffic flow is proposed. From a number of simulation examples, it can be seen that one-step real-time prediction of short duration traffic has the following advantages.

The ARIMA model is used to forecast hydroelectric power generation of Guangxi in 2020 as 5.3 billion kWh in May, 5 billion kWh in June, 41 kWh in July, 40 kWh in August, and 36 kWh in September. According to this forecast, it can be seen that Guangxi's hydroelectric power generation will continue to develop in a sustainable trend, and its quantitative forecast will provide useful references for production planning and industrial layout of Guangxi energy industry.

**Acknowledgements** Project: The project of improving the basic scientific research ability of young and middle-aged teachers in Guangxi universities, project No.2020ky64014

Project: Professor training project of Nanning College, project No.2016JSGC05

## References

1. Zhang, X., Yang, L., Li, Y., et al.: Impacts of lead/zinc mining and smelting on the environment and human health in China. Environ. Monitor. Assess. **184**(4), 2261–2273 (2012)
2. Seo, D.J., Koren, V., Cajina, N.: Real-time variational assimilation of hydrologic and hydrometeorological data into operational hydrologic forecasting. J. Hydrometeorol. **4**(3), 627–641 (2003)
3. Skamarock, W.C., Klemp, J.B.: A time-split nonhydrostatic atmospheric model for weather research and forecasting applications. J. Comput. Phys. **227**(7), 3465–3485 (2008)
4. Koster, R.D., Mahanama, S.P.P., Yamada, T.J., et al.: The second phase of the global land-atmosphere coupling experiment: soil moisture contributions to subseasonal forecast skill. J. Hydrometeorol. **12**(5), 805–822 (2011)
5. Brown, J.D., Demargne, J., Seo, D.J., et al.: The Ensemble verification system (EVS): a software tool for verifying ensemble forecasts of hydrometeorological and hydrologic variables at discrete locations. Environ. Model. Softw. **25**(7), 854–872 (2010)
6. Sanford, W.E., Pope, J.P.: Current challenges using models to forecast seawater intrusion: lessons from the Eastern Shore of Virginia, USA. Hydrogeol. J. **18**(1), 73–93 (2010)
7. China National Bureau of statistics. Production of major energy products. 10 June 2020. https://data.stats.gov.cn/easyquery.htm?cn=E0101

# Chapter 19
# Sockpuppet Detection in Social Network Based on Adaptive Multi-source Features

**Hang Yu, Feng Hu, Li Liu, Ziyang Li, Xiangpeng Li, and Zhimin Lin**

**Abstract** Social media, such as Facebook and Twitter, has become an indispensable part of the Internet. These online social platforms facilitate the information exchange between people. At the same time, some users create multiple accounts on the same social media, pretend to be others, and publish articles or comments through different accounts to influence online public opinion. We refer to multiple accounts created by the same user named as sockpuppet accounts. In the past, people's research on sockpuppet detection was mostly based on verbal features or non-verbal features. These kinds of methods have their unique advantages and drawbacks, most of them ignore the connection between the two methods and study separately. For this purpose, we propose an adaptive multi-source feature fusion method, which not only considers the importance of text in identifying the sockpuppet, but also considers the characteristics of users' online behavior which is not easy to disguise. The experimental results on real datasets show that our method has a better detection rate than compared with competitive algorithms.

## 19.1 Introduction

Social media has developed tremendously in the past few years. Many users have invested part of their personal lives in virtual space, which has promoted the prosperity and development of the Internet. However, sockpuppet accounts in social networks have brought some negative effects [1]. For example, in addition to the normal accounts used by a user in a social, create other accounts called sockpuppet accounts [2]. These accounts are used to perform various malicious activities, such as manipulating presidential elections, affecting financial markets, and spreading spam, phishing sites and malware [3], as well as the collection of private user data (privacy issues) [4], which affects the online experience of ordinary users [5].

H. Yu · F. Hu · L. Liu (✉) · Z. Li · X. Li · Z. Lin
Chongqing University of Posts and Telecommunications, Chongqing 400065, China
e-mail: liliu@cqupt.edu.cn

 
V. Chang et al. (eds.), *Modern Industrial IoT, Big Data and Supply Chain*,
Smart Innovation, Systems and Technologies 218,
https://doi.org/10.1007/978-981-33-6141-6_19

For the security of the Internet and the user experience, many methods have been specifically proposed by researchers in identifying sockpuppet accounts in social media. The current mainstream methods include verbal features-based solutions [6–8] and non-verbal features-based solutions [9–11]. These two types of methods have a certain effect on sockpuppet accounts detection. The solution based on verbal features pays more attention to text content, such as the number of words spoken by the account, sentence length, text similarity, etc. the detection method based on non-verbal features which pay more attention to user behavior, such as user login time, number of posts, and number of comments. However, the camouflage and concealment of sockpuppet accounts are more sophisticated for now. It is difficult to accurately identify sockpuppet accounts only through text content and simple user behavior analysis. Therefore, a more comprehensive method for sockpuppet accounts identification is required.

For this purpose, we propose a sockpuppet detection method based on adaptive multi-source feature fusion. Considering the importance of text in identifying sockpuppet and regarding the characteristics of users' online behavior that is not easy to disguise, a more comprehensive analysis of the text verbal and behavior characteristics of sockpuppet accounts is carried out. We convert the sockpuppet accounts identification problem into a classification problem. Using adaptive multi-source feature fusion, which can automatically integrate verbal and non-verbal features according to specific problems, select the best combination of features. Experiments show that our method can identify sockpuppet accounts effectively.

In this article, we adopt adaptive multi-source feature fusion, which improves the accuracy of identifying sockpuppet accounts. The main contributions are as follows:

1. Firstly, the verbal and non-verbal features of sockpuppet accounts can adaptively be integrated, language characteristics and user behavior are more comprehensively considered, and sockpuppet accounts identification is more reasonable.
2. In order to verify the effectiveness of the method, we collected public dataset from English Wikipedia as our experimental data. The accuracy of the experimental results reached 85%, indicating that the proposed method can be widely used and is better than a single verbal feature or non-verbal feature experiment.

## 19.2 Related Work

Sockpuppet detection is to detect whether a given single account [12, 13] or account pair [14–16] are owned by the same user. The existing methods of detecting sockpuppet accounts are mainly based on verbal features or non-verbal feature.

Solorio et al. [6] proposed the use of natural language processing technology, collected 77 ordinary users and sockpuppet users in Wikipedia, using the support vector machine (SVM) algorithm, the final experimental results showed that the accuracy reached 68.83%. Subsequently, the study extended the dataset to 623 instances, and the experimental results show that the optimal value of F-measure reached 73% [7].

Tsikerdekis et al. [9] proposed a way to identify sockpuppet accounts using non-verbal characteristics in social media. Some variables are expressed as non-verbal user behaviors, such as the number of modifications, the total number of bytes added, and the total number of deleted bytes at a specific time after the user registers on the website. The overall detection result accuracy rate reached 71.3%. Yang et al. [10] used the study of non-verbal user behavior to identify sockpuppet accounts in Renren. The accuracy of the model to identify sockpuppet accounts on Renren reached 99%, but the selected features are not attributes shared by social platforms, so the model cannot be extended to other social platforms.

## 19.3 Proposed Method

### *19.3.1 Verbal Features*

In the social media, all the comment texts of each user are gathered to form a new text, and a word vector matrix is trained through the word2vec model with fixed dimensions. Then use the trained word vector matrix to convert the new text of each user into a word vector, and then concatenate it into a sentence vector. Then extract verbal features such as text length, question frequency, and number of emotional words $N = \{n_1, n_2, \ldots, n_i\}$, here $N$ is verbal features.

### *19.3.2 Non-verbal Features*

Non-verbal features include non-verbal features such as the speaking time of each user, the number of comments per user per day, the number of interactions per user, and the number of online visits per user per day. $M = \{m_1, m_2, \ldots, m_j\}$, here $M$ is non-verbal features.

### *19.3.3 Adaptive Multi-source Feature Fusion*

Suppose $N = \{n_1, n_2, \ldots, n_i\}$, $M = \{m_1, m_2, \ldots, m_j\}$, initialize the weight matrix $W_1 = \left\{w_1^1, w_1^2, \ldots, w_1^i\right\}$, $W_2 = \left\{w_2^1, w_2^2, \ldots, w_2^j\right\}$, $Y = \{y_1, y_2, \ldots, y_k\}$, the verbal features and non-verbal features are combined horizontally as in formula (19.1):

$$Y = W_1 * N \oplus W_2 * M \tag{19.1}$$

Considering that verbal features are easy to disguise, so, it is necessary to make adaptive feature selection for this case. The linear weighting idea of using logistic regression algorithm is as formula (19.2) and formula (19.3):

$$f(x) = \frac{1}{e^x + 1} \tag{19.2}$$

$$h = \frac{1}{e^Y + 1} \tag{19.3}$$

where $h$ is the linear regression with the sigmoid function added. Use the maximum likelihood function to calculate the parameters in the model, such as formula (19.4) and formula (19.5):

$$L(w) = \prod h^Y * (1 - h)^{1-Y} \tag{19.4}$$

$$l = \log L(w) = \sum Y * \log h + (1 - Y) * \log 1 - h \tag{19.5}$$

For adaptive feature selection, add an $L1$ regularization, because $L1$ regularization can generate a sparse feature weight matrix, that is, a sparse model, which can be used for feature selection as in formula (19.6):

$$J = l + \alpha * ||W_1||_1 + \alpha * ||W_2||_1 \tag{19.6}$$

$\alpha$ is the regular term coefficient, $||W_1||_1$, $||W_2||_1$ is the sum of the absolute values of each element. The training model can adaptively select verbal and non-verbal features by iteratively updating the weights $W_1$ and $W_2$. Get more comprehensive sockpuppet detection dataset characteristics.

### 19.3.4 Sockpuppet Detection Model

Figure 19.1 is the architecture of our model, a simple and complete process of detecting sockpuppet accounts. The process is divided into three stages. The first stage is data collection and data preprocessing, the second stage is feature extraction and adaptive multi-source feature fusion, and the third stage is the use of machine learning algorithms to detect sockpuppet accounts and model evaluation.

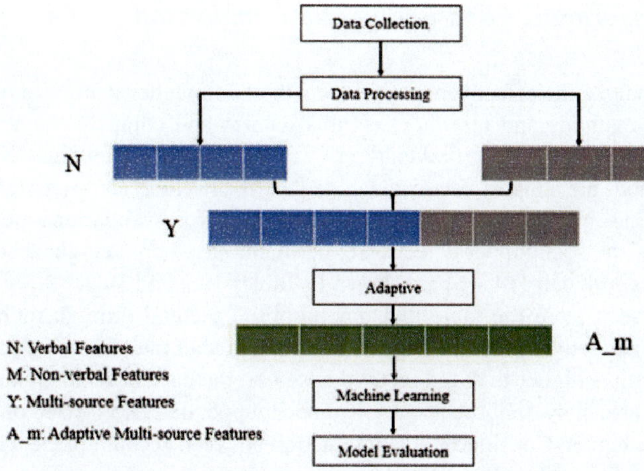


**Fig. 19.1** Sockpuppet detection model

## 19.4 Experiment and Result

### *19.4.1 Data Collection and Analysis*

In order to prove the effectiveness of our method, we collected public data from English Wikipedia as dataset. We collected user comments in the discussion area, user personal information, posting information, etc.

The accounts are sorted and filtered at beginning. A total of 370 sockpuppet accounts have been sorted out, sockpuppet users have at least two accounts or more, and 315 ordinary users. Then the accounts are randomly combined in pairs. Among them, there are 312 account pairs with sockpuppet relationship and 323 account pairs without sockpuppet relationship, as shown in Table 19.1.

**Table 19.1** Dataset information

| Datasets | Number of samples |
|---|---|
| #Sockpuppet accounts | 370 |
| #Ordinary accounts | 315 |
| # Account pair with sockpuppet relationship | 312 |
| # Ordinary user account pair | 323 |

### 19.4.2 Performance Evaluation and Comparison

In order to analyze the effectiveness of our method comprehensively, we reproduce the previous methods and test them on this dataset and compare them with our proposed method. Based on verbal features (VF) [6], non-verbal features (NVB) [9], and our method, the support vector machine (SVM) machine learning algorithm is used for training and classification testing, and fivefold cross-validation is performed. The results of the sockpuppet detection experiment based on a single account [12, 13] and an account pair [14–16] are shown in Tables 19.2 and 19.3.

It can be seen from the table that our proposed method through multi-source feature fusion adaptive selection, the accuracy rate, recall rate, F1-score value, and other indicators is higher than the method based on the current feature experiment. Experiments also show that the accuracy of sockpuppet detection based on account pairs is a little higher. Considering the interaction between account pairs, it has great help to identifying sockpuppet accounts.

In order to verify the effectiveness of this method to identifying sockpuppet accounts, we use different machine learning classification algorithms including support vector machine (SVM), random forest (RF), and adaptive increase (ADA) used above, as shown in Tables 19.4 and 19.5.

**Table 19.2** Experimental results of sockpuppet detection based on a single account

| Model | Precision | Recall | F1-score |
|---|---|---|---|
| VF | 0.71 | 0.73 | 0.74 |
| NVB | 0.73 | 0.74 | 0.76 |
| Our method | **0.85** | **0.76** | **0.80** |

The bold indicates the best experimental results, and their purpose is to be more eye-catching, in order to distinguish them from the data of other methods

**Table 19.3** Experimental results of sockpuppet detection based on account pairs

| Model | Precision | Recall | F1-score |
|---|---|---|---|
| VF | 0.73 | 0.69 | 0.75 |
| NVB | 0.76 | 0.71 | 0.77 |
| Our method | **0.79** | **0.91** | **0.84** |

The bold indicates the best experimental results, and their purpose is to be more eye-catching, in order to distinguish them from the data of other methods

**Table 19.4** Experiments on different classification algorithms based on a single account

| Machine learning | Precision | Recall | F1-score |
|---|---|---|---|
| SVM | 0.85 | 0.76 | 0.80 |
| RF | 0.80 | 0.79 | 0.81 |
| ADA | 0.72 | 0.90 | 0.78 |

**Table 19.5** Experiments on different classification algorithms based on account pairs

| Machine learning | Precision | Recall | F1-score |
|---|---|---|---|
| SVM | 0.79 | 0.91 | 0.84 |
| RF | 0.77 | 0.86 | 0.82 |
| ADA | 0.73 | 0.85 | 0.80 |

It can be seen from Tables 19.4 and 19.5 that the experimental effect based on account pairs are better than the effect of single account identification, indicating that the method based on account pairs is more suitable for distinguishing ordinary users from sockpuppet users.

## 19.5 Conclusion

This paper mainly studies the method of fusion of multi-source features to identify sockpuppet accounts, using adaptive multi-source feature fusion. Experimental results based on account pairs and single accounts show that our method is better than other methods. It has a better detection rate and can effectively identify sockpuppet accounts to achieve practical application effects.

In the future, the main research activities are not frequent and the sockpuppet accounts with low social participation. For this type of account, it is still the main direction of the current social media to detect sockpuppet accounts, and will be considered in the follow-up work.

**Acknowledgements** This work is partially supported by National Key R&D Program of China under Grant No. 2017YFB0802305 and NSFC Grant 61806031, and in part by the Science and Technology Research Program of Chongqing Municipal Education Commission under Grants KJQN201800638.

## References

1. Akram, W., Kumar, R.: A study on positive and negative effects of social media on society. Int. J. Comput. Sci. Eng. **5**(10), 351–354 (2017)
2. Jain, P., Kumaraguru, P., Joshi, A.: Identifying users across multiple online social media. In: Proceedings of the 22nd International Conference on World Wide Web, pp. 1259–1268. Association for Computing Machinery, New York (2013)
3. Thomas, K., Grier, C., Ma, J., et al.: Design and evaluation of a real-time url spam filtering service. In: 2011 IEEE Symposium on Security and Privacy, pp. 447–462. IEEE, California (2011)
4. Bonneau, J., Anderson, J., Danezis, G.: Prying data out of a social network. In: 2009 International Conference on Advances in Social Network Analysis and Mining, pp. 249–254. IEEE, Athens (2009)
5. Clauser, R.C.: Offline rules, online tools. J. Brand Manag. **8**(4), 270–287 (2001)

6. Solorio, T., Hasan, R., Mizan, M.: A case study of sockpuppet detection in Wikipedia. In: Proceedings of the Workshop on Language Analysis in Social Media, pp. 59–68. Association for Computational Linguistic, Atlanta (2013)
7. Solorio, T., Hasan, R., Mizan, M.: Sockpuppet detection in wikipedia: a corpus of real-world deceptive writing for linking identities. Comput. Lang. **131**(1), 677–679 (2013)
8. Bu, Z., Xia, Z., Wang, J.: A sock puppet detection algorithm on virtual spaces. Knowl.-Based Syst. **37**(1), 366–377 (2013)
9. Tsikerdekis, M., Zeadally, S.: Multiple account identity deception detection in social media using nonverbal behavior. IEEE Trans. Inf. Forensics Secur. **9**(8), 1311–1321 (2014)
10. Yang, Z., Wilson, C., Wang, X., et al.: Uncovering social network sybils in the wild. ACM Trans. Knowl. Discov. Data (TKDD) **8**(1), 2 (2014)
11. Yamak, Z., Saunier, J., Vercouter, L.: Detection of multiple identity manipulation in collaborative projects. In: Proceedings of the 25th International Conference Companion on World Wide Web, pp. 955–960. International World Wide Web Conferences Steering Committee, Switzerland (2016)
12. Kumar, S., Cheng, J., Leskovec, J., et al.: An army of me: Sockpuppets in online discussion communities. In: Proceedings of the 26th International Conference on World Wide Web, pp. 857–866. International World Wide Web Conferences Steering Committee, Switzerland (2017)
13. Maity, K., Chakraborty, A., Goyal, P., et al.: Detection of sockpuppets in social media. In: Companion of the 2017 ACM Conference on Computer Supported Cooperative Work and Social Computing, pp. 243–246. Association for Computing Machinery New York (2017)
14. Liu, D., Wu, Q., Han, W., et al.: Sockpuppet gang detection on social media sites. Front. Comput. Sci. **10**(1), 124–135 (2016)
15. Hosseinia, M., Mukherjee, A.: Detecting sockpuppets in deceptive opinion spam. In: International Conference on Computational Verbals and Intelligent Text Processing, pp. 252–272. Springer, Cham (2017)
16. Zheng, X., Lai, Y.M., Chow, K.P., et al.: Sockpuppet detection in online discussion forums. In: 2011 Seventh International Conference on Intelligent Information Hiding and Multimedia Signal Processing, pp. 374–377. IEEE, Dalian (2011)

# Chapter 20
# The Ethical Issues of Location-Based Services on Big Data and IoT

**Victor Chang, Yeqing Mou, and Qianwen Ariel Xu**

**Abstract** Both Internet of Things (IoT) and big data are hot topics in recent years. They indeed have brought about the change of business, promoted the progress of science and technology, and facilitated the lives of human beings. IoT creates the opportunity to connect every item to the Internet, and countless science and technology have supported the achievement of this goal. LBS is one of the indispensable technologies. It brings significant benefits to the business community, the individual, the society, and the national defense. However, at the same time, an individual's personal information is disclosed and even attacked by 'information thieves'. An inevitable reality is that the prerequisite of getting a location service is to expose your position first. Therefore, the privacy-related ethics issues are generated, and the danger is imminent, although there are corresponding protective measures.

## 20.1 Introduction

In the Internet of Things (IoT) environment, everything could be connected by the Internet. IoT refers to the kind of Internet-based expanded network. Besides, the extended client-side exchanges information with each other and communicates between any objects and items. The data and information could be exchanged among different devices, such as between different smartphones, which not only improves the convenience of life but also causes some data-related problems violating ethics. As an essential component of IoT, location-based services (LBS) become the services provider as well as the 'arch-criminal' of information leaks. Based on the Internet of Things, the LBS system can collect and convey all kinds of situational perceived data flexibly and efficiently through the perceived interaction between physical space and information space. The LBS system based on IoT will provide users with the

V. Chang (✉) · Q. A. Xu
Artificial Intelligence and Information Systems Research Group, School of Computing, Engineering and Digital Technologies, Teesside University, Middlesbrough, UK

Y. Mou · Q. A. Xu
International Business School of Suzhou, Xi'an Jiaotong-Liverpool University, Suzhou, China

 
V. Chang et al. (eds.), *Modern Industrial IoT, Big Data and Supply Chain*,
Smart Innovation, Systems and Technologies 218,
https://doi.org/10.1007/978-981-33-6141-6_20

customizable personalized real-time location service by processing the collected perceived information intelligently. It is undeniable that LBS, based on IoT, brings attractive services to people. However, at the same time, it also brings dangers about disclosure of personal information. This type of information often contains both the user's sensitive information, such as habits, health status, social relationships, and even identification. This paper will be divided into five parts to illustrate the situation of the application of LBS based on big data and IoT background and provide some solutions.

## 20.2 Big Data and IoT

### *20.2.1 Big Data*

Big data is an emerging term that describes the surprisingly rapid proliferation in the volume of data in structured and unstructured form Stergiou and Psannis [1]. According to Stergiou et al. [2], more trusted decisions are made based on accurate big data, greater operational efficiency cost reduction and reduced risk resulting from better decisions. In the Internet of Things environment, the data acquisition capability of various sensors, such as smartphones and wearable computing devices, has been improved significantly. With the development of IoT, cloud computing and other related technologies, every change in our behavior, location, and even body physiology data is sunk for the kind data which could be recorded and analyzed [3]. Besides, the concept of 'big data' is generated with the advent of the wide application of the Internet. The number of Internet users, including individuals, enterprises, and institutions increases sharply, making it relatively effortless to access and share data. Users can easily access the data by network, and massive data is created with user's Internet habits. For example, through intentional or unintentional sharing, clicks, and browsing, these operations quickly provide a large amount of data. Unlike the previous massive data concept, the emerging conception, 'big data', has four unique features, volume, variety, velocity, and veracity [4]. Variety refers to the multiform sources of data and various existing forms of information. For example, social media provide an enormous amount of users' location and life condition information. Vehicle navigation record a large number of driving trajectory information. Velocity means the value of the data will be depreciated—the development of a traffic query system dependent on the velocity [4]. The location of big data collected through LBS also has these four features.

### *20.2.2 IoT*

With the emerging IoT era, every object could be equipped with a smart device to collect data through the Internet. The challenge for personal privacy and data is generated with the development of IoT. Rivera and van der Meulen [5] predicted that a $14 billion economic influence is expected to be created due to IoT by 2022. The sensors, a crucial technology of IoT, are used to collect transmit data to the internet and achieve 'everything is connected'. According to Caron et al. [6], IoT might strongly impact personal privacy when more data with different sources are collected using these sensors.

## 20.3 LBS and LBS Applications

### *20.3.1 The Principle of LBS*

LBS, location-based service, refers to a collection of location-related services based on the location information of users. There are two technologies to access location information of mobile terminal users and provide location services for them. They are the radio communication network operated by telecom or mobile, such as GSM and CDMA and external positioning technology, such as GPS. Besides, a kind of value-added service provides the corresponding services for users with the support of the Geographic Information System (GIS) platform. In a word, LBS consists of three kinds of technologies, GPS, base station, and GIS.

GPS is just one of the most common components of LBS to provide location service for outdoor users. The application of GPS can be divided into two aspects, military and civil. In the context of the IoT, GPS is a vital component of Location-based services in the civil field. In terms of GIS, there are many elementary differences between LBS and GIS, although many of the functions in LBS are similar to those of traditional GIS systems. GIS system could usually make use of significant computing resources to provide professional geographic data analysis and processing for a small number of technicians.

On the contrary, LBS provides limited geographic data services to a large number of ordinary users and these services run on mobile terminals with limited resources. According to Zhou et al. [7], the LBS service providers usually possess these characteristics: high-performance, expansibility, and instantaneity, etc. These properties allow LBS to provide location services based on big data. Taking high-performance and expansibility as an example allows LBS to process the massive user query requests to decrease time waiting [8].

### 20.3.2 The Application of LBS

LBS has become the current application and research hotspot. Whether it is a public user or an industry user, there is a wide range of needs for access to locations and related services. The early LBS system was mainly used to quickly locate the caller's location in an emergency to implement rescue activities. At present, LBS has been widely used in military, transportation, logistics, medical, people's livelihood, and so forth.

The specific application of these location-based services can be divided into four categories, trigger services, information services, tracking service, and assistance [9]. The LBS information service and tracking focus on providing location services for users on social media platforms. Taking navigation software as an example, navigation software, such as Baidu mapping, can provide real-time information about road conditions and plan optimal travel routes for users. Additionally, drivers could find the nearest gas station with the help of these services. Besides, in a large museum (such as the Forbidden City Museum), visitors can enjoy the explanation of each collection through location-aware voice guider with the tracking service [7].

These applications of LBS created values to two groups, business and individuals, in the IoT environment. On the one hand, LBS changed how business works. As for the vehicle insurance company, their development benefits from LBS services because they could plan the insurance solutions that are most beneficial to the interests of the company with the help of LBS. It could record the trajectory information, such as the common driving locations and paths of every insured vehicle, analyze the risk levels, and then custom car insurance prices for different insurants [10]. As for the transport company, LBS applications help them plan the optimal driving route with the lowest fuel costs and breakdown cost [10]. Then, LBS realizes the accurate marketing of the service industry. For example, a restaurant could provide the best recommendation for users initiatively, accurately and duly by establishing the cooperation with map software because LBS records where users are and even analysis what they want through the users' query.

On the other hand, individuals indeed enjoy the kind of data-based position services. Firstly, customer experience on social media platforms is improved since LBS records user-generated locational data and transform it into the available information. Social media platforms of mobile Internet terminal offer a wide range of location services to users. It helps individuals create closer connections with the outside world and enhance the relevance between social networks and geographic locations. For example, individuals could share their location information with their friends on Weibo, WeChat, Facebook, and other social media platforms, bringing convenience to our lives. Then, vehicle navigation is one of the most widely used areas of LBS. With the application of vehicle navigation, self-driving tours meet individuals' needs for a journey. LBS not only provides navigation information services for drivers but also offers information about the nearby tourist attractions. Users could send a query about fining the nearby tourist attractions information, and then, they could receive a large number of traveling strategies. As for the bus and taxi,

LBS achieves the intelligence of transportation, including smart bus and smart taxi. For instance, the public transport dispatching monitoring and management system could generate the optimal driving route for buses base on the real-time location information. Besides, it could issue the scheduling instruction, such as acceleration, bypass and departure, etc., to these public vehicles according to the driving condition set in advance in the database. Finally, the wearable devices are also the kind of LBS application. The smartwatch, smart bracelet, and even the positioning chip embedded in sneakers are used to provide location services for customers. They can collect the outdoor sports route and generate images to record the exercise process.

## 20.4 Ethics Issues

### *20.4.1 Ethics Issues*

The research opportunities of LBS brought about a number of research challenges on theory, technology, and ethics. For example, the types of big data are usually diverse, unstructured, and multidimensional. Effective representation models and processing technologies are required to organize and analyze these data more efficiently and effectively [11]. In addition, all the services mentioned above are based on the location information from users. The ethics issue is related to the usage of this information. If used to analyze the query from users only, then the ethics problem is not worth raising concerns. However, if hackers or other people want to earn immoral and illegal gains from the data stream, there are serious ethical issues related to privacy for individuals. Location privacy is location information from the IoT users, one of the essential elements for IoT to perceive information and the prerequisite for IoT to provide location-based service with users. As for privacy, it is defined as the kind of information that the users are unwilling to expose. Therefore, it is unethical and even illegal to access and analyze user data without permission. Privacy is the right of citizens to their own privacy, the core of which is to control their own privacy according to their own will. Unauthorized positioning, location disclosure, mandatory commercial infringement, unreasonable use will harm the user's interests. For this reason, some users tend not to provide their information to the LBS providers, leading to insufficient data for LBS applications. Zhang et al. [12] tried to deal with this problem from a technology perspective. They proposed an algorithm established on the k-anonymity criterion to generate dummy locations and defined the trajectory entropy as a new variable to evaluate the performance of anonymity. Their simulation showed excellent performance. Different from them, this paper tends to focus on detailed ethical issues.

Figure 20.1 exemplifies the relationship between data analysis and ethical issues based on LBS. It is obvious that the source of data is used and the data could be analyzed by LBS providers to improve the location services. However, the data also could be attacked by both adversary and LBS providers. The former could attack

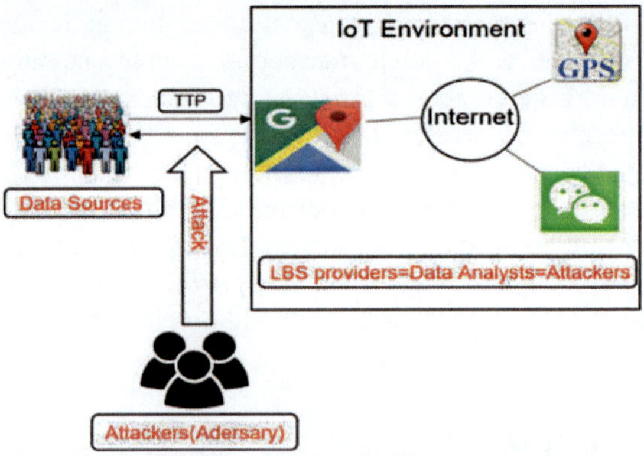


**Fig. 20.1** Ethical issues based on LBS

the defense mechanism, such as third trusted group (TTP), to access the database and analyze the user's personal information. As for the latter, they have the authority to obtain user data. Even though the data are not exact, LBS providers still could analyze more information by combining all queries or employing other methods.

### *20.4.2 Potential Privacy Disclosure Issues*

It is undeniable that the downside always arises with the benefits. From the view of ethics, the drawbacks are the kind of threats related to privacy for users. It is effortless to infer the privacy information and even sensitive information, such as social status and users' health condition by analyzing their location information since the prerequisite to enjoy location services is to deliver the location information.

Figure 20.2 shows the information used for data analysis and the benefits and privacy problems generated with the data analysis. The social media application is one of the most popular application areas of LBS and recent research outputs show that the privacy issues caused by social media cannot be ignored. Vicente et al. [13] revealed two kinds of threatens from the social media platform: sensitive privacy and re-identification. Besides, researchers, Li et al. explained that contextual information makes it is easier to achieve re-identification [14]. For example, an adversary could keep tracking users' trajectory in real-time through free software, such as Google Maps, and freely available data sources such as the Twitter stream data [14]. Then, they combine and analyze these data from two sources to identify an individual's life condition. Furthermore, Wang et al. [15] mentioned that when the users issue the location information on the social media platform, metadata tends to be revealed without the user's permission, and they even may be unaware of it if the location-based service on users' phone is in the active condition. The data thieves can access to these metadata to identify individuals and their moving tracks directly.

In fact, these metadata exposed individuals' information and used to improve the service provided for themselves. Taking Twitter as an example, there are more than

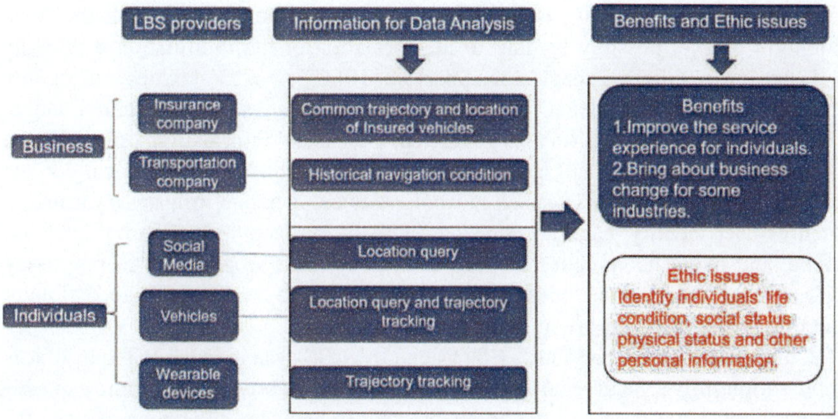


**Fig. 20.2** Benefits and privacy problems with data analysis

500 million tweets on twitter platform, and they are used to enrich a wide range of geospatial target analysis. This is why users could receive recommendations for restaurants or the prompt about nearby friends [15]. Additionally, there is a similar danger to privacy in the domestic market. Taking WeChat as an example, its built-in programs, such as nearby people, shake it up and WeChat sports also involve infringement of privacy through location services. User avatars, personal data, and other information will be seen by people nearby through the 'People nearby' program. Furthermore, the exact distance from the nearby person to your location will also be displayed. Zhao [16] also revealed that WeChat accesses user location information indirectly through the leaderboard on WeChat sports. Although users could empty the location information so that the information is not available to the other users, backstage will still record it for big data analytics [16]. Secondly, as for the vehicles with the LBS, it provides the optimal route for drivers.

However, at the same time, users' related information is recorded by LBS. This information includes the life condition of users and their identification as well as social status. For example, LBS could infer the individuals' addresses of home and company by analyzing the most common query on the navigation software. Combined with other query information, such as the grade of the most frequented restaurant, the LBS can even infer the standard of living and consumptionability of the person who sends a query. This information could be accessed by car compass as well as navigation software on smartphones. Finally, as for the wearable devices, there are also some ethical issues related to owners' privacy. Take smart bracelets as an example; these bracelets record runners' physical condition information by analyzing their heart rate and other indexes. It is undeniable that the authorities can analyze bracelets' physical status based on the health information from their smart bracelet records; however, if they are trafficked to the healthcare industry, customers will receive unnecessary medical marketing ads.

What is more worrying is that the physical condition is more valuable than common people, possibly leading to the formation of the information trafficking industry if the authorities neglect to supervise [10]. These ethical issues are not only immoral but also even illegal. The privacy trafficking industry is a potential danger caused by LBS. Once this trade is generated, then users' information, especially the celebrity's information, will be the source of illegal gains. The situational context data of LBS service in IoT space not only contains personal privacy information such as user identity, bank account, current location, activity trajectory, behavior mode, and living habits, but also may involve business secrets such as marketing plan, product information, customer information, and so on. Lu and Liao [17] point out that business secrets are more likely to cause crime.

Compared with the previous LBS system, the LBS system in the IoT space faces more serious privacy and security problems. Suppose the privacy and security of LBS services in the IoT space cannot be guaranteed. In that case, the popularity of LBS services in personal applications and the large-scale promotion of the commercial sector will be seriously affected [17].

## 20.5 Recommendation and Discussion

### *20.5.1 Existing Protection Method and the Weakness*

There is a collection of methods and techniques to solve these location-related privacy issues while they are defective. Most location privacy protection solutions can be categorized into two architectures, TTP-based schemes and (ii) TTP-free schemes. Figure 20.3 illustrates that, under the TTP-based scheme, TTP is responsible for receiving and transforming user location information and query, and then sending it to lbs. LBS can only deal with the query based on imprecise location information and then return the results to the user via TTP. For example, the most common method is the cloaking technique, which is based on the K-anonymity model. K ensures that the query from one user is mixed with at least K candidates and K is the most significant risk information disclosure. The queries will be sent into the centralized location anonymizer, and then the queries will be generated into a bigger cloaking region to mix these queries. Finally, these cloaking location queries rather than the exact one will be sent from anonymizer to the location-based services devices [18], which guarantee that the attacker cannot identify private information belongs to which specific individuals. As for this technique, the most important as well as the weakest

**Fig. 20.3** TTP-based scheme

segment is the cloaking area. If the adversary attacks this area, all information will be revealed because all locations are anonymized [19]. A relatively more advanced technique is 'dummy location'. However, the side information of this technique is neglected, and it is easy to be attacked by adversaries, then it is neither not functional [18]. Furthermore, the protocol about privacy preserving is also a popular method [20]. However, the drawback of it is that the service provider cannot predict all purposes.

What is more, it is meaningless to tell the uses of what they will do with this location information because it will make the location data become biased and not truth [3]. Apart from these vulnerabilities, there also other doubts about these protection methods. He et al. [21] revealed that the surrounding physical environment fundamentally limits privacy protection. Jiang et al. [22] found that the continuous LBS queries are ignored because the previous protection method only deals with snapshot user locations, one-time queries for the K-anonymity. In conclusion, the trusted third party is a bottleneck in computing and communication in the whole system. The low success rate for anonymity, the high cost of communication, and the weak ability are also drawbacks to resisting attacks from adversaries.

### *20.5.2 Recommendation*

The drawbacks of these existing methods are innumerable. 'The biggest challenge is how to balance individual privacy while still making the individual's location data publicly available for mining and analysis' [15]. Users' information is not only helpful for the business and individuals themselves but also practical for authorities. From real-time universal health monitoring for people to future urban planning, large amounts of data are needed. It would generate great benefit to users if researchers and government agencies access this information and use it for research. Therefore, it is crucial to balance their relationship.

When the user wants to get the efficient and convenient location service, their clear text information can easily become the attacker's attack objects. According to Yan et al. [8], today, many LBS providers abroad pay the growing attention to user data confidentiality, such as

- binding users to hardware uniquely to avoid situations where multiple users are on the same device,
- increasing user behavior management, identifying suspicious user groups or user behavior, and
- providing real-time alerts based on customized policies.

However, few of these app developers or operators at the domestic are concerned. Therefore, supervisors need to do more to strengthen the regulation of consulting privacy on a social level. From a technical point of view, Yan et al. [8] point out that two kinds of recognition abilities should be improved for the LBS providers. On the one hand, it is necessary to enhance user recognition ability and make it is difficult for non-real users to access real data. On the other hand, improving the hardware recognition capabilities for mobile devices and developing corresponding security policies for relevant tools such as simulators are essential. Besides, since the principle of location-based services, the biggest contradiction is that the higher the quality of location-based service, the higher the risk of privacy leakage.

According to Zhou et al. [7], the goal of most current research revolves around the 'relative privacy' protection of the location, or 'controllable privacy,' where location information is released in a form that is relatively difficult for attackers to identify. This form could meet the user's personalized privacy needs and ensure that LBS can obtain the location even though it cannot access the location information [7]. Besides, the three biggest challenges to break the existing methods are as following. The first one is about the measurement standards of the extent of user privacy leaks. Then, the second one is how to be responsible for the comprehensive protection for data-based location privacy. The last one is how to trade-off between privacy protection and the availability of these location services. The traditional technique, such as K-anonymity, does not consider how to measure the degree of a privacy breach or protect users' privation completely, let alone balance them.

## 20.6 Conclusion

Along with the development of big data mining and analysis technology, in the IoT environment, all groups are experiencing and enjoying the convenience that LBS brought to all aspects of social development, such as military, commercial, and personal life. In the business world, LBS has brought about substantial business changes. For example, the transportation industry could compute cost-optimal routes for the future transportation plan by analyzing historic navigation records. For individuals, everyone lives in the services provided by LBS since all query content sent by the user are the best source of data analysis for the LBS providers. However, the contradiction is that the high quality of LBS is based on the extent of the exposed location information and fragile protection mechanisms. Therefore, whether it is for LBS providers, government regulators, or individual users, it is essential to raise personal information protection awareness. For the first two groups, it is urgent to find a breakthrough to protect user privacy information and find a definition of privacy.

**Acknowledgements** This work is supported by VC Research (VCR 0000082).

## References


1. Stergiou, C., Psannis, K.E.: Efficient and secure BIG data delivery in Cloud Computing. Multimed. Tools Appl. **4**, 1–20 (2017)
2. Stergiou, C., Psannis, K.E., Gupta, B.B., Ishibashi, Y.: Security, privacy and efficiency of sustainable cloud computing for big data and IoT. Sustainable Computing: Informatics and Systems (2018)
3. Mayer-Schönberger, V., Cukier, K.: The Age of Big Data. Zhe Jiang People Publishing House, s.l. (2013)
4. Ma, H.: Haixiang Ma Blog. [Online] (2014). Available at: https://www.mahaixiang.cn/sjfx/803.html. Accessed 10 Dec 2018
5. Rivera, J., van der Meulen, R.: Gartner (2014). [Online] Available at: https://www.gartner.com/newsroom/id/2819918. Accessed 6 Dec 2018
6. Caron, X., Bosua, R., Maynard, S.B., Ahmad, A.: The Internet of Things (IoT) and its impact on individual privacy: an Australian perspective. Comput. Law Secur. Rev., 4–16 (2016)
7. Zhou, A., Yang, B., Jin, C., Ma, Q.: Location-based services: architecture and progress. Chin. J. Comput. **34**(7), 1155–1571 (2011)
8. Yan, D., Wang, Y., Xue, Z., Hu, L.: Research on the security risk of APP data for LBS service. Commun. Technol. **49**(12), 1702–1708 (2016)
9. Sheng, X.: LBS—Application of technology in social media. Friend Sci. **2012**(2), 159–161 (2012)
10. Wang, L., Meng, X.: Location privacy preservation in big data era: a survey. Ruan Jian Xue Bao/J. Softw. **25**(4), 639–712 (2014)
11. Yao, X.A., Huang, H., Jiang, B., Krisp, J.M.: Representation and analytical models for location-based big data. Int. J. Geogr. Inf. Sci. **33**(4), 707–713 (2019). https://doi.org/10.1080/13658816.2018.1562068
12. Zhang, L., Qian, Y., Ding, M., Ma, C., Li, J., Shaham, S.: Location privacy preservation based on continuous queries for location-based services. In: IEEE INFOCOM 2019—IEEE Conference on Computer Communications Workshops (INFOCOM WKSHPS), Paris, France, pp. 1–6 (2019)
13. Vicente, C.R., Freni, D., Bettini, C., Jensen, C.S.: Location-related privacy in GEO-social networks. Internet Comput. **15**(3), 20–27 (2011)
14. Ajayakumar, J., Ghazinour, K.: I am at home: spatial privacy concerns with social media check-ins. ScienceDirect **08**, 551–558 (2017)
15. Wang, S., Sinnott, R., Nepal, S.: Protecting the location privacy of mobile social media users. In: International Conference on Big Data (Big Data), pp. 1143–1150 (2016)
16. Zhao, C.: The privacy protection dilemma of social media in the new media era-taking WeChat as an example. Theory Research **2017**(12), 5–6 (2017)
17. Lu, H., Liao, L.: Privacy-preserving model of LBS in Internet of Things. Comput. Eng. **50**(15), 91–96 (2014)
18. Sun, G. et al.: Efficient location privacy algorithm for Internet of Things (IoT) services and. J. Netw. Comput. Appl., pp. 3–13 (2017)
19. Zhao, H., Yi, X., Wan, J.: Privacy-area aware all-dummy-based location privacy algorithms. In: Proceedings of the IEEE First International Conference on Mobile Service, pp. 9–16 (2016)
20. Li, X., Jung, T.: Search me if you can: privacy-preserving location query service. In: 2013 Proceedings IEEE INFOCOM, pp. 2760–2768 (2013)
21. He, X., Jin, R., Dai, H.: Leveraging Spatial diversity for privacy-aware. IEEE Trans. Inf. Forensics Secur. **13**(6), 1524–1534 (2018)
22. Jiang, H., Zhao, P., Wang, C.: RobLoP: towards robust privacy preserving against location dependent attacks in continuous LBS queries. IEEE/ACM Trans. Netw. (n.d.)

# Chapter 21
# A Grammar-Aware Pointer Network for Abstractive Summarization

**Yinghua Zhou, Fang Cao, Yangyang Cao, Ning Yang, and Zhen Li**

**Abstract** Pointer network (PN) has achieved breakthrough in recent years of text summarization research. But it only focuses on semantic relevance of the source sequence; in fact, the text should also comply with explicit grammar rules. The semantics and syntax are the two main granularities to research in this paper. In more detail, we proposed a grammar-aware pointer network (GPAN) for abstractive summarization, which not only tracks the key semantics of the original text, but also observes the syntax rules. To enforce the syntactic constraints, we get each word attached with the part-of-speech (POS) tag and syntactic dependency (DEP) tag and input them into the recurrent network when training the network. Then, we predict the POS and DEP at each decoder time step; by this way, we trying to let the model learn to track the grammar information of the ground truth. We evaluate our model on the benchmark dataset CNN/Daily Mail and GiGaword. The experimental results show that our model leads to significant improvements.

## 21.1 Introduction

The research work [1–4] for artificial intelligence mainly involves the theory, method, technology, and application system, which is embodied in the simulation and extension of human intelligence from the algorithm. Text summarization is a sub-process study that purifies redundant information, rephrasing a shorter one reserving the trunk. According to the form of generation, it is divided into two categories: extractive and abstractive. With explicit formulas, the former directly extracts sentences or subsequences from the source text to form into the summarization, keeping syntactic correct, but lacking of coherence and logic. The latter generates by the algorithm model itself possessing of more association and reasoning. In recent years, deep learning has gotten a critical application in this field. The basic approach is using an encoder–decoder framework [5, 6], in which the encoder transforms the global

Y. Zhou · F. Cao (✉) · Y. Cao · N. Yang · Z. Li
Chongqing University of Posts and Telecommunications, Chongqing, China
e-mail: s180201020@stu.cqupt.edu.cn



V. Chang et al. (eds.), *Modern Industrial IoT, Big Data and Supply Chain*,
Smart Innovation, Systems and Technologies 218,
https://doi.org/10.1007/978-981-33-6141-6_21

information of the input sequence to a latent space, and the decoder infers the corresponding output sequence. In this paper, we major in abstractive summarization. Rush [7] is the first time to apply deep learning technology to the research of generalized automatic summarization. The innovative point of the article is that it proposed a fully data-driven approach to calculate the relevance between the source and abstract. Zhou [8] and Lin [9] proposed a selective coding model which is equivalent to "selecting" relatively important content so that the decoder can read the source text in a targeted manner. Tan [10] combines the TextRank algorithm with the Attention mechanism, improving the saliency of document representation. Li [11] conducted a linguistic analysis of the artificial abstracts, using a variational auto-encoder on the decoder side enables the model to learn vectors containing potential structural information. Ling [12] select the block with the largest attention weight and use the word attention distribution in the block to directly calculate the context vector. Li [13] use key information to guide generation. Homoplastically, Wang [14] refer to the latent topics of source text to do it. Furthermore, Li [15] introduce an novel method to produce attention distribution, generating more abstractive summarization instead of simple repetition.

In the previous text summary generation work, most of the basic research objects are words, to explore the relationship between each word in the source text, and each word in the abstract, at the time of decoding, is also to calculate the probability of each word being correctly predicted. The contributions of our work are as follows:

1. The motivation of this research is to explore the essence of the auto-summarization generation by simulating people's thinking logic in the process of writing abstracts, giving deep learning a certain degree of interpretability and reasoning.
2. There are a lot of repetition problems in the abstracts generated in the previous related studies. Our model does not purposefully deal with it, instead thinking about the essential factors behind the problem.
3. Our model has greatly improved the logic and smoothness of generated sequence.

## 21.2 Baseline Model

Donate the source text as $X = [x_1, x_2, ..., x_J]$, $J$ is total numbers of the words (fine-grained division, it also could be chars) of $X$. Then, there is a fixed enough large vocabulary for the corpus. In the embedding layer of the neural network, there will be trainable vectors to represent corresponding words (chars) called word2vec as input for the recurrent network (LSTM, GRU). Then, the information of each words (chars) in the $X$ is represented by the hidden state $h_i$, $i \in (0, J)$. Due to the limitation of the vocabulary size, some low-frequency words cannot be included in the vocabulary. These words are out of vocabulary (OOV). They will be uniformly replaced and displayed by UNK, and their semantic information will also be discarded. The main idea of pointer network [16–19] is that the output element comes from the input. For abstractive summarization, directly copying OOV of the source forms into abstract when necessary. The model has to solve two problems: at each time step of decoding, whether to choose to generate a word from the vocabulary or copy a word from the source text; if you choose to copy, where to copy from the source text.

Since the attention distribution [20, 21] and position distribution of lexical items are basically the same, the model directly uses the attention distribution as the position distribution, and the complexity of the model is reduced through parameter reuse.

$$a_t = softmax(f(h_i, d_t)) \tag{21.1}$$

$f$ is a score function to measure the alignment between $h_i$ and $d_t$.The context vector $c_t$ and the probability of the next token are computed as below:

$$\begin{aligned} c_t &= \sum_i h_i a_{t,i} \\ y_t^* &= [d_t; c_t]L \\ P_{vocab} &= softmax(y_t^* W) \end{aligned} \tag{21.2}$$

[;] means concatenation and $L$. We are trainable parameters. $P_{vocab}$ is the probability distribution over the fixed vocabulary.

The model sets up a switch network base on a multilayer perceptron and inputs the semantic vector and the decoder hidden layer St into the network to obtain a switch variable $P_{gen} \in [0; 1]$ for time step $t$ is calculated from the context vector $h_t^*$, the decoder state $s_t$ and the decoder input $x_t$:

$$p_{gen} = \sigma(W_{h^*} h_t^* + W_s s_t + W_x x_t + E + b) \tag{21.3}$$

The copy probability is $1 - P_{gen}$. For each source text, at each decoding time step, the words in the text are dynamically added to the vocabulary to get expanded vocabulary, and the expanded vocabulary distribution is:

$$p(w_t) = p_{gen} p_{vocab}(w_t) + (1 - p_{gen})\alpha_t \tag{21.4}$$

When the word $w$ is not in the vocabulary, $P_{vocab}(w) = 0$, $P_{gen}$ is equivalent to a "soft switch," learning to generate a word from the vocabulary distribution or copy a word from the input sequence, which improve the training accuracy of the rare terms.

## 21.3 Grammar-Aware Pointer Network

The previous research work predicts words by sampling from vocabulary probability distributions. Although with the support of some tricks and model structure, the

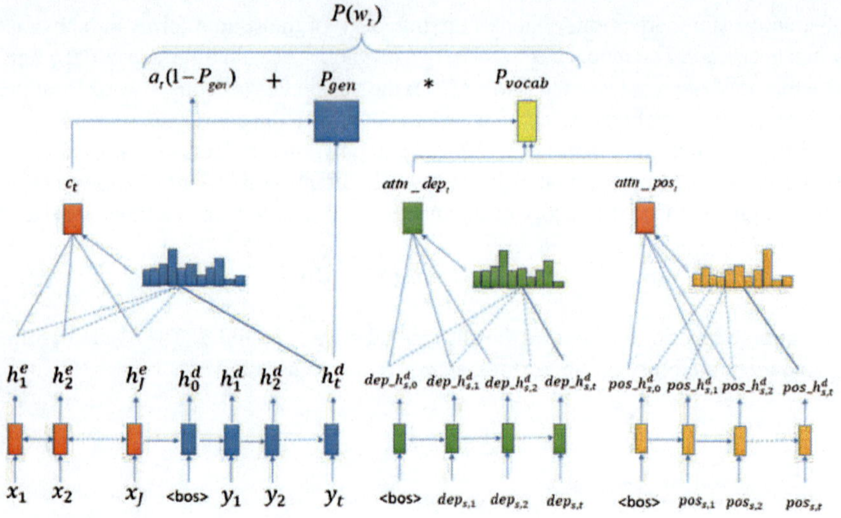


**Fig. 21.1** Grammar-aware pointer network

effect has been greatly improved, and this is not the original intention of cognitive learning. For example, the beginning of the text tends to be the noun and the subject in the sentence. There must only be one verb and only one predicate in an English sentence. In our research, tracking on this information for the model, explaining and studying this phenomenon, then combining with specific in our experiments verify its effectiveness. In the previous research work, words were the most basic research object.

There is a strong semantic relationship between text sequences. For example, verbs are usually followed by nouns, and subjects are followed by predicates, etc. Some scholars have begun to study the role of grammar in text summarization by syntactic tree. But, in deep learning, it is probability distribution that can model these rules, what is exactly the main contribution of this paper (Fig. 21.1).

In the previous research work, words were the most basic research object. As shown in the figure, we put DEP and POS together with words as the basic objects of research.

### *21.3.1 Grammar Embedding*

Among the existing research results, there are already many tools, such as NLTK, Spacy, and Stanford CoreNLP which provide analysis functions for English text, such as POS, DEP, and named entity recognition. In addition to the fixed vocabulary list, we also created a part of speech table of size 49 to represent the part of speech of each word (such as nouns, verbs, and adverbs) and a dependency table of size 21 to represent each word. The syntactic components (such as subject, predicate, and object) assumed in the sentence. These two vocabularies, like the fixed vocabulary, are used as parameters to participate the model training.

### 21.3.2 Multi-task Learning

In the training phase, each term generated by the decoder has a standard abstract and DEP, POS for reference and backpropagation of errors to correct model parameters. But, in testing phase, we could only take the last time step model generation as the input of the decoder at this time step, so we use the teacher forcing mechanism to alleviate the gap between these two stages. In addition, in the training process of the model, we also let the model predict the DEP and POS at each moment and the components assumed in the sentence and use this part as a loss function for training.

$$\begin{aligned} dep\_dist &= f_1(\text{out}) \\ pos\_dist &= f_2(\text{out}) \end{aligned} \tag{21.5}$$

where $f_1$, $f_2$ is a linear function, followed by softmax, which output a probability of distributions, and we choose the max as current generated word and

$$out = [e; s_t; c_t^e; attn\_dep; attn\_pos; dep_t; pos_t] \tag{21.6}$$

$s_t$ is the decoder hidden state at time step $t$, and $c_t^e$ is the source context vector, $att\_dep$ is an attention about the DEP information with weighted, which indicate the current DEP relevance about the previous. We calculate it by the following:

$$attn\_dep = deps \cdot softmax(V \tanh(W \cdot deps + b)) \tag{21.7}$$

$deps$ is the previous decoder hidden state of the DEP. $W$, $V$, $b$ are learnable parameters. Following a similar approach, we also get the $att\_pos$. At each decoder time step, we calculate negative log probability loss of predicted word, DEP, and POS:

$$\begin{aligned} loss_{dep} &= NLLoss(dep\_dist, dep) \\ loss_{pos} &= NLLoss(pos\_dist, pos) \\ loss_{word} &= NLLoss(word\_dist, word) \\ loss &= \alpha loss_{dep} + \beta loss_{pos} + \gamma loss_{word} \end{aligned} \tag{21.8}$$

where $\alpha, \beta, \lambda$ are hyperparameters for our model. NLLoss is defined as follows:

$$NLLoss = -\frac{1}{T}\sum_{t=0}^{T}\log \mathrm{P}(\mathrm{w}_t^*) \tag{21.9}$$

Following [22], we take the evaluation metric as part of loss and define the following loss function:

$$L_{rl} = (r(\hat{y}) - r(y^*))\sum_{t=1}^{T}\log\ p(y_t^*|y_1^*, ..., y_{t-1}^*, x) \tag{21.10}$$

where $r$ is the average of ROUGE-1 F1 of the results, $\hat{y}$ is sampled from output probability distribution by argmax, and $y^*$ is sampled by corresponding probability distribution at each decoder time step. Then, the loss of model is defined as follows:

$$L = \mu loss + (1 - \mu)L_{rl} \tag{21.11}$$

## 21.4 Experiments

### *21.4.1 Experiment Setting*

According to word frequency from large to small, take the top 50,000 as a fixed vocabulary. The hidden size in the recurrent neural network is 512. The embedding dimension of word, DEP, and POS both are 200, initialed by truncated normal distribution. Set batch size=80 for dataset GiGaword and batch size=10 for dataset CNN/Daily Mail, training on GPU of GTX2080Ti. In the training process, Adam is used as the optimizer, and the learning rate is set to 0.001. During the experiment, the validation set data is used for verification every 1000 iterations, and the model parameters are saved if the f-score of ROUGE-1 is greater than 0.38. During the experiment, we found that the model is more volatile when using argmax (pick the word with the highest probability as the generated word) to sample the generated words, though it can get a good score when the batch number is relatively small. We use beam search as sampling strategy. Unlike argmax, beam search is a topK algorithm, which means that at each decoding moment, take the top K words with the highest probability as the words generated at the current moment, and finally, the probability of each word corresponding to the K abstracts is averaged, and then, the one with the highest probability is taken as the abstract generated by the current batch model. In addition, due to the exposure bias between training phase and the test phase, we set a teacher force rate as 0.75 to alleviate the gap of the two phrases.

### 21.4.2 Experiment Result

We have tried to hierarchically get the representation of source text from sentence to word [23, 24]. But experimental results show that ROUGE has not been significantly improved, even getting worse as training iteration. Perhaps, the hierarchical attention segmented the text, causing semantic incoherence. Therefore, we finally choose the most traditional global attention to calculate the source context vector. As it is depicted in the table, there are obvious advantages compared with these mentioned models (Tables 21.1 and 21.2).

**Table 21.1** F1 score of ROUGE on CNN/Daily Mail dataset

| Model | ROUGE-1 | ROUGE-2 | ROUGE-L |
|---|---|---|---|
| Words-lvt2k [24] | 32.5 | 11.8 | 29.5 |
| Words-lvt2k-ptr [24] | 32.5 | 11.7 | 29.2 |
| Words-lvt2k-hie [24] | 31.8 | 11.6 | 28.7 |
| Baseline [19] | 36.44 | 15.66 | 33.42 |
| Pointer coverage [19] | 39.53 | 17.28 | 36.90 |
| C2F [12] | 31.10 | 15.40 | 28.80 |
| GBA [10] | 38.10 | 13.90 | 34.00 |
| AES [25] | 35.80 | 13.60 | 33.40 |
| DivCNN Seq2Seq [15] | 32.28 | 12.75 | 24.30 |
| KIGN + prediction-guide [13] | 38.95 | 17.12 | 35.68 |
| **GAPN** | **39.74** | **16.98** | **38.45** |

**Table 21.2** F1-Score of ROUGE on GiGaword dataset

| Model | ROUGE-1 | ROUGE-2 | ROUGE-L |
|---|---|---|---|
| ABS [7] | 29.76 | 11.30 | 26.40 |
| ABS+ [7] | 29.8 | 11.90 | 27.0 |
| DRGD [11] | 36.27 | 17.57 | 33.62 |
| SEASS [8] | 36.15 | 17.54 | 33.63 |
| BASE + E3T [26] | 36.89 | 16.86 | 34.74 |
| CGU [9] | 36.3 | 18.7 | 33.8 |
| Re3Sum [27] | 37.04 | 19.03 | 34.46 |
| AC-ABS [28] | 36.05 | 17.35 | 33.49 |
| Filtered + pseudo [29] | 35.85 | 17.94 | 33.72 |
| **GAPN** | **38.14** | **17.89** | **36.28** |

### *21.4.3 Experiment Analysis*

When processing natural language tasks in deep learning, the input is a tensor of three dimensions, [batch_size; num_words; embedding_dim], in which batch_size is the number of samples and num words is the number of words in each sample and embedding dim is the dimension of the word vector of each word. Grammar is originally a conception based on sentences in English. That is to say when it comes to grammar, the domain is a complete and standardized sentence. Based on this principle, For the DataSet CNN/DailyMail, there are four sentences on average of its summary. More specifically, the input tensor is four dimensions [batch_size, num_sents, num_words, embedding_dim]. We use NLTK's sent_tokenize and word_tokenize to divide the summary into sentences according to the ending punctuation (e.g. '.', '!', '?') and then divide each sentence into words. For grammar attention, we use an intra-sentence attention to capture the grammar information, which separately calculate the alignment weight of DEP and POS in each sentence. We do this because the grammatical relationship only exists in a single complete sentence. For example, they are more semantic relationships between the subject in the first sentence and the predicate in the second sentence instead of grammatical relevance. However, it does not work well as what we look forward to. The main reasoning is that the number of sentences in the abstract is different, and the length of the sentence is also different, but for the program, the dimensions must be unified, which means that there are many truncation and padding operations in the abstract, which seriously damages the data semantic information and grammatical rules of the source.

But for GiGaword, the abstract is actually a headline, only one sentence. In other words, there is no need to add another dimension. There are few truncation and padding operations. Relatively speaking, the data source information is preserved well, resulting in an excellent score.

## 21.5 Conclusion

Our model generates text abstracts from the perspective of syntax and semantics, imitating people's logical thinking activities when dealing with this type of task in the process of text summary writing, exploring feasible solutions based on the essence of this task. We **firstly** should focus on the essential parts of the source text. **Then** before writing, we should grasp correct grammar knowledge. We are trying to use algorithms to simulate human brain thinking, although the specific implementation details are still very different from human beings, which is also the most common trend of deep learning research at this stage.

## References


1. Xia, S., Liu, Y., Ding, X., Wang, G., Yu, H., Luo, Y.: Granular ball computing classifiers for efficient, scalable and robust learning. Inf. Sci. **483**, 136–152 (2019)
2. Xia, S., Peng, D., Meng, D., Zhang, C., Wang, G., Giem, E., Wei, W., Chen, Z.: A fast adaptive k-means with no bounds. IEEE Trans. Pattern Anal. Mach. Intell. (2020)
3. Xia, S., Wang, G., Chen, Z., Duan, Y., et al.: Complete random forest based class noise filtering learning for improving the generalizability of classifiers. IEEE Trans. Knowl. Data Eng. **31**(11), 2063–2078 (2018)
4. Xia, S., Zhang, Z., Li, W., Wang, G., Giem, E., Chen, Z.: Gbnrs: a novel rough set algorithm for fast adaptive attribute reduction in classification. IEEE Trans. Knowl. Data Eng. (2020)
5. Cho, K., van Merrienboer, B., Gulcehre, C., Bahdanau, D., Bougares, F., Schwenk, H., Bengio, Y.: Learning phrase representations using rnn encoder–decoder for statistical machine translation. In: Proceedings of the 2014 Conference on Empirical Methods in Natural Language Processing (EMNLP), pp. 1724–1734 (2014)
6. Sutskever, I., Vinyals, O., Le, Q.V.: Sequence to sequence learning with neural networks. In: Advances in Neural Information Processing Systems, pp. 3104–3112 (2014)
7. Rush, A.M., Chopra, S., Weston, J.: A neural attention model for abstractive sentence summarization. In: Proceedings of the 2015 Conference on Empirical Methods in Natural Language Processing, pp. 379–389 (2015)
8. Zhou, Q., Yang, N., Wei, F., Zhou, M.: Selective encoding for abstractive sentence summarization. In: Proceedings of the 55th Annual Meeting of the Association for Computational Linguistics (Volume 1: Long Papers), pp. 1095–1104 (2017)
9. Lin, J., Sun, X., Ma, S., Su, Q.: Global encoding for abstractive summarization. In: Proceedings of the 56th Annual Meeting of the Association for Computational Linguistics (Volume 2: Short Papers), pp. 163–169 (2018)
10. Tan, J., Wan, X., Xiao, J.: Abstractive document summarization with a graph-based attentional neural model. In: Proceedings of the 55th Annual Meeting of the Association for Computational Linguistics (Volume 1: Long Papers), pp. 1171–1181 (2017)
11. Li, P., Lam, W., Bing, L., Wang, Z.: Deep recurrent generative decoder for abstractive text summarization. arXiv preprint arXiv: 1708.00625 (2017)
12. Ling, J., Rush, A.: Coarse-to-fine attention models for document summarization. In: Proceedings of the Workshop on New Frontiers in Summarization (2017)
13. Li, C., Xu, W., Li, S., Gao, S.: Guiding generation for abstractive text summarization based on key information guide network. In: Proceedings of the 2018 Conference of the North American Chapter of the Association for Computational Linguistics: Human Language Technologies, Volume 2 (Short Papers), pp. 55–60 (2018)
14. Wang, L., Yao, J., Tao, Y., Zhong, L., Liu, W., Du, Q.: A reinforced topic-aware convolutional sequence-to-sequence model for abstractive text summarization. In: Proceedings of the 27th International Joint Conference on Artificial Intelligence, pp. 4453–4460 (2018)
15. Li, L., Liu, W., Litvak, M., Vanetik, N., Huang, Z.: In conclusion not repetition: Comprehensive abstractive summarization with diversified attention based on determinantal point processes. In: Proceedings of the 23rd Conference on Computational Natural Language Learning (CoNLL), pp. 822–832 (2019)
16. Vinyals, O., Fortunato, M., Jaitly, N.: Pointer networks. In: Advances in Neural Information Processing Systems, pp. 2692–2700 (2015)
17. Gulcehre, C., Ahn, S., Nallapati, R., Zhou, B., Bengio, Y.: Pointing the unknown words. In: Proceedings of the 54th Annual Meeting of the Association for Computational Linguistics (Volume 1: Long Papers), pp. 140–149 (2016)
18. Gu, J., Lu, Z., Li, H., Li, V.O.K.: Incorporating copying mechanism in sequence-to-sequence learning. In: Proceedings of the 54th Annual Meeting of the Association for Computational Linguistics (Volume 1: Long Papers), pp. 1631–1640 (2016)

19. See, A., Liu, P.J., Manning, C.D.: Get to the point: summarization with pointer generator networks. In: Proceedings of the 55th Annual Meeting of the Association for Computational Linguistics (Volume 1: Long Papers), pp. 1073–1083 (2017)
20. Bahdanau, D., Cho, K., Bengio, Y.: Neural machine translation by jointly learning to align and translate. In: 3rd International Conference on Learning Representations, ICLR 2015 (2015)
21. Luong, M.-T., Pham, H., Manning, C.D.: Effective approaches to attention-based neural machine translation. In: Proceedings of the 2015 Conference on Empirical Methods in Natural Language Processing, pp. 1412–1421 (2015)
22. Paulus, R., Xiong, C., Socher, R.: A deep reinforced model for abstractive summarization. In: International Conference on Learning Representations (2018)
23. Yang, Z., Yang, D., Dyer, C., He, X., Smola, A., Hovy, E.: Hierarchical attention networks for document classification. In: Proceedings of the 2016 Conference of the North American Chapter of the Association for Computational Linguistics: Human Language Technologies, pp. 1480–1489 (2016)
24. Nallapati, R., Zhou, B., dos Santos, C., Gulcehre, C., Xiang, B.: Abstractive text summarization using sequence-to-sequence rnns and beyond. In: CoNLL 2016, p. 280 (2016)
25. Chen, Y.B., Ma, Y., Mao, X.D., et al.: Abstractive summarization with the aid of extractive summarization. In: Proceedings of Asia-Pacific Web and Web-Age Information Management Joint International Conference on Web and Big Data, vol. 10987, pp. 3–15. Springer, Cham (2018)
26. Amplayo, R.K., Lim, S., Hwang, S.W.: Entity commonsense representation for neural abstractive summarization. In: 2018 Conference of the North American Chapter of the Association for Computational Linguistics: Human Language Technologies, NAACL HLT 2018, pp. 697–707. Association for Computational Linguistics (ACL) (2018)
27. Cao, Z., Li, W., Li, S., Wei, F.: Retrieve, rerank and rewrite: Soft template based neural summarization. In: Proceedings of the 56th Annual Meeting of the Association for Computational Linguistics (Volume 1: Long Papers), pp. 152–161 (2018)
28. Li, P.J., Bing, L.D., Lam, W.: Actor-critic based training framework for abstractive summarization. (2018–8–15) [2018–9–15].https://arxiv.org/abs/1803.11070.
29. Matsumaru, K., Takase, S., Okazaki, N.: Improving Truthfulness of Headline Generation. arXiv, 2020: arXiv: 2005.00882

# Chapter 22
# Single-Phase Photovoltaic Inverter Control Based on Quasi-Prand Big Data Analysis

**Yujia Zhang, Xinlong Zheng, Peixiao Fan, and Yong Xiang**

**Abstract** In this paper, a micro-grid system based on single-phase photovoltaic inverter double closed-loop feedback is established. The double closed-loop feedback model is discussed, and the advantages and disadvantages of PI, PR and quasi-PR control in terms of dynamic performance and steady-state error are compared according to the amplitude–frequency characteristics. First, the advantages of the quasi-PR control strategy are expressed from the theoretical level, which can realize tracking without static errors and enhance the anti-interference ability. After that, the combination of pole configuration and big data analysis is used to obtain the optimal solution of PR and quasi-PR controller parameters, so as to more strictly carry out the impact of both on the performance of the double closed-loop system in the actual system. Finally, a simulation model platform was established based on MATLAB/Simulink, which verified the superior performance of the system.

## 22.1 Introduction

At present, in the control of single-phase photovoltaic inverters, the double closed-loop control scheme of voltage outer loop current inner loop is one of the development directions of high-performance inverter power supplies. In the development process, the controller in the double closed-loop system has been further developed. Now, the main application of proportional integral controller, namely PI regulator. The PI control structure is simple and easy to implement, which enables the system to obtain better dynamic and steady-state performance, but it cannot achieve static error-free tracking of sinusoidal signals and has the disadvantages of steady-state error and poor anti-interference ability. Further, scholars have proposed a PR control strategy. The method is also easy to implement. Because of the advantages of eliminating steady-state errors and improving anti-jamming capabilities, PR control has also been further applied [1–3].

Y. Zhang (✉) · X. Zheng · P. Fan · Y. Xiang
Automated Institute Central South University, Changsha 410006, Hunan, China



V. Chang et al. (eds.), *Modern Industrial IoT, Big Data and Supply Chain*,
Smart Innovation, Systems and Technologies 218,
https://doi.org/10.1007/978-981-33-6141-6_22

The relevant concepts of quasi-PR controllers were first proposed by Professor Chen Hongsheng, demonstrating the advantages of quasi-PR controllers over PI and PR controllers under single closed-loop feedback control [4]. In other scholar articles, there are also double closed-loop systems based on quasi-PR control [5], but there are some problems: In the double closed-loop model, the transfer function and parameter selection are complex high-order systems, the parameters, etc., which are all estimated values. On the other hand, when comparing the quasi-PR controller with other controllers, it is only a simple analysis on the structure of the controller itself, because it is impossible to get the optimal parameter solution and accurately represent the control parameter pairs. The impact of performance, so there is a certain degree of rigor in the comparison on the actual system.

This paper further verifies the advantages of quasi-PR control through the analysis of actual systems and the combination of pole placement and big data analysis.

## 22.2 Single-Phase Inverter Model Structure and Control Strategy

## 22.3 Inverter Double Closed-Loop Control Strategy

We first established the structural block diagram of the double closed-loop system. The instantaneous current reference value is generated by $G_v(s)$, which interferes with the filter inductor current signal and the load feedforward. The current signal is compared to form an instantaneous current error signal. The modulation signal is generated through $G_i(s)$, and the SPWM switching signal is generated after calculation with the carrier signal to control the switching device to form a modulated output voltage on the AC side.

The transfer function during feedforward with load disturbance is:

$$U_o(s) = \frac{G_v(s)G_i(s)K_{pwm}U_{\text{ref}} - [(r + Ls)]I_o(s)}{LCs^2 + \left[G_i(s)K_{\text{pwm}} + r\right]Cs + G_v(s)G_i(s)K_{\text{pwm}} + 1} \tag{22.1}$$

For the outer loop voltage control, we have proposed three control methods and conducted amplitude–frequency analysis, respectively, to obtain theoretical stability and dynamic performance. First, we analyze the simplest PI control when $G_v(s)$ selects PI control:

$$G(s) = K_{ip} + \frac{K_{ii}}{s} \tag{22.2}$$

At the grid fundamental frequency $\omega_0$, the amplitude of PI control is:

$$A_{PI}(\omega_0) = \sqrt{K_p^2 + \left(\frac{K_i}{\omega}\right)^2} \tag{22.3}$$

Therefore, both of the transfer functions are finite, the output voltage $U_o$ is smaller than the reference voltage $U_{ref}$, and there is a steady-state error. In order to realize the system without steady-state error. Just set the system input to a given reference value, then the command can be tracked without static error. When $G_v(s)$ selects PR control:

$$G_v(s) = K_p + \frac{K_r s}{s^2 + \omega_0^2} \tag{22.4}$$

At the grid fundamental frequency $\omega_0$, the amplitude of PR control:

$$A_{PR}(\omega) = \sqrt{\frac{K_p^2 + (K_r\omega)^2}{(-\omega^2 + \omega_0^2)^2}} \tag{22.5}$$

Substituting $\omega_0$ into the transfer function, the first term is equal to $U_{ref}$, and the second term tends to 0. Compared with PI control, after introducing PR control, the output at the fundamental frequency is basically equal to the reference voltage and closed loop at other frequencies. The gains are both small and less than 0. However, it can be seen in subsequent simulations that when the load is disturbed, the response speed is slower and the dynamic response time is longer. Due to the problems caused by the above PR controller, we intend to propose a quasi-PR control that is easier to implement.

When $G_v(s)$ selects quasi-PR control:

$$G_{PR^*}(s) = K_p + \frac{2K_R\omega_c s}{s^2 + 2s\omega_c + \omega_0^2} \tag{22.6}$$

At the grid fundamental frequency $\omega_0$, the amplitude of quasi-PR control:

$$A_{PR^*}(\omega_0) = \sqrt{K_p^2 + \left(\frac{2K_R\omega_c\omega_0}{-\omega_0^2 + 2\omega_c\omega_0 + \omega_0^2}\right)^2} \tag{22.7}$$

$K_p$, $K_r$ are proportional and resonance parameter. Output image is shown in Figs. 22.1 and 22.2.

Using the control variable method and analyzing the law through the amplitude–frequency characteristics of the controller under different parameters: $K_r$, $K_p$. Therefore, we conclude that quasi-PR not only has a sufficiently large gain, but also increases the bandwidth, which significantly increases the anti-interference ability of

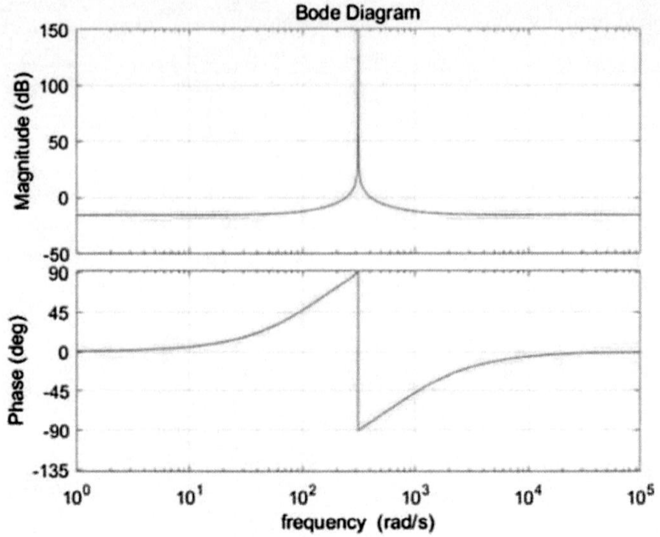


**Fig. 22.1** Amplitude frequency of ideal PR controller

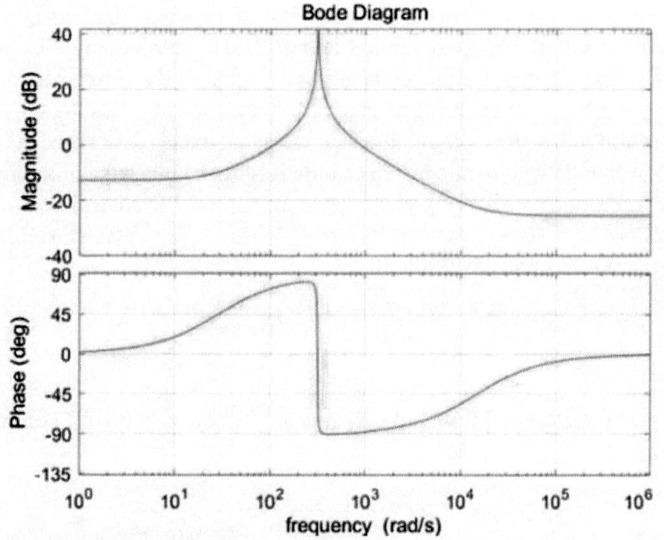


**Fig. 22.2** Amplitude frequency of ideal QPR controller

the system. Based on the analysis of a large number of papers, ω is generally greater than or equal to 3.14. Here, in order to facilitate calculation, we take $\omega = 3.14$.

## 22.4 Optimal Solution of PR Controller and Quasi-PR Controller Parameters

The closed-loop characteristic equation after order reduction is as follows:

$$D(s) = s^4 + B_0 s^3 + B_1 s^2 + B_2 s^1 + B_0 \tag{22.8}$$

According to the closed-loop characteristic equation before order reduction:

$$K_p = \frac{B_1 LC - 1 - K_{ii} K_{\mathrm{pwm}} C}{K_{\mathrm{pwm}} K_{pi}} \tag{22.9}$$

$$K_r = \frac{B_3 LC}{K_{\mathrm{pwm}} K_{ii}} \tag{22.10}$$

Substitute the relevant values of this system and finally obtain the optimal solution of the PR controller under the pole placement method as: $K_p = 0.166$, $K_r = 160.56$.

The voltage deviation percentage is calculated by the following formula:

$$\Delta U\% = \left| \frac{U_o - U_{\mathrm{ref}}}{U_{\mathrm{ref}}} \right| \times 100\% \tag{22.11}$$

Because the system steady-state error is less than 0.1% under most parameters when the system reaches 0.02 s, only the average value of output voltage deviation percentage under the previous 0.02 s is used to represent dynamic response performance of the system. After applying various models for numerical analysis, the following conclusions are drawn (Figs. 22.3 and 22.4).

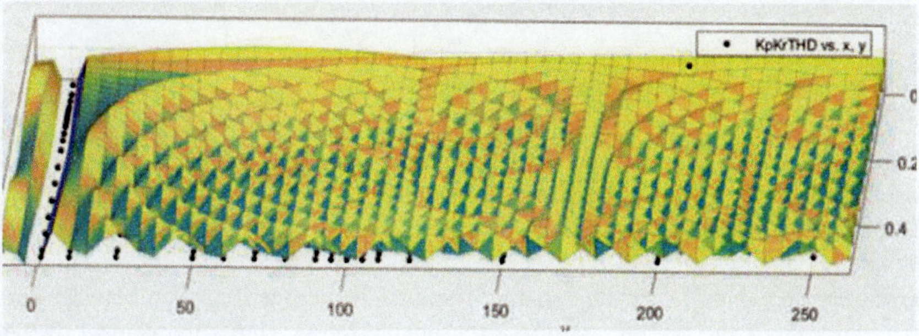


**Fig. 22.3** Simulation diagram of harmonic distortion function

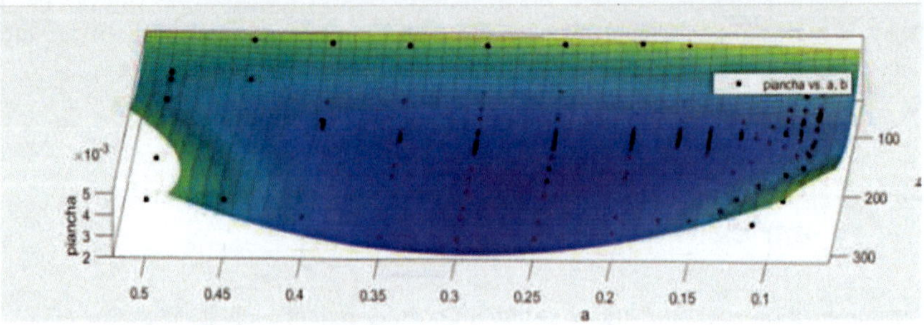


**Fig. 22.4** Output voltage deviation percentage

We are applying a variety of data analysis functions to simulate the THD and 0.02 s transmission voltage deviation percentages within a 95% confidence range.

The original model of THD harmonic distortion rate by custom equation is as follows:

$$f(x, y) = a + b\sin(m\pi xy) + ce^{-\frac{wy}{2}} \tag{22.12}$$

In the end, the optimal range we achieved, taking into account some margin, $K_p = 0.18$–$0.24$, $K_r = 120$–$140$.

The output voltage percentage of the first 0.02 s in Fig. 22.1 is through polynomial, which sets $x$ and $y$ to be third-order (high-order simulation effect is equivalent to third-order). The original model is as follows:

$$\begin{aligned} f(x, y) = p00 + p10y + p20x^2 + p11xy + p02y^2 + p30x^3 \\ + p21x^2y + p12xy^2 + p03y^3 \end{aligned} \tag{22.13}$$

The final scope is $K_p = 0.22$–$0.3$, $K_r = 120$–$200$.

Finally, combined with the anti-jamming capability and dynamic response of the system, $K_p = 0.23$ and $K_r = 125$ are taken to determine the optimal parameter solution for double closed-loop feedback quasi-PR + PI control.

## 22.5 Simulation Model and Results

Using the above model parameters, a simulation model is established based on the MATLAB/Simulink platform (Table 22.1).

It can be seen that the voltage deviation between the two is very small. From the data in the table, it can be clearly seen that quasi-PR control the smaller the voltage deviation rate is, the faster the dynamic response is, but the deviation between the two is not large and both are within a reasonable range. Under the control of the current inner loop PI and the voltage outer loop PR, the total THD of the output

**Table 22.1** Model parameter

| Name | Controller parameters |
|---|---|
| DC voltage source E | 130 V |
| Reference voltage $U_{\text{ref}}$ | 110 V |
| Filtering inductance L | 937.6 μH |
| Filter capacitor C | 33.6 μF |
| Resistance r | 0.1 Ω |
| Voltage outer loop quasi-PR control $K_{p1}$ | 0.23 |
| Voltage outer loop quasi-PR control $K_{r1}$ | 125 |
| Voltage outer loop PR control $K_{p2}$ | 0.166 |
| Voltage outer loop PR control $K_{r2}$ | 160.56 |
| Current inner loop PI control $K_{pi}$ | 0.11 |
| Current inner loop PI control $K_{ri1}$ | 328.8 |

**Table 22.2** Comparison of the rate

| Controller type | THD (%) | | |
|---|---|---|---|
| | 0–0.02 s | 0.02–0.08 s | 0.08–0.1 s |
| Quasi-PR | 0.32 | 0.18 | 0.13 |
| PR | 0.92 | 0.69 | 0.61 |

**Table 22.3** Voltage deviation (%)

| Controller type | THD (%) | | |
|---|---|---|---|
| | 0–0.02 s | 0.02–0.08 s | 0.08–0.1 s |
| Quasi-PR | 0.263 | 0 | 0 |
| PR | 0.471 | 0 | 0 |

voltage harmonic distortion rate of the system is 2.22%, while under the quasi-PR control, the total THD is 0.63%. Both controls are less than 5%, which meets the grid connection standard, but it is obvious that the anti-interference ability of the quasi-PR control is significantly enhanced (Tables 22.2 and 22.3).

## 22.6 Conclusion

This paper establishes a system model based on the double closed-loop feedback of a single-phase inverter and compares the advantages and disadvantages of PI, PR, and quasi-PR control in terms of dynamic performance and steady-state error according to the amplitude–frequency characteristics. In theory, the amplitude–frequency characteristic curve It can be seen that the advantages of quasi-PR control. Through the actual system, the impact of the two on the performance of the double closed-loop system can be carried out more rigorously. The final conclusion is drawn: Both PR and quasi-PR control can achieve no static error tracking of sinusoidal signals, but quasi-PR control can greatly enhance the anti-jamming capability, combined with double closed-loop feedback, and has better steady-state performance and dynamic performance.

## References

1. Zongxiang, L., Wang, C., Min, Y., Zhou, S., Lv, J., Wang, Y.: A review of microgrid research. Power Syst. Autom. **31**(19), 100–107 (2007)
2. Zheng, Z., Ai, Q.: Research status of microgrid and its application prospects in my country. Grid Technol. **32**(16), 27–58 (2008)
3. Chen, J., Xie, Y., Chen, B.: Inverter circuit control technology and strategy. Electr. Appl. **25**(9), 102–106
4. Wang, F., Wu, J., Wang, H.: Research on single-phase grid-connected inverter based on quasi-proportional resonance control. Electrical Switch, 3 (2017)
5. Zhang, Y.: Design of Single-Phase Photovoltaic Inverter Based on Double Closed-Loop PI and Quasi-PR Control. Northeast Agricultural University (2019)

# Chapter 23
# New Thoughts on the Course Teaching of Computer Specialty Under the Background of Big Data

**Zhang Xingyun**

**Abstract** This article analyzes the current problems in the teaching of computer majors in the era of big data and proposes that teachers should change their teaching methods. Teachers should implement personalized teaching reforms with the help of student learning resource data and combine teaching with big data massive teaching resources. To improve teaching methods, guide students to independently formulate learning programs based on their own knowledge of knowledge, promote students' autonomous learning ability, and achieve the goal of improving teaching quality. Colleagues suggested that teaching in the era of big data needs to strengthen the construction of resource platforms as a strong support.

In the era of big data with the rapid development of the Internet, its knowledge and technology are rapidly updated. Computer technology is particularly prominent. Under this situation, the computer professional courses of various universities have not kept pace with the times. Similarly, computer graduates The actual hands-on ability has not reached the current requirements of social enterprises in the era of big data, and it has become increasingly significant. In order to cultivate talents that meet the needs of society and keep up with the pace of the big data era, some new thoughts are put forward to the teaching of computer courses.

## 23.1 Thinking on the Problems Existing in the Teaching Mode of Computer Major Courses

The talent training mode of computer major courses requires students to master the knowledge from both theoretical and practical hands-on skills. At present, the common problems in our computer teaching are: passive learning ability of students,

Z. Xingyun (✉)
School of Artificial Intelligence, Hubei Business College, No. 634 Xiongchu Road, Hongshan District, Wuhan, Hubei Province, China

 
V. Chang et al. (eds.), *Modern Industrial IoT, Big Data and Supply Chain*, Smart Innovation, Systems and Technologies 218,
https://doi.org/10.1007/978-981-33-6141-6_23

classroom teaching focuses on the teacher's "teaching" as the leading role. In the process of teaching organization, there is less interaction between teachers and students, especially in the computer professional course class is dull and boring [1]. The teacher's grasp of the teaching rhythm can only be based on the teacher's limited judgment in the classroom, that is, through assignments and classroom performance; the traditional teaching mode is mainly based on the study of students' theoretical knowledge, focusing on the level of student test results, and ignoring student innovation Cultivation of abilities, improvement of hands-on abilities, comprehensive abilities and guidance of autonomous learning; knowledge content in the field of computer science is updated very quickly, while the curriculum system of traditional computer courses is updated slowly and textbooks cannot keep up to date. So that the computer application graduate's practical ability cannot meet the needs of social enterprises.

## 23.2 Thoughts on Teaching Methods of Computer Major Courses Under the Impact of Big Data

### *23.2.1 Change Teaching Ideas*

Education must return from the pursuit of unilateral learning "scores" to the educational nature of "educating people." Teachers need to change their teaching concept so that their roles can be transformed from learners to learners, promoters and helpers of student learning, researchers in education and teaching, and developers and builders of courses. The use of educational information technology to vigorously promote interactive discussions, blended teaching-classroom plays small videos and other student-oriented classroom teaching. Combined with the quality of students doing questions in class and watching the trajectory of network resources after class to study the status of students' grasp of knowledge and the needs of students, the classroom teaching methods are further improved based on the data analysis [2]. Introduce digitization in classroom teaching, change the teacher's teaching thoughts, and transform unified teaching into personalized education targeted by people, as shown in Fig. 23.1, the classroom teaching module.

### *23.2.2 Transform Students' Learning Methods and Guide Students to Learn Independently*

Teachers and students should gradually transition from "textbook learning" to "network learning." The study of computer majors should be more closely integrated with the Internet, encourage students to change to a personalized, decentralized, and remote educational organization, and guide students to learn independently through

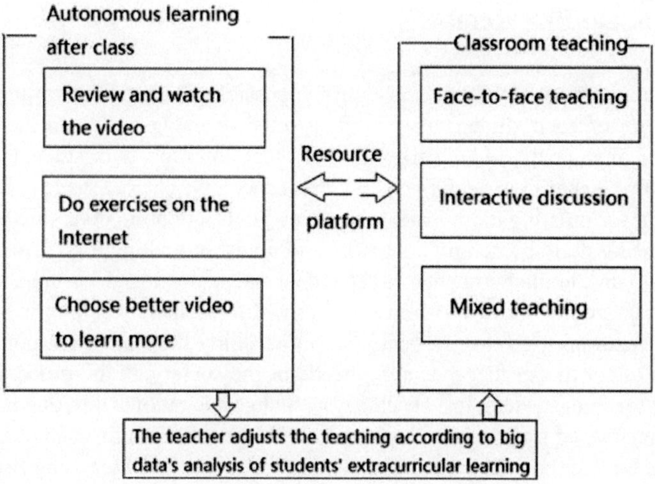


**Fig. 23.1** Teaching transformation model diagram of computer major courses

the resource platform: students choose the lesson according to their own knowledge of the knowledge review or watch videos, or do online questions to deepen or consolidate, or choose better resources for further study. Teachers can obtain a large amount of data on the resource platform—student learning trajectories to assist teaching, as shown in Fig. 23.1, the students' learning module.

### *23.2.3 Strengthening the Construction of Resource Platforms*

In the era of big data, there are huge amounts of resources, and these resources come from the Internet. In order to quickly and conveniently use Internet resources, schools also need to establish their own resource platforms. Create your own high-quality courses, upload high-quality network resources, test questions, resource courseware, etc. on the resource platform so that students can independently arrange learning plans, and break through the limitations of time and space to truly achieve independent and active learning [3]. Therefore, the construction and enrichment of the resource library is an important part of computer teaching [4]. Each course builds an appropriate resource library and platform according to the characteristics of the course to improve the quality of teaching.

## 23.3 Concluding Remarks

With the support of big data, it is entirely possible for teachers to analyze the learning data of each student, carry out targeted teaching based on the learning weaknesses of students, and digitize the entire learning process of students, thereby matching the teacher's curriculum characteristics with the characteristics of students. Combining, formulating more flexible teaching goals, adopting diversified teaching methods, using digital teaching behaviors and decision-making are more scientific and regular, can stimulate students "independent and personalized learning, and make teachers" classrooms more vivid. Interestingly, the teacher's teaching is no longer limited to textbooks, so that students can more enjoy the fun of learning [5]. To cultivate a talent model that meets the needs of the society, in the process of optimizing the teaching system, it is necessary to change educational thinking in a timely manner, innovate educational concepts, and use advanced teaching technologies such as multimedia and the Internet under the guidance of advanced teaching concepts to stimulate students' active learning initiative, to improve the application ability and comprehensive quality of computer students.

## References

1. Rong, Hu: Research on computer teaching in colleges and universities under the background of big data. Technol. Outlook **24**, 227 (2016)
2. Lihua, Zhang: Talking about the application of MOOC and big data technology in college computer basic teaching. Contemp. Educ. Pract. Teach. Res. **7**, 6 (2018)
3. Jiang, X., Wang, H.: Analysis on the optimization of the teaching system of computer majors in colleges and universities in the Internet + era. Wirel. Internet Technol. **10**, 75–76 (2017)
4. Jinling. Big data and the reform of information-based teaching. Audio-Vis. Educ. China **10**, 8–13 (2013)
5. Liang, W.: Big data era-classroom teaching will usher in a real change. Beijing Inst. Educ. Learn. J. (Nat. Sci. Ed.) **03**, 14–16

# Chapter 24
# Design and Research on Automatic Launch and Retrieval Righting Mechanism of Traffic Road Cone

**Linsen Du and Haiyang Zheng**

**Abstract** The righting mechanism of the traffic road cone automatic launch and retrieval system is installed on the frame. During normal work, the righting mechanism mainly contacts and interacts with the traffic cone by the movement of the vehicle to right or invert road cone. The study used the basic theory of mechanics to design and check the structure of the righting mechanism and carried out parameter design, selection and simulation model check calculation on the linear motor of the control system of the righting mechanism, so as to complete the design and research on automatic launch and retrieval righting mechanism of traffic road cone.

During the operation of the traffic road cone automatic launch and retrieval system car, the righting mechanism could automatically right or topple the road cone through the interaction with the traffic road cone. Therefore, the structure of the righting mechanism and the rationality of the system control design directly affect whether the highway traffic road cone could be launched and retrieved automatically. Based on the basic data of the shape of the highway traffic road cone, this paper analyzes the shape parameters of the righting mechanism and designs the control system.

## 24.1 Basic Theory

During the placement of the highway traffic road cone automatic launch and retrieval system, the retracting mechanism topples the road cone on road surface, the vehicle moves forward, the righting mechanism interacts with the falling road cones to right the traffic road cone. During the retrieval of the highway traffic road cone automatic

L. Du (✉)
Research Institute of Highway Ministry of Transport, 100088 Beijing, China
e-mail: ls.du@Rioh.com

H. Zheng
Shenyang University of Technology, 110870 Shenyang, China

 
V. Chang et al. (eds.), *Modern Industrial IoT, Big Data and Supply Chain*,
Smart Innovation, Systems and Technologies 218,
https://doi.org/10.1007/978-981-33-6141-6_24

launch and retrieval system, the vehicle moves backward, the righting mechanism interacts with the standing road cones to topple the traffic road cone. According to the principle of the action of the moment relationship, the theoretical conditions for the road cone to be righted or toppled are:

$$\begin{cases} F_f \leq f = mg\mu \\ F_f l_1 > mgl_2 \end{cases} \tag{1}$$

$F_f$ is the force acting on the road cone by the righting mechanism, N; $f$ is the ground static friction force, N; $m$ is the road cone quality, kg; $g$ is the gravitational acceleration, 9.8 m/s$^2$; $l_1$ is the height of road cone stress point from ground, mm; $l_2$ is the horizontal distance of the center of gravity relative to the flip side of the road cone, mm; $\mu$ is the road cone and ground friction coefficient, it is 0.7 here.

From (1), the theoretical condition that the road cone can be brought down is:

$$\frac{l_2}{l_1} < \mu \tag{2}$$

The righting process is as in Fig. 24.1.

The toppling process is as in Fig. 24.2.

**Fig. 24.1** Schematic diagram of the road cone righting process

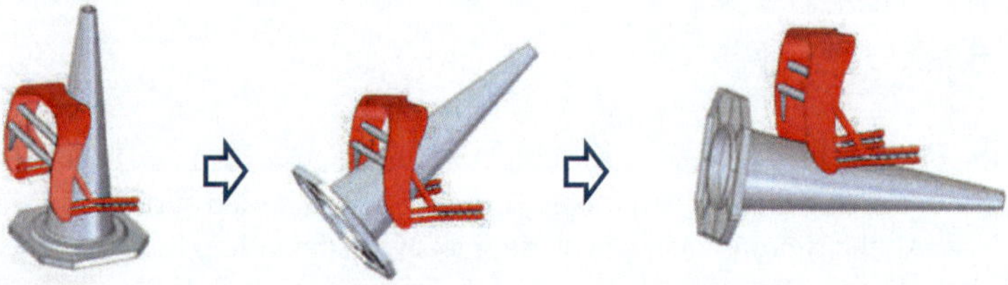

**Fig. 24.2** Schematic diagram of road cone toppling process

**Table 24.1** Parameters of traffic cones

| Traffic road cone | Product specifications | | Use required parameters (highway) | |
|---|---|---|---|---|
| | Shape | Conical cone, regular octagonal base | Type | H90 |
| | Size | Top circle inner diameter is 55 mm<br>Outer edge thickness of base is 28 mm<br>Inscribed circle diameter of outer edge of base is 440 mm | Placement distance | 10–20 m |
| | Material | plastic | Working surface size | 1000 mm |
| | Performance | Stability, drop resistance, low temperature impact resistance, weather resistance | Working length | 3–5 km |
| | Weight | 4.5 kg | | |

## 24.2 Structural Design of Righting Mechanism

### *24.2.1 Traffic Road Cone*

The road cone used by the highway traffic road cone automatic retractable system vehicle is a H90 reflective rubber cone. The road cone has a size of 440 × 440 × 900 mm and a mass of 4.5 kg. The detailed parameters are shown in Table 24.1.

### *24.2.2 Structural Design of Righting Mechanism*

The working principle of the righting mechanism is that when the traffic road cone automatic launch and retrieval system is ready for work, the righting mechanism slides out of the preset position by a lateral movement device when the righting is in a working state. The preset distance is 370 mm. Interaction to make it right; when it is in working state, the righting mechanism interacts with the falling road cone by means of the vehicle's backward movement to make it right (Fig. 24.3).

1. **Checking the Righting (toppling) Process**

*Righting check.* According to the size of the road cone in Fig. 24.4, the height $l_1$ of the road cone's stress point from the ground is 399.5 mm; the horizontal distance $l_2$ of the road cone's center of gravity relative to the road cone's flip side is 204 mm, $\mu$ is the coefficient of friction between the road cone and the ground, here is 0.7, then (Fig. 24.5):

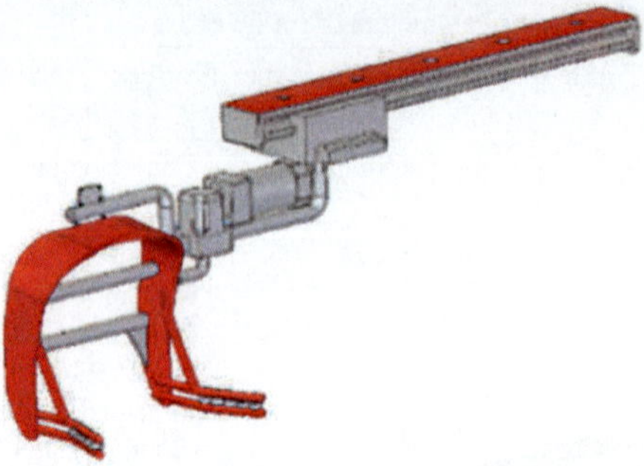

**Fig. 24.3** Overall three-dimensional structure of the righting mechanism

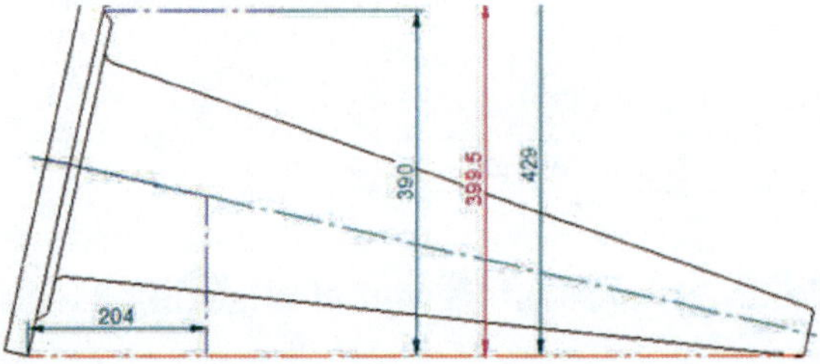


**Fig. 24.4** Road cone centering action size

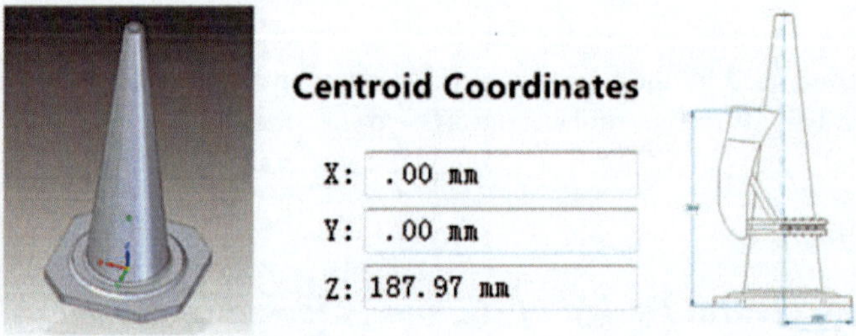


**Fig. 24.5** Road cone downside action size

$$\frac{l_2}{l_1} = \frac{204}{399.5} = 0.51 < \mu \tag{3}$$

*Toppling check.* As can be seen from the size of the road cone in Fig. 24.4, the height $l_1$ of the road cone's stress point from the ground is 564 mm; the horizontal distance $l_2$ of the center of gravity of the road cone with respect to the flip side of the road cone is 220 mm, $\mu$ is the coefficient of friction between the road cone and the ground, here is 0.7, then:

$$\frac{l_2}{l_1} = \frac{220}{564} = 0.39 < \mu \tag{4}$$

2. **Righting Guard Plate Check**

Considering the design of the contact between the righting-toppling guard and the standing road cone, it can be known that the road cone has a contact height of 564 mm and a road cone contact diameter of 131.19 mm. The arc length of the contact surface includes two arcs with an angle of 33.21° and a radius of 80 mm. An arc length of 58.8° and a radius of 145 mm. As shown in Fig. 24.6, the corresponding angle of the arc length of the green contact surface is 58.8°, and the radius is 145 mm. From this, the arc length $L_1$:

$$L_1 = \frac{n\pi R}{180} = 58.8 \times 3.14 \times 145 \div 180 = 148.73\,\text{mm} \tag{5}$$

As shown in Fig. 24.6, the arc lengths of the red and yellow contact surfaces are equal, both at an angle of 33.21° and a radius of 80 mm. From this, the arc length $L_2$:

$$L_2 = \frac{n\pi R}{180} = 33.21 \times 3.14 \times 80 \div 180 = 46.35\,\text{mm} \tag{6}$$

So, the total arc length:

$$L_m = L_1 + 2L_2 = 148.73 + 2 \times 46.35 = 241.43\,\text{mm} \tag{7}$$

Because the diameter of the contact surface of the standing road cone and the righting guard is 131.19 mm, it can be known that the contact surface circumference $C$ is:

$$C = \pi D = 3.14 \times 131.19 = 411.94\,\text{mm} \tag{8}$$

So,

$$L_m = 241.43 > \frac{C}{2} = 205.97\,\text{mm} \tag{9}$$

**Fig. 24.6** Contact arc length of tilting righting frame

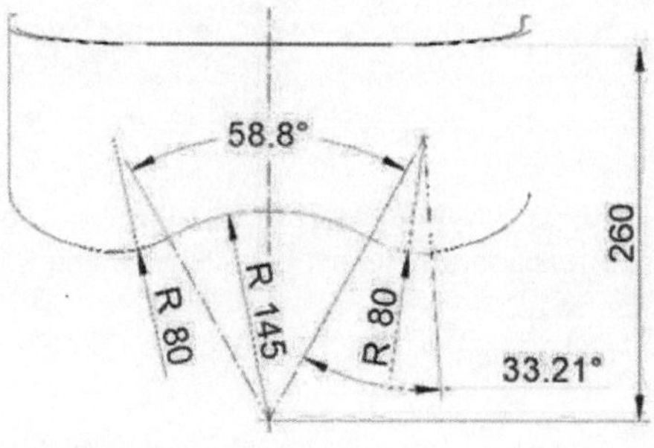

The design of the contact surface of the righting-toppling shield takes into account a radius of 145 mm and two radian surfaces with a radius of 80 mm. The total arc length is greater than half of the circumference of the contact surface of the standing road cone, which can "envelop" the halfway. The cone can play the role of handstand more stably, and it will not be stuck.

## 24.3 System Control Design of Righting Mechanism

### *24.3.1 Design Parameters and Selection of Righting Mechanism Components*

The righting mechanism system is controlled by a linear motor to control its lateral movement. The selection parameters of the linear motor are mainly determined by the rated driving power and rated voltage. The size of the driving power is determined by the linear motor slider and the resistance load of the centering device. Calculate the lateral driving force and driving power of linear motor 2.

It is known that the mass $m_1$ of the lying centering device is 23.67 kg, the mass $m^2$ of the linear motor slider is 17 kg, and the mass of the road cone $m_3 = 4.5$ kg. According to the overall performance index of the system, the device operating speed $v$ is 7.7 m/s.

Driving power $P_A$:

$$P_A = F_U \cdot v = m_1 + m_2 g f \cdot v \tag{10}$$

$P_A$ is the load power demand, W; $F_U$ is the load force, N; $v$ is the running speed, m/s; $m_1$ is the mass of the righting and toppling device, kg; $m_2$ is the mass of the linear motor slider, kg; gravity acceleration $g$ is taken as 9.8 m/s$^2$.

Calculated:

$$P_A = 613.8\,\mathrm{W} \tag{11}$$

When selecting a motor, considering the power loss and the appropriate safety margin, the power $P_M$ of the motor is calculated by the following formula:

$$P_M > P_A/\eta \cdot S \tag{12}$$

$P_M$ is the output power of the motor, W; $P_A$ is the driving power, W; $\eta$ is the transmission efficiency, between 0.8 and 0.95, here is 0.95; S is the safety margin, here is 1.2.

Calculated:

$$P_M > 775.32\,\mathrm{W} \tag{13}$$

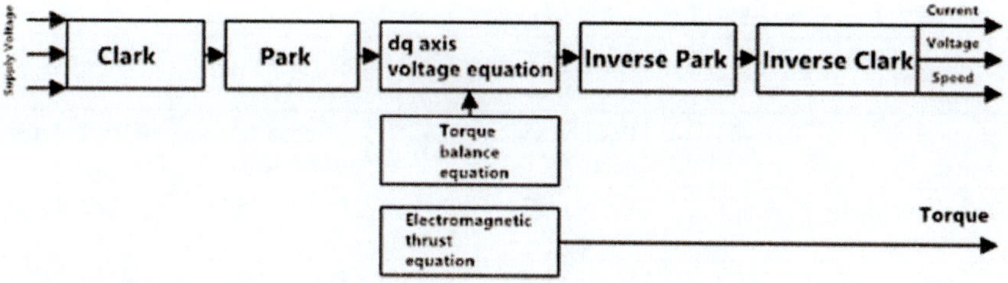


Fig. 24.7 Body model of a permanent magnet synchronous linear motor

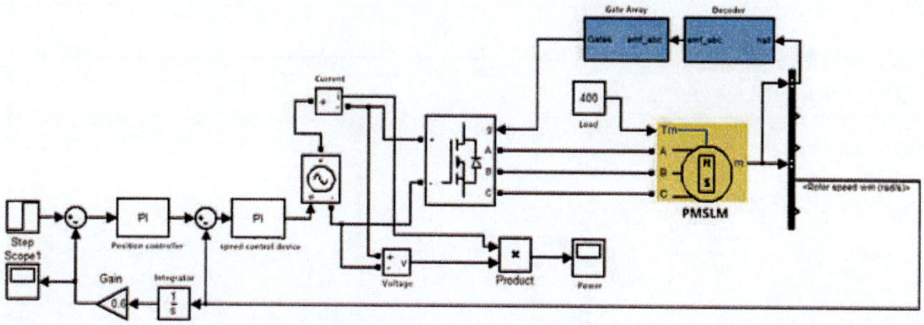


Fig. 24.8 Simulink simulation model

Therefore, the actual recommended power of the motor is 900 W, and the recommended voltage is 220 V or 550 V.

## 24.3.2 *Linear Motor Selection Simulation Calculation Check*

### 24.3.2.1 Building a Simulation Model

By establishing the body model of the permanent magnet synchronous linear motor (as shown in Fig. 24.7) and the Simulink simulation model (as shown in Fig. 24.8), the linear motor selection simulation calculation is checked.

The linear motor parameter settings are shown in Table 24.2.

### 24.3.2.2 Simulation Calculation Check Result

Figures 24.9 and 24.10 show the dynamic response of the control system when the supply voltage changes. It can be clearly seen that the lower the supply voltage, the worse the dynamic performance of the control system. When the power supply voltage is 48 V, the motor will be forced to work in the fourth quadrant of the

**Table 24.2** Parameter table of mathematical model of linear motor

| | | | | | |
|---|---|---|---|---|---|
| Controlled DC voltage | 220 V | Primary resistance. | 2.875 Ω | Primary inductance | 0.0085H |
| Moment of inertia | 0.0008 kg m$^2$ | Position sensor gain | 0.6 | Viscous damping coefficient | 0.0001 N m s |
| Viscous damping coefficient | 1 N m/s | Position PI | P (160) I (1.9) | Position PI Limit | [−220, 220] |
| Load resistance torque | 5 N | Speed PI | P (110) I (1.6) | Position PI Limit | [−220, 220] |
| Static friction moment | 4 N | Moment constant | 1.4 N/A | Phase | 3 |

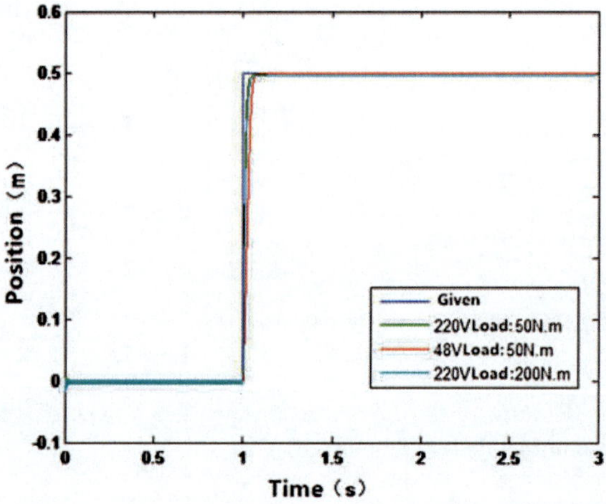


**Fig. 24.9** Effect of motor supply voltage and load on the dynamic response performance of the control system

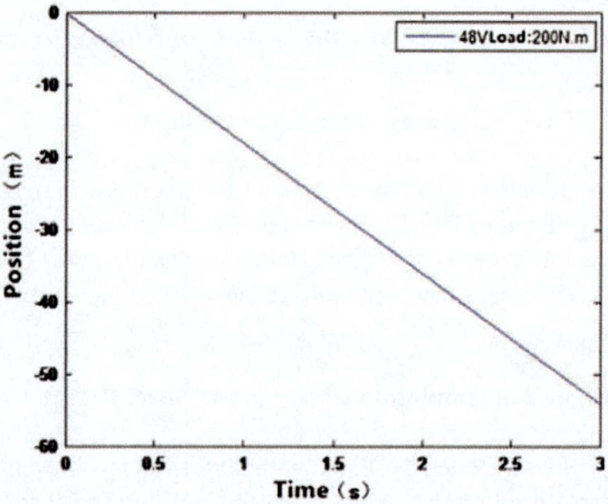


**Fig. 24.10** Dynamic response of the control system when the supply voltage is 48 V

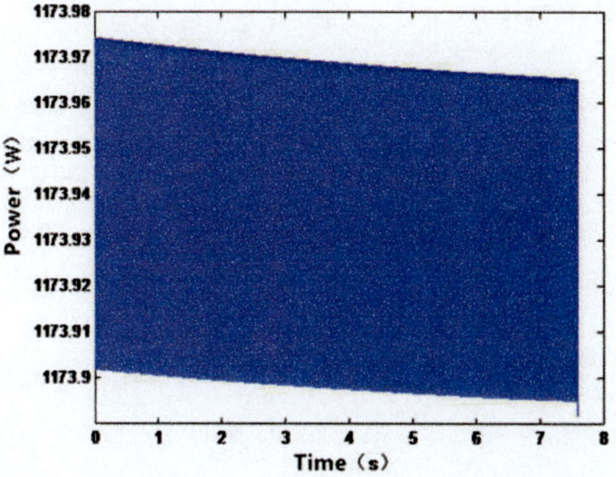


**Fig. 24.11** Steady-state input power change of the motor when the voltage is 220 V with a load

mechanical characteristics and the phenomenon of "flying" will occur. If the control voltage is maintained at 220 V, the change of the load basically does not have much influence on the dynamic performance of the control system. Note that the starting performance of the motor at the moment of power-on at 220 V is slightly worse. If a higher supply voltage to the motor (such as 550 V) will significantly improve the above performance. Figure 24.11 shows the power obtained when the motor is running stably, and its average power is 1173.95 W. Therefore, the simulation check result of the motor is that the rated voltage is 220 V or 550 V and the rated power is 1174 W.

## 24.4 Conclusion

In this paper, through the design and study of the traffic road cone automatic launch and retrieval righting mechanism, the conclusions are as follows:

1. The structure design of the righting mechanism meets the basic requirements of righting or toppling the traffic road cone, and the length of the contact surface of the righting mechanism is longer than the length of the contact surface of the road cone.
2. The parameter design and selection of the linear motor of the righting mechanism are reasonable, and the simulation verification of the system control model is passed, which meets the basic requirements of system control design.
3. The selection parameters of the linear motor are mainly determined by the rated driving power and rated voltage [1]. It can be clearly seen that the lower the supply voltage, the worse the dynamic performance of the control system [2], the power obtained when the motor is running stably [3].

## References

1. Xu, P., Song, J., Shen, G.: Research on the characteristics of electric vehicle motor drive system. **40**(11), 43–50 (2007)
2. Dkarimi, M., Mokhtari, H., Iravani, M.R.: Wavelet based on-line disturbance detection for power quality application. Power Delivery **15**(14), 1212–1220 (2000)
3. Wang Yu cold: Powertrain Parameter Matching of Pure Electric Vehicle Considering Cycle Conditions, p. 27. Chongqing University, Chongqing (2013)

# Chapter 25
# The Performance Study of Database Full-Text Retrieval

**Daiwen Wu**

**Abstract** In the paper, it implements the full-text retrieval in MYSQL Database by Lucene API and compares the performance and recall–precision of Lucene full-text retrieval with that of database full-text retrieval and like clauses retrieval. By comparing, we found Lucene is close to the database in full-text retrieval performance, and the time consuming of Lucene is 1/5 that of like clauses retrieval. What is more, the Lucene full-text retrieval can achieve the 77.2% recall ratio in Chinese retrieval aspects that is substantially higher than database full-text retrieval. Therefore, as for MYSQL database with larger quantity data, the Lucene full-text retrieval will be a good choice.

## 25.1 Introductions

In recent years, with the emergence of an open-source, high-performance full-text search engine Lucene, applications based on the Lucene and secondary development increased. As long as the software system using full-text search, it can be integrated with Lucene full-text search capabilities. However, we all know that MYSQL database can achieve string searching through SQL like clauses, and MYSQL database has full-text function itself. So how are the performance of Lucene full-text search, the MYSQL database full-text retrieval and like clauses retrieval? The article is about to carry out experiments in this area.

D. Wu (✉)
Department of Education, Weinan Normal University, Weinan 714099, Shanxi, China
e-mail: wudaiwen@foxmail.com



V. Chang et al. (eds.), *Modern Industrial IoT, Big Data and Supply Chain*,
Smart Innovation, Systems and Technologies 218,
https://doi.org/10.1007/978-981-33-6141-6_25

**Table 25.1** Pdf books table structure

| Field name | Field type | Null | Remark |
|---|---|---|---|
| id | int | No | Primary key |
| bookName | varchar(40) | No | bookName |
| page | varchar(20) | No | Page number |
| pagetext | varchar(40) | Yes | Page context |

## 25.2 Design of the Experiment

The database quoted in this article was MYSQL, because MYAQL supports for full-text search function. It can achieve the full-text retrieval of database using SQL statements like "select * from [Table name] where Match (field name) Against (keyword)". At this moment, the database table engine type is required to be "MyISAM" [1], because this is the only engine to support full-text search. While with SQL statement "select * from table name where field name Like '% keyword%'", we can achieve database string retrieval [2]. A data table will be designed in the database in this article, and import part of the text data through program. Then write corresponding programs with Java to achieve retrieval of the three ways and distinguish the pros and cons of the three retrieval methods by comparing time-consuming and recall of three kinds of retrievals.

## 25.3 Realization of the Experiment

### *25.3.1 Database Table Design*

First, design a simple data table called pdf books, which will be used to store each page of text of pdf ebook, the data Table 25.1 shows the structure.

### *25.3.2 Experiment Code*

#### 25.3.2.1 Data Import

In this work, we process PDF documents with importing PDFBOX API which helps to abstract the text form PDF by pages and then save to database. Core code [3]:

```
public class drawtext {
 public static void main(String[] argv) {
        COSDocument cosdoc = null;
        PDDocument document = null;
        PDFTextStripper stripper = null;
        String filepath = "e:/pdfdocs/bookname.pdf";
        String restr = new String();
        try {
              File f = new File(filepath);
              PDFParser parser = null;
              parser = new PDFParser(new FileInputStream(f));
              parser.parse();
              cosdoc = parser.getDocument();
              stripper = new PDFTextStripper();
              document = new PDDocument(cosdoc);
              String SQL = "";
              String filename=filepath.substring(filepath.lastIndexOf("/")+1);
              int pagesum = document.getNumberOfPages();
              for (int i = 1; i <= pagesum; i++) {
                    boolean sort = false;    // sort or not
                    int startPage = i;        // start extracting pages
                    int endPage = i;          // end extracting pages
                    stripper.setSortByPosition(sort);   // set sort or not
                    stripper.setStartPage(startPage);   // set the start page
                    stripper.setEndPage(endPage);       // set the end of page
                    restr = stripper.getText(document);
                    ConnDb cdb = new ConnDb();
                    SQL = "insert into pdfbooks(bookName,page,pagetext)
values('"+filename+"',"+ i + ",'" + restr + "')";
                    cdb.executeUpdate(SQL);//insert records to the database
              }
        } catch (Exception e) {
              e.printStackTrace();
        }
    }
}
```

#### 25.3.2.2 Full-Text Retrieval in Database

In this work, we do the full-text retrieval by using the SQL sentence "select * from tableName where Match(fieldName) Against('KeyWord')". Following is a segment of JAVA code as an example [4]:

```
public static void QueryByFulltext(String keywords) {
        try {
                Date start = new Date();
                GetConn(user,password);//connect to the database
                String SQL = "select * from pdfbooks where Match(pagetext)
Against('"+keywords+"')";
                record = getResultSet(SQL);// get record set
                int count=0;
                try {
                        while(record.next()){
                                count++;
                        }
                } catch (SQLException e) {
                                e.printStackTrace();
                }
                Date end = new Date();
                long time_index = end.getTime() - start.getTime();
                System.out.println("Database full text retrieval time: (MS)");
                System.out.println(time_index);
                System.out.println("Hit results of database full text retrieval:"+count);

        } catch (IOException e) {
                e.printStackTrace();
        }
}
```

#### 25.3.2.3 Retrieval by Using "Like"

Through the SQL sentence "select * from tableName where fieldName like '%keyWord%'", we can do work about matching string. Following is an example implements by JAVA, the core code [5]:

```
public static void QueryByLike(String keywords) {
    try {
        Date start = new Date();
        GetConn(user,password); // Connect to the database
        String SQL = "select * from pdfbooks where pagetext like
'%"+keywords+"%'";
        record = getResultSet(SQL); //get record set
        int count=0;
        try {
            while(record.next()){
                count++;
            }
        } catch (SQLException e) {
            e.printStackTrace();
        }
        Date end = new Date();
        long time_index = end.getTime() - start.getTime();
        System.out.println("Database like retrieval time: (MS)");
        System.out.println(time_index);
        System.out.println("Hit results of database like retrieval:"+count);
    } catch (IOException e) {
        e.printStackTrace();
    }
}
```

#### 25.3.2.4 Indexing and Retrieval Based on Lucene

There are two steps to implement this function. First we use Lucene to index the records needed in the full-text retrieval through iterating over all the records in the database. Then we can do the retrieval work in the local indexing files without accessing the database. The core code [6, 7]:

```
public static void indexdb() throws SQLException {
    Date start = new Date();
    try {
        File indexpath = new File(Dest_Index_Path);
        IndexWriter writer = new IndexWriter(indexpath,
                new StandardAnalyzer(), true);
        GetConn(user,password);// Connect to the database
        String SQL = "select * from pdfbooks";
        record = getResultSet(SQL);// get record set
        // Traverse the record set, do index for fields which need to be
retrieved
         DbIndexBuilder(writer);
         writer.optimize();
         writer.close();
     } catch (IOException e) {
         e.printStackTrace();
     }
     Date end = new Date();
     long time_index = end.getTime() - start.getTime();
     System.out.println("Indexing time: (seconds)");
     System.out.println(time_index / 1000);
 }
 public static void QueryIndex() {
     try {
         Date start = new Date();
         IndexSearcher searcher = new IndexSearcher(Dest_Index_Path);
         Term term = new Term("pagetext", "keyword");
         Query query = new TermQuery(term);
         Hits hits = searcher.search(query);
         int count=0;
         for (int i = 0; i < hits.length(); i++) {
             count++;
         }
         Date end = new Date();
         long time_index = end.getTime() - start.getTime();
         System.out.println("Lucene Retrieval time: (MS) ");
         System.out.print(time_index);
         System.out.println();
         System.out.println("Hit results of Lucene full text retrieval:"+count);
     } catch (IOException e) {
         e.printStackTrace();
     }
     System.out.println("Search success!");
     System.out.println("");
     }
```

## 25.4 Experimental Analysis

### 25.4.1 Experimental Environment

Program running environment in this experiment:

Hardware platform: HP 520 laptop
CPU: Core Duo T2600 (2.16G)
Memory size: 2 G
Database: MYSQL
Software environment: Eclipse 3.4

### 25.4.2 Experimental Data and Analysis

In this work, we imported three PDF documents with 1474 pages and 1474 records. Following is the time-consuming table about searching six randomly picked key words by the three methods [8].

As can be seen from Table 25.2, the performance of full-text retrieval based on Lucene is close to MYSQL database full-text retrievals, but much faster than like clause, which consumes 1/5 of time by comparing with like clause retrieval. It can meet the requirements of the speed when dealing with a large number of data retrieval in performance. The shortcoming is that full-text retrieval based on Lucene need to do index for fields in the database in advance, which may consume a certain amount of time and hard disk storage space. However with the development of computer hardware technology, these shortcomings can be easily ignored, and it is worthwhile to get a better retrieval performance by losing some time and storage space.

There are several norms of algorithm evaluation of text information retrieval and the field of search engine recall, precision, retrieval time and sort of results and so on. Only recall will be discussed in this thesis. Recall is the ratio of the number of related documents retrieved and the number of all the documents in documentation

**Table 25.2** Time consuming by three methods

| Keyword | Time consuming (ms) | | |
|---|---|---|---|
| | Lucene | DB like | DB full-text |
| Computers | 156 | 406 | 79 |
| Action | 78 | 391 | 47 |
| Manning | 63 | 406 | 31 |
| Development | 78 | 453 | 47 |
| Software | 93 | 469 | 47 |
| Economy | 79 | 453 | 47 |
| Average time | 87.9 | 429.7 | 51 |

**Table 25.3** Recall of three ways of retrieval

| Keyword | Hit number | | | Recall | |
|---|---|---|---|---|---|
| | Lucene | DB like | DB full text | Lucene | DB full text |
| Computers | 3 | 3 | 3 | 100% | 100% |
| Action | 7 | 7 | 7 | 100% | 100% |
| Manning | 0 | 3 | 3 | 0 | 100% |
| | Average recall | | | 66.7% | 100% |
| Development | 810 | 948 | 0 | 85.4% | 0 |
| Software | 110 | 137 | 0 | 80.3% | 0 |
| Economy | 603 | 914 | 0 | 66.0% | 0 |
| | Average recall | | | 77.2% | 0 |

set, which measure recall of the retrieval system (search engine), and the formula shows the calculation of recall:

$$\text{Recall} = \text{Numbers of right answer returned/Numbers in backup database} \quad (25.1)$$

Table 25.3 shows the recall of three ways of retrieval. The number of right answer in backup database gets the value from the column named "DB Like" when calculating recall.

As can be seen from Table 25.3, the recall of database full-text retrieval of English can reach 100%, but barely achieve for Chinese keyword, which can be retrieved are only some sentences separated by punctuations. For example, we can get results by running "select * from pdf books where Match (pagetext) Against (updating method not provided)", but get no result by running "select * from pdf books where Match (pagetext) Against (provided)". This example shows that database full-text retrieval does less better in Chinese word segmentation [9]. On contrary, the recall of retrieval on Chinese keyword based on Lucene does better at a ratio of 77.2%. The recall on Chinese keyword is related to segmentation algorithm. JE Chinese word segmentation je-analysis-1.5.1.jar was adopted in this thesis. If a better segmentation algorithm will be used, there could be a further improvement on the recall [10].

## 25.5 Summaries

Summing up the experimental analysis above, we can see that if we are going to do full-text retrieval to English, it is wise to choose full-text retrieval in MYSQL database to get a better retrieval performance and recall and choose both full-text retrieval based on Lucene and database like clause match retrieval to retrieve Chinese documents. Taking a further consideration of the retrieval performance, even though the performance of like clause match retrieval will be less good than it of full-text retrieval based on Lucene when dealing with few data, the recall can be better, and when there are plenty of data, the performance of retrieval will be greatly improved by using full-text retrieval based on Lucene.

**Funded Project** The integrated management system of Weinan City Science and Technology Business (2017CXPT-3). Weinan City Science and Technology Business integrated management system optimization promotion and management operation (2020FWNL-8), University research business integrated management system application research (WXQY002-012).

## References

1. Forta, B., Liu, X., Zhong, M.: MySQL Be Sure to Know. People's Posts and Telecommunications Press, Beijing (2017)
2. Liu, Z.: SQL2005 Baodian. Electronic Industry Press, Beijing (2016)
3. Wang, X.: Lucene+nutch Developing Search Engines. People's Posts and Telecommunications Press, Beijing (2017)
4. Rongqin Science and Technology: JSP Dynamic Web Site Development and Example. Tsinghua University Press, Beijing (2017)
5. Feisi Technology: JSP Application Development Explanation. Electronic Industry Press, Beijing (2016)
6. Qiu, Z., Fu, T.: Develop Your Own Search Engine—Lucene 2.0+Heriterx. People's Posts and Telecommunications Press, Beijing (2015)
7. Gospodnetic, O., Hatcher, E.: Lucene in Action. Electronic Industry Press, Beijing (2017)
8. Li, Z., Yu, Q., Chen, S.: Application research of information retrieval technology based on full-text retrieval engine. Comput. Digit. Eng. **36**(9), 81–85 (2015)
9. Bell, C.A.: Deep Understanding MySQL. People's Posts and Telecommunications Press, Beijing (2016)
10. Li, G., Song, W., Qiu, Z.: Conquer Ajax+Lucene Building Search Engine. People's Posts and Telecommunications Press, Beijing (2016)

# Chapter 26
# The Importance of Data Mining Technology to University Research in the Big Data Environment

**Deng Meiling and Lei Guiping**

**Abstract** Under the background of the rapid development of information technology, the application of Internet+ mode and big data technology has gradually penetrated in various industries, greatly changed the way of social production and management and improved the efficiency of all walks of life in society. The use of big data in my country's universities can better help improve the management model of my country's universities and improve management efficiency. At the same time, it also provides a new direction for the future university management model. At present, optimizing the management mode of universities can be achieved through data mining technology. This article mainly analyzes the application of data mining technology and focuses on the mainstream direction of efficient management methods.

## 26.1 Technical Brief

### *26.1.1 Big Data Technology*

Big data is an effective means of storing data through computers. It is the primary method of data management. Through large-scale and complex data collection and processing, a comprehensive, three-dimensional data collection of school research can be realized, so as to have a more comprehensive and systematic understanding of the various factors of university research development, so that industry decision-makers can more easily find problems and solve problems [1]. The use of big data technology greatly reduces the difficulty of data processing in the process of industry management, save resources and greatly improves the speed of management work. From a sense, big data relies on information and Internet+ related technologies to develop. Through massive information processing, individual problems are abstracted into conceptual overall problems, and laws are found in massive individual data to achieve accurate and concrete presentation of industry data. In terms

D. Meiling (✉) · L. Guiping
Guangdong University of Science & Technology, Dongguan 523083, China



V. Chang et al. (eds.), *Modern Industrial IoT, Big Data and Supply Chain*,
Smart Innovation, Systems and Technologies 218,
https://doi.org/10.1007/978-981-33-6141-6_26

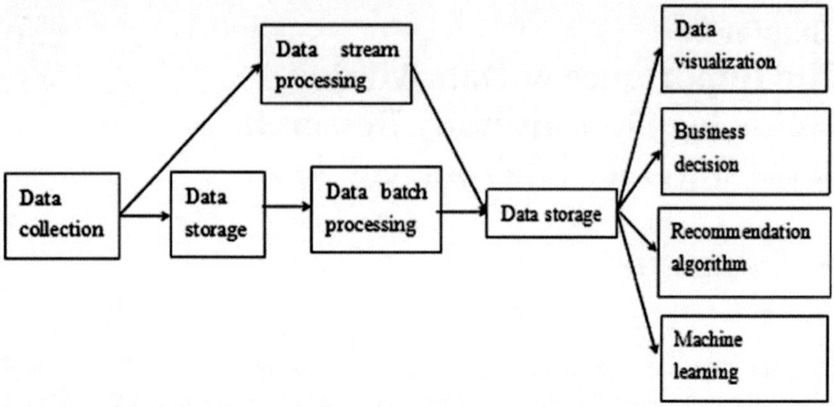


**Fig. 26.1** Examples of big data technology application patterns

of university research management, the scientific research management in colleges and universities is an intuitive embodiment of the change management mode of scientific and technological progress. Big data technology can change the disadvantages of traditional scientific research management that is not targeted and too general. The application process of big data technology is shown in Fig. 26.1.

### *26.1.2 Data Mining*

Information sources for data mining may include data warehouses, databases, online public data and other forms of dynamic or system data storage [2]. At present, data warehouse and data mining technology have become the key information technology for business applications, academic research and administrative management. At present, data mining technology is a key technology in the analysis of various industries at home and abroad. Besides large-scale commerce, the application of data mining technology in finance, post and telecommunications, industry and scientific research has been relatively common. Introducing data mining technology into the scientific research management of universities is a more effective technical means to solve the large amount of scientific research data in universities, but it is difficult to integrate it into effective information. Effective management of research in universities [3].

### *26.1.3 Information Management*

Information management clearly divides the responsibilities of each position in team production, so that everyone can carry out repeated and single work in their own position, they can achieve the production goal with the minimum cost and greatly reduce errors in the work process.

First of all, the methodology defines information management as a method and tool in modern management theory and holds that information management as a method is a tool for enterprises to be applied to market operation, product production and service promotion in order to improve management efficiency and economic benefits. And the stage theory holds that information management is the inevitable stage of the development of modern management theory, which is a new management mode derived from the development of high-tech development of the times. This is only a part of the historical evolution of management theory. Finally, the process theory holds that information management is one of the management means to further standardize the management system under the condition of meeting the basic management needs and quality standards, so as to facilitate the enterprise leaders to improve the production efficiency.

## 26.2 Design of University Scientific Research Management System

### *26.2.1 Data Standardization*

The scientific research projects of universities and colleges have developed rapidly, with more and more data generated, and the storage capacity is increasing. The rational use of data mining technology can provide effective technical means for healthy development university scientific research. In order to function better, it is necessary to standardize the processing of scientific research data and set up relevant-related parameters in advance. The name of the scientific research project, the time of establishment of the project, the funds, the key words and the time of completion or termination should be arranged in advance, which can facilitate the computer to collect, input and calculate the data [4].

Data mining technology-driven scientific research management in colleges and universities can analyze the data chain of individual big data to big data management through the index parameters reflected by students and schools' scientific research data. Using big data to collect and process individual project data, it is easy to get the progress of each research project, the required resources, and occupy the starting time of resources. It can divide the specific resources for each research project and reasonably use the limited resources of colleges and universities to support more scientific research projects scientific research projects. These data can also provide a reference for the school to evaluate the value of scientific research projects. In the scientific research management system of colleges and universities constructed by data mining technology, the work that schools need to do to establish big data processing system includes data collection, data integration, data analysis, data application and data visualization. This work requires professional personnel to establish a data processing platform and data repository, students and teachers to upload personal project information and related data, together to improve the system and regular system maintenance. Only in this way can the system platform be supported by big data provide support for the development of scientific research in colleges and universities and realize the information management of scientific research.

### 26.2.2 Main Functions of the System

The research system that has been built in institutions of higher learning often has the following basic modules, including project declaration, project management, achievement display, award-winning record, school society exchange, data mining and system management. The application of this project is mainly for the application, review, evaluation and establishment of scientific research projects. The project management module manages the basic information of the project, including the input of scientific research project information, query and status adjustment, such as in progress, suspension, suspension and completion. The project management module also supports data analysis and statistics, including the management of funds. In addition to the normal budget allocation, the withdrawal of funds, expenditure and carryover of funds, additional or early freezing of funds also exists. Results display module is mainly aimed at completed scientific research projects, and scientific research results are the main pursuit of scientific research projects in colleges and universities and can reflect the level of scientific research and the highest technical strength of a university. The result display module shows the input, query, management and data analysis of the papers, works, patents and appraisal results of college teachers and students. The winning record is mainly to record the information of the scientific research achievements and the winning information, which is responsible for the query, management, statistics and analysis of the relevant winning information. The academic exchange module mainly involves four aspects: academic conference, teaching, visiting and studying. The data mining module provides information support for making important decisions in scientific research management of universities. The final system management module is mainly for the system itself, to achieve normal system maintenance, user rights management, set system parameters, monitoring system running state, check unsafe access, database maintenance and management, and so on [5]. The design function diagram of university scientific research project management system is shown in Fig. 26.2.

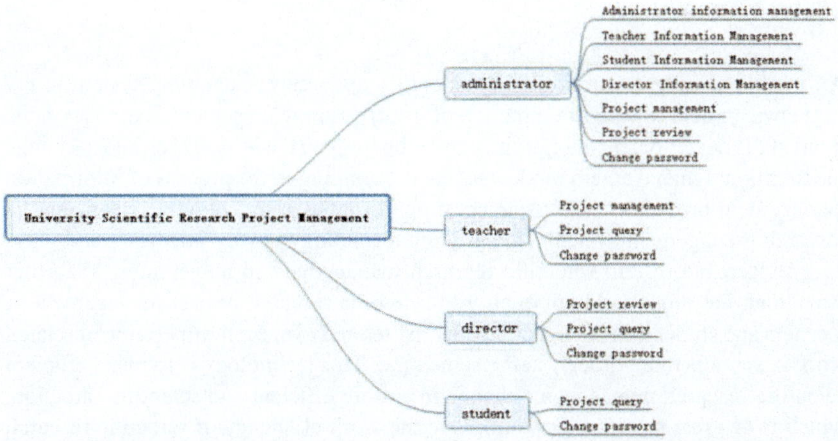


**Fig. 26.2** Function diagram of management system of scientific research projects in colleges and universities

## *26.2.3 Key Technologies*

The university scientific research management system uses the B/S mode of Microsoft.NET technology. Users can import and modify scientific research data with corresponding permissions. The Web interface of the front desk of the system can be implemented in C++ or Java advanced language, using Eclipse or Visual Studio series of products to create development space. Data storage in the background uses database-related knowledge, at present in the university scientific research management system using MySQL technology is relatively common, data in the database to achieve storage and management, through the front-end code call, back-end code input, modification, deletion and calculation. The system collates the data through the data mining technology, analyzes, obtains the effective information and displays and finally forms the system data analysis report.

Firstly, it is the a priori algorithm, and Zhuge is an algorithm to mine Boolean association rules, which belong to single-dimensional, single-layer, Boolean association rules in classification. Finally, Bayesian classification is based on probabilistic reasoning, which is the process of reasoning and making decisions under various uncertain conditions. Through the above algorithms, data mining technology can realize the relevance analysis of key data in university scientific research management system and the depth mining of data and provide guarantee for university research projects.

## 26.3 Conclusion

All in all, the progress of university scientific research management system in big data environment toward the direction of information management is the inevitable trend of the development of high and new technology in the information age to change the traditional management mode, and the data mining in the process of information management provides strong technical support for the design of university scientific research management system. Data mining technology is very important and plays an irreplaceable role in scientific research management in universities, to further strengthen the order and high quality of scientific research project management of teachers and students in colleges and universities and to make the operation of related work in schools more orderly at the same time. This technology is to make efficient scientific research management change in a more efficient and scientific direction, which is of great significance to improve the work efficiency of scientific research in colleges and universities and enhance the scientific research strength of schools.

## References

1. Zijing, S.: Application of computer data mining technology in university scientific research management system. Technol. Inf. **15**(17), 195–196 (2017)
2. Yue, X.: Application. Research on NET and data mining technology for higher vocational scientific research management system. Comput. Knowl. Technol. **14**(24), 48–50 (2018)
3. Chenzhi, Z.: Design and research of university scientific research management system based on data mining technology in big data environment. Netw. Secur. Technol. Appl. **05**, 29–30 (2018)
4. Zhanbo, L., Liwei, W., Xiaoli, W.: Design of university scientific research management system based on data mining technology in big data environment. Electron. Test. **02**, 44–45 (2017)
5. Pastoral: Analysis and design of educational administration system based on data mining technology. Electron. Technol. Softw. Eng. **22**, 194 (2015)

# Chapter 27
# The Importance of Big Data in the Automotive Industry

**Ying-shun Wang**

**Abstract** In the process of social and economic development in my country, the number of private cars is increasing. When a car is driving, it will not only produce noise and exhaust, but also cause traffic jams, which is not conducive to the rapid development of the city. In this context, new energy vehicles came into being. This paper analyzes the application prospects of big data technology in the new energy vehicle industry, clarifies the important concepts of big data analysis technology for new energy vehicles, and analyzes the application of big data technology in the field of new energy vehicles to ensure that new energy vehicles are large. The quality of data application has been comprehensively improved.

## 27.1 Introduction

As the most popular research field, big data has attracted the attention of tens of thousands of people. Big data mainly means the data and information that people obtain every day has increased significantly [1]. Through big data, it cannot only bring new development opportunities to people's lives, but also change the life production mode, accompanied by artificial intelligence cloud computing. As well as the rapid development of 5G technology, the level of intelligent and electrified vehicles has been continuously improved [2]. The application of big data technology in the field of new energy vehicles has very significant value, which can effectively reduce the consumption of energy resources and improve the overall utilization efficiency of resources.

Y. Wang (✉)
Guangdong University of Science & Technology, Dongguan 523083, China



V. Chang et al. (eds.), *Modern Industrial IoT, Big Data and Supply Chain*,
Smart Innovation, Systems and Technologies 218,
https://doi.org/10.1007/978-981-33-6141-6_27

**Table 27.1** Comparison of main new energy vehicle technology routes

| Type drive way | Flint battery car motor driven | Pure electric vehicle motor driven | Hybrid electric vehicle motor drive + internal combustion engine drive |
|---|---|---|---|
| Energy system | The fuel cell | Battery | Battery internal combustion engine power generation unit |
| Key points of technological development | Fuel cell engine technology, fuel life, drive system | Battery system integration control technology, charging technology | Power battery system integration and control technology, drive system matching and control |
| Technology maturity | In development | Mature and have sales | Most mature, have sales |

## 27.2 New Energy Vehicle Big Data Analysis Technology

Using intelligent information network as the carrier can ensure to increase the savings of new energy vehicles. For example, in order to ensure that new energy vehicles drive more efficiently and safely, a monitoring and management platform must be established to comprehensively collect data and information generated during the driving process of new energy vehicles. It is necessary to integrate relevant information such as car files, weather, and road conditions, so as to build a rationalized news new energy vehicle database, ensure the organic combination of new energy vehicle quantity resources and big data technology, make sure that each piece of data has different values to dig deeper, optimize the hidden value of various data information; thus, the problem of safe driving of new energy vehicles can be improved, as shown in Table 27.1. In addition, the use of big data technology can also ensure new energy vehicle battery monitoring and accurate judgment, ensure timely diagnosis and elimination of new energy vehicle faults, better improve the operating quality of new energy vehicles [3].

## 27.3 The Specific Operation of New Energy Vehicle Big Data Technology

### *27.3.1 Big Data Information Collection*

When using big data to collect new energy vehicle data, the most commonly used is to use the Internet to collect data. By collecting the logs and files of new energy vehicles, it can ensure a wider range of information sources. With the continuous development of new energy vehicles, the number of various sensors onboard is also increasing, traditional methods can no longer meet the current increasing demand for

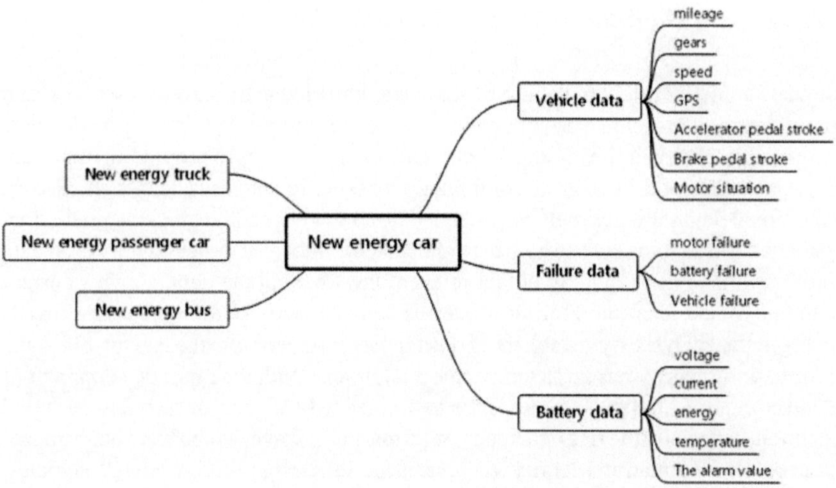


**Fig. 27.1** Platform of the national monitoring and management center for new energy vehicles

data information. For this reason, it is necessary to use big data collection technology to standardize the communication of new energy vehicles [4]. At the same time, it is necessary to collect statistics on new energy vehicle driving data and process fault information quickly to ensure more accurate data transmission, as shown in Fig. 27.1.

### *27.3.2 Big Data Processing and Storage*

In order to enhance the management level of new energy vehicle hardware and software, it is necessary to actively make good use of big data processing technology to ensure that data quality is improved. Data processing can be divided into two parts [5]. The first part is to organize the data, summarize them according to different categories, and use the onboard terminal program or other applications to accurately estimate the positive and negative accuracy of the data. The second part is to use the real-time computing module to accurately judge the data information when the data is transmitted to the platform, so as to find out the problems such as dropped frames and wrong frames in time. During data analysis, accurate processing must be performed according to the actual situation, and missing data values must be selected to fill or clean the data to ensure that the production efficiency of new energy vehicles is greatly improved. At present, most new energy vehicles in my country adopt self-built or leased methods to enhance the effect of data storage.

### 27.3.3 Big Data Statistics and Visualization

Big data statistics refers to the use of statistical knowledge to analyze resources, and all information must have clear connotations. By refining and extracting hidden data information, discover the internal laws of data changes. So as to provide an important reference basis for the design and optimization of subsequent new energy vehicles [6]. Traditional data statistics methods include cluster analysis and regression prediction. Visualization technology can help data miners make more accurate judgments on data information, so as to quickly obtain relevant laws behind the data, China's current automotive and local monitoring platforms can also achieve the processing results of big data analysis by establishing visual large screens. In the era of big data, information security has attracted people's attention. With the rapid development of cloud computing and the Internet, information security has also attracted widespread attention, especially the risk of personal information leakage, which has a huge impact on people's information security. At present, in the design of new energy vehicles, massive data storage, and data transmission also have some loopholes [7]. At present, problems such as weak data supervision and lax law enforcement in cities also lead to data information leakage, which significantly increases the risk.

### 27.3.4 Big Data Information Mining

For big data information mining, we must first pay attention to the close integration of big data technology and new energy technology, improve the processing quality and processing level of data analysis, and use big data to design data information models for new energy vehicles [8]. At the same time, it can improve the accuracy of the algorithm and ensure the development of repeated data. It is a specific application of big data analysis technology in the new energy automobile industry. In the process of driving a new energy vehicle, it is necessary to use a variety of sensors to monitor the vehicle status in real time, so as to obtain a large amount of data information, including vehicle location information, system component information input information, driver operation information, vehicle peripheral information, and so on. Their own driving behavior is also closely related, users through driving behavior can ensure that the driving energy consumption of new energy vehicles decline. According to the relevant research results, it is shown that the driver's speed will be affected if the driver fails to operate, and the energy consumption of the new energy vehicle will be increased. Therefore, it is necessary to judge the driver's historical driving behavior through neural network and nonlinear data regression equation, and to form a unique data model to optimize the bms of the new energy vehicle, so as to provide an important auxiliary basis for the driver's operation.

### *27.3.5 Enhanced Data Access*

At present, the problem of isolated data island seriously affects the ability of big data information processing. Therefore, government departments should play a leading role and improve the construction of information platforms.

### *27.3.6 Construction of Information Security System*

In the design of new energy vehicles, we must strengthen the sharing of information resources. We must improve the information security system and avoid being leaked by illegal users. There are still many backward situations in law and law enforcement in our country at present. Only by strengthening industry, self-discipline can we ensure the protection of personal information security. We should formulate corresponding industry norms according to the actual situation of the industry. We should actively encourage the broad user groups to clarify the focus of personal information protection. In order to ensure the self-discipline of industry members or abide by the law, it is necessary to introduce punishment mechanism or disclosure mechanism, perfect self-discipline standard, as the form contract clause between new energy automobile group and consumers, in order to ensure the consistency of consumers' right to obtain information. The establishment of data collection and management system, which can accurately divide sensitive information and general information, ensure the reasonable distribution of data, and in legal practice, it is necessary to reserve space for legal interpretation as far as possible, so as to effectively release sensitive information and guarantee the rights of information subjects. For example, different disclosure standards should be set for sensitive information and general information with high sensitivity. Highly sensitive information should not be disclosed, while low-sensitive information can be desensitized and disclosed, and general information can be disclosed.

## 27.4 Conclusion

With the intelligent development of new energy vehicles, computing power is also more powerful. Compared with traditional vehicles, the electric and intelligent level of new energy vehicles is higher. In the process of rapid development of data in the future, we must actively increase the level of analysis and processing of new energy data, ensure the overall quality of data transmission greatly improved, and ensure the overall effect of data design.

## References

1. Min, L.: Research and practice on teaching reform of database system course. Wirel. Internet Technol. **16**(07), 88–89+104 (2019)
2. Jia, L.: Research on the teaching mode of software engineering in colleges and universities based on big data analysis. Econ. Outlook Bohai Sea (03), 176 (2019)
3. Wenjun, Y., Lin, T.: The exploration and practice of multi-dimensional teaching mode in the teaching of software engineering in colleges and universities. Sci. Educ. J. (Mid-Term) **12**, 73–75 (2016)
4. Tingting, H.: Bilingual teaching model of software engineering for software service outsourcing in colleges and universities. Comput. Educ. **13**, 67–69 (2010)
5. Dawkins, S., Tian, A.W., Newman, A., Martin, A.: Psychological ownership: a review and research agenda. J. Organ. Behav. **2**, 163–183 (2017)
6. Alhazmi, Y.A., Salama, M.M.A.: Economical staging plan for implementing electric vehicle charging stations. Sustain. Energy Grids Netw. **2**, 12–25 (2017)
7. Li, Y., Abdel-Monem, M., Gopalakrishnan, R., Berecibar, M., Nanini-Maury, E., Omar, N., van den Bossche, P., Van Mierlo, J.: A quick on-line state of health estimation method for Li-ion battery with incremental capacity curves processed by Gaussian filter. J. Power Sources **10**, 40–53 (2018)
8. Panchal, S., Dincer, I., Agelin-Chaab, M., Fraser, R., Fowler, M.: Experimental and theoretical investigations of heat generation rates for a water cooled $LiFePO_4$ battery. Int. J. Heat Mass Transf. **5**, 1093–1102 (2016)

# Part III
# Supply Chain Fundamentals

# Chapter 28
# Competitive Decision Policies in Supply Chains Considering Investment of Low-Carbon Technology

**Ruidong Zhao, Huafei Huang, and Bo Li**

**Abstract** This paper considers a supply chain with two competitive suppliers and a retailer, in which no supplier or one of them decides to invest the low-carbon technology. Their game models are built and the comparisons are performed. The conclusions show that when one of the suppliers invests, compared with the case under no investment in low-carbon technology, although the supplier sets the higher wholesale price and the retailer also correspondingly makes a higher sales price, he also wins the higher profit, and the other supplier and the retailer accordingly achieve more profits as the free riders.

## 28.1 Introduction

With the continuous introduction of the low-carbon emission reduction policies and the continuous enhancement of consumers' environmental awareness, many manufacturers have considered the investment of low-carbon technologies in their production operations in order to achieve energy conservation and emission reduction goals. For example, clothing brands such as Zara have begun to use sustainable and environmentally friendly organic cotton in their production operations and H&M has strengthened its control over its raw materials to produce sustainable apparel products and reduce carbon emissions [1]. In addition, some competitive manufacturers, like Baowu Group and Hebei Iron and Steel Group, are also investing in low-carbon innovative technologies in their corresponding producing processes in order to improve the product structure and achieve sustainable development. Furthermore, it is verified that more and more consumers are beginning to pay attention to the sustainable development of products manufactured by enterprises, and the purchase

R. Zhao (✉)
School of Environment, Hohai University, Nanjing 210098, China

H. Huang · B. Li
College of Management and Economics, Tianjin University, Tianjin 300072, China

 
V. Chang et al. (eds.), *Modern Industrial IoT, Big Data and Supply Chain*,
Smart Innovation, Systems and Technologies 218,
https://doi.org/10.1007/978-981-33-6141-6_28

decisions of consumers have also prompted manufacturing companies to further increase investment in low-carbon technologies to satisfy consumers' demand [2].

The literature review related to the topics in this paper includes two aspects: one is the researches on the decision-making strategy in supply chains when the enterprise invests in low-carbon technologies, the other is the studies about the decision-making strategy in the competitive supply chain under the low-carbon environment.

First of all, reducing carbon emissions of products through low-carbon technology investment is an effective measure for energy saving and emission reduction in various enterprises. For example, Zhang et al. considered a manufacturer's decisions to invest in two types of products, namely traditional products and green products, and to sell through retailers when consumers are environmentally conscious [3]. Zhu and He studied the influences of retailer competition on the decision of suppliers' green investment level in the supply chain [4]. Du et al. focused the consumer's preference for low-carbon products produced by manufacturers, and designed a game model in which both manufacturers and retailers invested in low-carbon efforts [5]. Chen et al. discussed the pricing strategy and decision-making issues on low-carbon investment in supply chains [6]. Furthermore, Zhou and Ye analyzed the issue in which the manufacturer invested the low-carbon effort [7]. Further, Liu et al. investigated the cooperation problem of horizontal emission reduction input in a two-tier supply chain with a single supplier and multiple retailers [8]. He et al. compared the pricing strategies of firms under different carbon limits and trading policies for green technology investment [9]. Zhi et al. analyzed the impact mechanism of the cost-sharing contract in the supply chain [10].

However, the above papers are limited to the single-channel or dual-channel structure. Concerning with a competitive supply chain, Chen et al. studied the impacts of different production efficiency of two competing firms on their optimal pricing and production strategies when considering carbon emission taxes [11]. Giri et al. constructed a supply chain composed of two competing manufacturers and a retailer, and analyzed the government subsidies with environmentally conscious or the punishment measures [12]. Shi et al. analyzed the value of sustainable investment from the perspective of incentive analysis in the supply chain [13].

Different from the above results, this paper builds a supply chain consisting of two suppliers and one retailer, where the supplier considers whether to invest in low-carbon technology. Two scenarios are designed, and the game models are constructed and compared. It is concluded that when one supplier invests in low-carbon technology, in addition to its own benefits, the other supplier and retailer will also take a free ride to obtain higher profits, indicating that compared with the supplying competition between both suppliers, the impact of low-carbon technology investment is even greater.

## 28.2 Problem Description

The supply chain is composed of two suppliers and one retailer. The two suppliers sell two homogeneous products to consumers through the retailer. For simplicity, it is assumed that the two suppliers' production costs are the same, denoted as $c$, while the wholesale prices of the two suppliers are, respectively, $w_i, i = 1, 2$ and the retail price of the two products sold by the retailer is $p_i, i = 1, 2$.

Let $g_i, i = 1, 2$ indicate the suppliers $i$'s low-carbon technology investment level, based on the result from [13]. Meanwhile, we suppose that suppliers can save environmental taxes from the reduction of carbon emissions and reduce production costs. The definition $C(g_i) = \lambda g_i$ is the reduction of the supplier's production environmental tax, which $\lambda > 0$ is the elastic coefficient of the production environment tax reduction after the supplier's low-carbon technology investment. In addition, the fixed cost of the supplier's investment of low-carbon technology can be expressed as: $V(g_i) = \frac{\theta}{2} g_i^2$, where $\theta > 0$ is the investment cost coefficient [1, 13]. Based on Li et al. [14], we assume that suppliers' low-carbon technology investment can stimulate product demand. That is, if $q_i$ is represented as the product demand of suppliers, then the demand function is defined as $q_i = a - p_i + bp_{3-i} + \rho g_i$, where $a$ indicates the potential market size of the supplier's product, $\rho > 0$ indicates the sensitivity coefficient of the low-carbon input level in the supplier's demand. $b$ indicates the product substitution coefficient in price competition and $0 < b < 1$. All notations are shown in Table 28.1.

**Table 28.1** Parameters and decision variables in this paper

| Notations | Meanings | Notations | Meanings |
|---|---|---|---|
| $a$ | Potential market demand | $w_i$ | The wholesale price of the supplier, $i = 1, 2$ |
| $c$ | Unit production cost | $p_i$ | The retail price of the retailer's product, $i = 1, 2$ |
| $\theta$ | Fixed cost coefficient of low-carbon technology input | $\rho$ | Consumer demand sensitivity coefficient caused by low-carbon technology input |
| $\lambda$ | Cost reduction coefficient of the environment tax | $q_i$ | The supplier's product demand, $i = 1, 2$ |
| $b$ | Product replacement rate affected by price | $g_i$ | The supplier's low-carbon technology input level, $i = 1, 2$ |
| $\pi_{s,i}^j, \pi_r^j$ | The supplier's or the retailer's profit without/with low-carbon technology investment scenario, $i = 1, 2$, $j = NN, I$ | | |

## 28.3 Model Establishment and Analysis

### 28.3.1 Benchmark (NN)

For the benchmark NN scenario, we suppose that a Nash game between two suppliers first determines the wholesale prices to the retailer. Then, the retailer determines the retail price. From this, the optimal equilibrium solution can be derived as:

$$\begin{aligned} w_i^{NN*} &= (a+c)/(2-b), \\ p_i^{NN*} &= ((1-b)(2a+c)+a)/2(1-b)(2-b), \\ q_i^{NN*} &= Y/2(2-b), \end{aligned} \tag{28.1}$$

and the profits of the three members are: $\pi_{s,i}^{NN*} = Y^2/2(2-b)^2, i = 1, 2$.
Where $Y = a - c(1-b)$.

### 28.3.2 Under One Supplier's Investment (I)

This section considers a supplier's investment in low-carbon technology. The sequence of the game is as follows: first, we consider that the first supplier's investment in low-carbon technology belongs to a company's long-term decision, so he firstly determines the low-carbon investment level. Then, the two suppliers decide the wholesale prices to the retailer. Finally, the retailer decides the retail prices of the suppliers' two products.

Then, the profits of the two suppliers and the profit of the retailer can be given as:

$$\begin{aligned} \pi_{s,1}^I &= (w_1^I - (c - C(g_1^I)))q_1^I - \frac{1}{2}\theta(g_1^I)^2 \\ &= (w_1^I - (c - C(g_1^I)))(a - p_1^I + bp_2^I + \rho g_1^I) - \frac{1}{2}\theta(g_1^I)^2, \end{aligned} \tag{28.2}$$

$$\pi_{s,2}^I = (w_2^I - c)q_2^I = (w_2^I - c)(a - p_2^I + bp_1^I), \tag{28.3}$$

$$\pi_r^I = (p_1^I - w_1^I)q_1^I + (p_2^I - w_2^I)q_2^I. \tag{28.4}$$

The inverse method is used to solve the problem, and the following equilibrium solution is obtained, that is, in the first, when a supplier makes a decision to invest in low-carbon technology, the optimal equilibrium solutions of the two suppliers and the retailer in the supply chain is:

$$g_1^{I*} = \frac{(2+b)YZ}{X}, \quad w_1^{I*} = \frac{1}{(2-b)} \cdot \left( \frac{2(\rho-\lambda)YZ}{X} + a + c \right),$$
$$w_2^{I*} = \frac{1}{2-b}\left( \frac{(\rho-\lambda)YZ}{X} + \frac{(a+c)(\theta+2)}{2+b} \right),$$
$$p_1^{I*} = \frac{((2b^2+b-4)a\lambda - b\rho)Z - \theta(2-b)(2+b)^2(Y-2a(2-b))}{2(1-b)X} - \frac{c((3b+4)\rho + (b+1)b\lambda)}{2(1+b)X},$$
$$p_2^{I*} = \frac{a(\rho+b\lambda+\lambda)(2b^2-3) + (2\rho(b+2)b + \rho + (b+1)\lambda)}{2(1-b^2)X} Z - \frac{\theta(2-b)(2+b)^2(Y-2a(2-b))}{2(1-b)X},$$

and the product requirements of the two suppliers are

$$q_1^{I*} = \frac{\theta(2-b)(b+2)^2Y}{2X},$$
$$q_2^{I*} = \frac{Y}{2X}((2-b)(b+2)^2\theta - 2\rho^2 - \lambda(Z+\lambda(2-b^2)+\rho(4-b^2)).$$

The profits of the three members in the supply chain are

$$\pi_{s,1}^{I*} = \frac{\theta(b+2)^2Y^2}{2X},$$
$$\pi_{s,2}^{I*} = \frac{Y^2((2-b)(b+2)^2\theta - 2\rho^2 - \lambda(Z+\lambda(2-b^2)+\rho(4-b^2)))^2}{2X^2},$$
$$\pi_r^{I*} = \frac{Y^2(2(2-b)^2(b+2)^4(b+1)\theta^2 - 2(2-b)(b+1)(b+2)^2(\rho+(1+b)\lambda)\theta Z + (\rho+(1+b)\lambda)^2Z^2)}{2(1-b^2)X^2}.$$

where $X = \theta(4-b^2)^2 - (b^2\lambda - 2(\rho+\lambda))^2$, $Y = a - c(1-b)$, $Z = 2(\rho+\lambda) - b^2\lambda$.

As $\Delta w^{I*} = w_1^{I*} - w_2^{I*} > 0$, $\Delta p^{I*} = p_1^{I*} - p_2^{I*} > 0$, and computing $\Delta\overline{\pi}_s^{I*} = \pi_{s,1}^{I*} - \pi_{s,2}^{I*}$, it is easy to know that there exists a threshold $\overline{\theta}$, when $\theta > \overline{\theta}$, there is $\Delta\pi^{I*} > 0$, that is

**Proposition 1** *In a supply chain with two competing suppliers and a retailer, when the first supplier makes a low-carbon technology investment, through comparing the pricing strategies and profits of both suppliers, there are*

1. $\Delta w^{I*} > 0$, $\Delta p^{I*} > 0$, $\Delta p^{I*} > \Delta w^{I*}$.
2. *There is a threshold* $\overline{\theta}$ *for the first supplier's fixed investment cost coefficient* $\theta$, *when* $\theta > \max(\overline{\theta}, \theta_1, \theta_2)$, $\Delta\overline{\pi}_s^{I*} > 0$.

It can be seen from Proposition 1 (i) that because the first supplier invests in low-carbon technology, his wholesale price is higher than that under no investment in low-carbon technology. Then, the retailer also correspondingly set a higher sales price for this product, and the retailer's increased sales price is much higher than the difference between the wholesale prices of the two suppliers. That is, the suppliers who invest in low-carbon technologies have greatly improved the market demand and the price of the retailer, which may bring benefits to the retailer. In addition, from Proposition 1 (ii), it can be known that there is a threshold $\overline{\theta}$ for the fixed investment cost coefficient of the supplier, that is $\Delta\overline{\pi}_s^{I*} > 0$, when $\theta > \max(\overline{\theta}, \theta_1, \theta_2)$. This result is unexpected, because the fixed investment cost coefficient $\theta$ is large, indicating that the supplier's investment in low-carbon technology increases, the cost increases, but the benefits it brings will be higher than that without the low-carbon technology investment, indicating that when the investment in carbon technology is increased to a certain level, the level of low-carbon technology can greatly increase market demand and have a significant impact on the reduction of production costs, so as to ensure that suppliers can obtain greater profits.

Further, to ensure existence of the optimal solutions [1, 14], the following conditions are need to be satisfied: $c < a/(1-b)$, $\theta > \max(\theta_1, \theta_2)$ and $\rho > \lambda$ where $\theta_1 = ((1-b)\lambda + \rho)^2/2(1-b)(2-b)^2$, $\theta_2 = (2\rho^2 + (4 - b(b+2))\lambda\rho + (1-b)(2-b^2)\lambda^2)/(2+b)(2-b)^2$. By comparing the optimal solutions in the above two scenarios, the following Proposition 2 can be obtained.

**Proposition 2** *When comparing the benchmark scenario and one supplier investment scenario, the following properties are satisfied*:

1. *For each supplier and retailer, there are* $w_1^{I*} \geq w_2^{I*} \geq w_i^{NN*}$, $p_1^{I*} \geq p_2^{I*} \geq p_i^{NN*}$, $i = 1, 2$;
2. *For the retailer's order quantity*, $q_1^{I*} \geq q_2^{I*} \geq q_i^{NN*}$, $i = 1, 2$.

From Proposition 2 (i), one supplier also charges a higher wholesale price and the retailer set higher retail prices for products with higher levels of low-carbon technology investment. From Proposition 2 (ii), we can see that the market demand is also the highest when one supplier invests in low-carbon technologies. This is somewhat counterintuitive, generally speaking, when the sales price increases, it will damage the enthusiasm of consumers and lead to a decrease in sales. It can be seen that if suppliers are actively engaged in low-carbon technological innovation under a competitive environment, a green and environmentally friendly consumption environment will be formed for the entire market, which is beneficial to the enterprises participating in the supply chain.

Further, we compare the profits in these two scenarios, as already known $\pi_{s,1}^{I*} - \pi_{s,2}^{I*} > 0$ from Proposition 1 above. Similarly, we have Proposition 3.

**Proposition 3** When comparing the benchmark scenario and one supplier investment scenario, there are (i) $\pi_{s,1}^{I*} \geq \pi_{s,2}^{I*} \geq \pi_{s,i}^{NN*}$; (ii) $\pi_r^{I*} \geq \pi_r^{NN*}$.

Proposition 3 gives the profit comparison among suppliers and retailers under both scenarios. When one of the suppliers makes a low-carbon technology investment, it is obvious that the profit of this supplier is higher than that of the other supplier who does not make any investment in technology. As for the profit of the retailer, it can be seen that if the supplier invests in low-carbon technology, the retailer will benefit. That is, the retailer can obtain the spillover effect from the supplier's low-carbon innovation. The profit under the baseline scenario is the lowest.

## 28.4 Conclusion

This paper studies the supply chain composed of two suppliers and one retailer, in which two scenarios are designed, i.e., one of them or none of them will invest low-carbon technology. Through the establishment of a game model, some interesting conclusions can be drawn. Compared with the benchmark, if the supplier invests in low-carbon technology, then all members of the supply chain will profit from the investment. Thus, the supplier is encouraged to continue the investment process of carbon technology. Further, when a supplier invests in low-carbon technology, it can be seen that the suppliers or retailer who do not invest will enjoy a free-riding effect. This shows that the competitive suppliers are encouraged to invest in low-carbon technologies, then this investment has a significant effect on the overall performance improvement in the entire supply chain.

## References


1. Dong, C.W., Shen, B., Chow, P.S., Yang, L., Ng, C.T.: Sustainability investment under cap-and-trade regulation. Ann. Oper. Res. **240**(2), 509–531 (2016)
2. Chitra, K.: In search of the green consumers: a perceptual study. J. Serv. Res. **7**(1), 173–191 (2017)
3. Zhang, L.: Consumer environmental awareness and channel coordination with two substitutable products. Eur. J. Oper. Res. **241**(1), 63–73 (2015)
4. Zhu, W.: Green product design in supply chains under competition. Eur. J. Oper. Res. **258**(1), 165–180 (2017)
5. Du, S.: Low-carbon supply policies and supply chain performance with carbon concerned demand. Ann. Oper. Res. **255**(1–2), 569–590 (2017)
6. Chen, X.: Manufacturer and retailer coordination for environmental and economic competitiveness: A power perspective. Transp. Res. Part E: Logist. Transp. Rev. **97**, 268–281 (2017)
7. Zhou, Y.: Differential game model of joint emission reduction strategies and contract design in a dual-channel supply chain. J. Clean. Prod. **190**, 592–607 (2018)
8. Mingwu, L., et al.: Cooperation research on supply chain horizontal emission reduction under carbon trading policy. Ind. Eng. Manage. **3**, 28–35 (2015)
9. He, H.: Carbon limits and trading policies research on pricing strategy considering green technology investment. China Manage. Sci. **24**(5), 74–84 (2016)
10. Zhi, B.: Cost-based contract under carbon cap and transaction mechanism. Two-level supply chain coordination strategy. China Manage. Sci. **25**(07), 48–56 (2017)
11. Chen, X.: Sustainable pricing and production policies for two competing firms with carbon emissions tax. Int. J. Prod. Res. **53**(21), 6408–6420 (2015)

12. Giri, R.N.: Government intervention on a competing supply chain with two green manufacturers and a retailer. Comput. Ind. Eng. **128**, 104–121 (2019)
13. Shi, X.: Value of bargaining contract in a supply chain system with sustainability investment: an incentive analysis. IEEE Trans. Syst., Man, Cybern.: Syst. **50**(4), 1622–1634 (2020)
14. Li, B., Zhu, M., Jiang, Y., Li, Z.: Pricing policies of a competitive dual-channel green supply chain. J. Clean. Prod. **112**, 2029–2042 (2016)

# Chapter 29
# The Investigation of Housing Price in Yangtze-River Delta Region

**Yijie Chen and Youlin Huang**

**Abstract** The fluctuation of house price in a region is affected by the local government debt and local land rent, which produces the spillover effect of house price. In this paper, three provincial cities in Jiangsu, Zhejiang, and Shanghai are taken as the representatives, and the data from 2009 to 2019 is taken as the samples, and the spatial distribution pattern and evolution trend of housing prices in Jiangsu, Zhejiang, and Shanghai are analyzed by using exploratory spatial data analysis (ESDA). After that, we use the SAR, SEM, and SDM model to analyze the influencing factors of housing price change in Jiangsu, Zhejiang, and Shanghai, and study the spillover effect of housing price according to the geographical factors and economic factors. Finally, it analyzes the spillover effect and the difference and comparison of influence of local government debt, local land rent on housing prices. Furthermore, we could predict whether other real estate types, such as commercial and residential house, will have the same situation.

## 29.1 Introduction

In the last dozen years, the real estate industry has been a major driver of China's economic growth. Stimulated by economic policies, the real estate industry directly contributed 11% of the GDP in 2009. From 2009 to 2018, the real estate industry basically maintained a proportion of 10–12% of the GDP, which is a vital factor for China's economic recovery from the global crisis. The Yangtze River delta is one of the most representative urban development clusters of real estate in China. The total investment in real estate development has increased to 9503.561 billion RMB in 2014 from 1590.925 billion RMB in 2005, which accounted for 18.53% of total fixed investment in 2014. Moreover, according to the data of the National Bureau of Statistics, the investment in real estate in China increased to 492.15% from

Y. Chen (✉) · Y. Huang
Xi'an Jiaotong Liverpool University, Suzhou 215123, Jiangsu, China
e-mail: yijie.chen19@student.xjtlu.edu.cn



V. Chang et al. (eds.), *Modern Industrial IoT, Big Data and Supply Chain*,
Smart Innovation, Systems and Technologies 218,
https://doi.org/10.1007/978-981-33-6141-6_29

1086.093 billion RMB in 2005 to 6435.215 billion RMB in 2014, which shows that the residential market has been the most important part of the real estate market.

Chinese housing price is affected by various factors, among which local government debt and local land price are the two main factors. On the one hand, land price is the monetary reflection of land economic value. Land price is the sum of present value of multi-year land rent paid at one time. Real estate developers will determine the price of the house according to the fees paid when leasing the land. Therefore, local rent determines the price of the house. When the price of the house rises too fast, the price of the house can be limited first when the land is sold. If the price is higher, the price will be limited. If the price is lower, the price will be limited. The government to adjust the price of a means often uses this. All in all, the local land price almost determines the trend of house price all the time.

On the other hand, while China's urbanization is advancing rapidly, urban housing prices continue to rise rapidly. The rise in housing prices is not only driven by a variety of non-policy factors such as supply, demand, and speculation, but also influenced by the strategic actions of local governments, especially the long-term land fiscal strategy pursued by local governments. Debt financing based on land finance means "land mortgage—loan—construction—land sale—repayment". High housing price is conducive to local governments' acquisition of huge land fiscal revenue. In order to obtain more land fiscal revenue, local governments will maintain and promote the rise of housing price. It can be seen that local government debt generates land finance, and land finance interacts with housing prices.

Yangtze River delta area (Nanjing, Zhejiang, and Shanghai) depends on its good infrastructure developed the convenient location advantage of and interests in the education of science and technology as well as the improvement of investment environment, increasingly become the one of the fastest economic growth, the most rapid urbanization area of Yangtze river delta economic take off at the same time also led to the increase of urban housing demand, and the rising of housing price. Statistics show that in the past ten years between the Yangtze river delta city presents the overall rise in house prices and accelerate the differentiation between different cities, such as trends, especially the region at the center of the city of Shanghai and the deputy center city of Hangzhou and Nanjing, house prices to rise significantly, other cities within the region have appeared in the different level of housing prices. The reason why Yangtze River delta area is selected as the research object is that, on the one hand, it's housing market development ranks among the top in China, which can provide lots of valuable experience for other regions and is highly representative. On the other hand, because it has existed as a whole for a long time, internal members have high synergy and recognition, close relationship and frequent interaction, which are not influenced by geographical environment.

### *29.1.1 Aims and Research*

This paper studies the housing price in the Yangtze River delta region (Nanjing, Hangzhou, Shanghai), which is affected by local government debt and land price, and whether there is spillover effect between these regions.

Geographically, the Yangtze River delta region is quantified into three provinces: Jiangsu, Zhejiang, and Shanghai. In different provinces, provincial capitals are taken as examples to explore the impact of the above two factors on the housing price (mainly commercial housing). Then, after comparing the difference of influence, a spatial econometric model is established to explore the spillover effect of housing prices in these regions.

The paper is organized as follows. Section 29.2 gives an overview of my methodology and empirical approach.

## 29.2 Literature Review

Few years ago, a large number of scholars have done a lot of research on the impact of local government debt and land price on housing prices. Zhou and Du [1] found that the change of land price has a strong ability to explain the change of housing price, and land price promotes the rise of housing price, which is based on the granger causality test of land price and housing price. In the following years, most studies adopted the regression method of panel data to empirically study the positive impact of land price on local real estate price. On the other hand, the other study used the panel data of "urban investment bond" of 34 large and medium-sized cities in China from 2008 to 2012 to empirically analyze the relationship between local government debt and commercial housing price, and the results showed that there was a significant positive relationship between the two economic variables. Therefore, we speculate that both local government debt and land price have a positive relationship with the housing price. However, in terms of research methods and research perspectives, most domestic and foreign scholars adopt traditional time series and panel data models, all of which are based on the assumption that housing price fluctuations in various regions are independent of each other, and ignore the interaction caused by spatial proximity, which leads to deviations in the measurement model. James LeSage introduced the externality between objects into spatial econometrics and defined it as "spillover effect". Foreign scholars have conducted extensive studies on the spatial spillover effect between things. Kelejian and Piras [2] analyzed the spillover effect of cigarette price changes in American states. Ertur, Koch, Elhorst, and others [3] discussed the economic growth link between countries. Study [4] found that the spillover effect was not very significant between different countries, but mainly reflected in the interaction between different cities in the country. Therefore, this paper aims to study whether there is still a positive relationship between the spillover effect of local government debt and land price and housing price, as well as the difference between them. The research objects are cities in the Yangtze River delta region (Zhejiang, Hangzhou, Shanghai, etc.). The selection of cities in the Yangtze River economic belt as the research object can fill the gap in the existing relevant studies on the spatial spillover effect of housing prices in urban agglomerations, and the estimation results have certain universality, that is to the east, central, western cities have certain reference significance. It is also believed that regardless of real estate price, local government debt or other socio-economic indicators, the panel data of prefecture-level cities could better represent the real level of the research area than the panel data of provincial-level cities, and the empirical results obtained were also of more reference value.

## 29.3 Data and Method

To analyze the spatial relevance of housing prices, we collected the house price data in annually base which from 2009 till 2019 for three cities in the region to calculate the Moran's $I$ value. It is a measure of spatial autocorrelation which is characterized by a correlation in signal among locations in space. The following is the formula of Moran's $I$:

$$I = \frac{n}{\sum_{i=1}^{n}\sum_{j=1}^{n} w_{ij}} \frac{\sum_{i=1}^{n}\sum_{j=1}^{n} w_{ij}[x_i - \overline{x}][x_j - \overline{x}]}{\sum_{i=1}^{n}[x_i - \overline{x}]^2} (\forall j \neq i), \tag{29.1}$$

We use Moran's $I$ model to estimate the evolution of spatial correlation of housing prices in an area. When Moran's $I$ proved the existence of spatial correlation, three different time series or panel data regression were established, taking local government debt, local rent, income and population as the control variables, and the distance of different places as the spatial weight variable. After comparing the influence differences, a spatial measurement model was established to explore the spillover effect of housing prices. Spatial Durbin model (SDM), spatial lag model (SAR), and spatial error model (SEM) would be compared. Here, we construct a spatial panel model to study the driving factors of housing price. Here is the model formula:

$$\text{SAR}: Y_{it} = \rho \sum_{j \neq i} W_{ij} Y_{jt} + \sum_{k} X_{kit}\beta_k + \alpha_{it} + \varepsilon_{it}, \tag{29.2}$$

$$\text{SEM}: Y_{it} = \sum_{k} X_{kit}\beta_k + \alpha_{it} + \xi_{it};\ \xi_{it} = \lambda \sum_{j \neq i} W_{ij}\xi_{jt} + u_{it}, \tag{29.3}$$

$$\text{SDM}: \text{Y}_{it} = \rho \sum_{j \neq i} W_{ij} Y_{jt} + \sum_{k} \beta_k X_{kit} + \sum_{k} \left( \sum_{j \neq i} W_{ij} X_{kit} \right) \theta_k + \alpha_{i(t)} + \mu_{it} \tag{29.4}$$

$Y_{it}$ stands for the dependent variable, that is housing price in each spatial unit ($i = 1, \ldots, N$) at time $t$ ($t = 1, \ldots, T$); $X_{kit}$ is the $k$th independent variable, including local government debt, land rent, income and population; $\beta_k$ is the $k$-dimensional parameter to be estimated; $\alpha_{i(t)}$ is the intercept term, where it is normally due to examination of the fixed/random effects test; $\rho$ represents the coefficient of spatial autocorrelation in the SAR model, and $\lambda$ is the coefficient; In the SDM model, $\rho$ and $\theta_k$ stands for the coefficients of spatial autocorrelation which are independent variables and dependent, respectively. After that, LIN test and Wald test were performed on spatial lag and spatial error to determine whether SAR or SEM or SDM was used. Finally, the certain model would estimate the direct, indirect, and total effects on housing price.

## References

1. Zhou, B., Du, L.: Land finance and rise of house prices: a theoretical analysis and empirical research. Financ. Trade Econ. **8**, 109–116 (2010)
2. Kelejian, H.H., Piras, G.: Estimation of spatial models with dangerous weighting matrices and an application to demand model for cigarettes. RASI, Canada, Ottawa (2012)
3. Ertur, C., Koch, W.: Technological inter-denpendence and spatial externalities: theory and evidence. J. Appl. Economet. **22**(6), 1033–1062 (2007)
4. Vansteenkiste, I., Hiebert, P.: Do house price developments spillover across Euro area countries? Evidence from a global VAR. J. Hous. Econ. **20**(4), 299–314 (2011)

# Chapter 30
# Analysis of Laos' Macroeconomic Development

**Qiming Tang and Meijuan Li**

**Abstract** The national economy of Laos has maintained a steady growth trend, with a total GDP of us $18.4 billion and a per capita GDP of us $2720.3 in 2018, with a GDP growth rate of 6.5%. Laos' overall unemployment rate remains low at around 0.66%. Since 2002, Laos' total imports of goods have been greater than its total exports, resulting in a long-term trade deficit. To boost its economy, Laos welcomes foreign investment. Foreign direct investment in Laos is mainly concentrated in vientiane and the areas along the Mekong River, among which the investment from China plays an important role in the foreign direct investment in Laos, and the investment amount increases year by year. The main areas attracting foreign direct investment in Laos are mineral development and power infrastructure. Laos has a chronic current account deficit, which reached us $2.277 billion in 2017 in particular.

## 30.1 Introduction

Lao people's democratic republic is a landlocked country on indo-china peninsula, bordering China in the north, Dongpu village in the south, Vietnam in the east, Myanmar in the northwest and Thailand in the southwest. 80% of the territory is mountainous and plateau, and is mostly covered by forest. In 1986, Laos implemented reform and opening up and introduced a market economy. Laos has fully implemented the policy of economic restructuring and opening up to the outside world. This paper mainly analyzes the macroeconomic development of Cambodia in recent years from the aspects of economic growth, unemployment rate, price level, import and export trade, foreign investment and international balance of payments.

Q. Tang · M. Li (✉)
Yunnan Normal University, Kunming, China



V. Chang et al. (eds.), *Modern Industrial IoT, Big Data and Supply Chain*,
Smart Innovation, Systems and Technologies 218,
https://doi.org/10.1007/978-981-33-6141-6_30

## 30.2 Economic Growth

The national economy of Laos maintained steady growth [1]. From the perspective of gross domestic product, since the 1990s, except for the Asian financial crisis in 1997, which slowed down the economic growth of Laos, its gross domestic product has maintained steady growth in other years. As can be seen from Fig. 30.1, the total GDP of Laos was only 5.9 billion us dollars in 2008, which increased to 18.4 billion us dollars in 2018, and 12.5 billion us dollars in 10 years. The imf-weo forecasts that Laos' GDP will total $20.2 billion in 2019 and $30.9 billion in 2024.

As Fig. 30.2 shows, since 2002, Laos' GDP per capita has also increased year by year, from $345 in 2002 to $2720 in 2019. In 2008, Laos's GDP per head exceeded $1000, reaching $1014; By 2014, GDP per capita had exceeded $2000, or $2017.

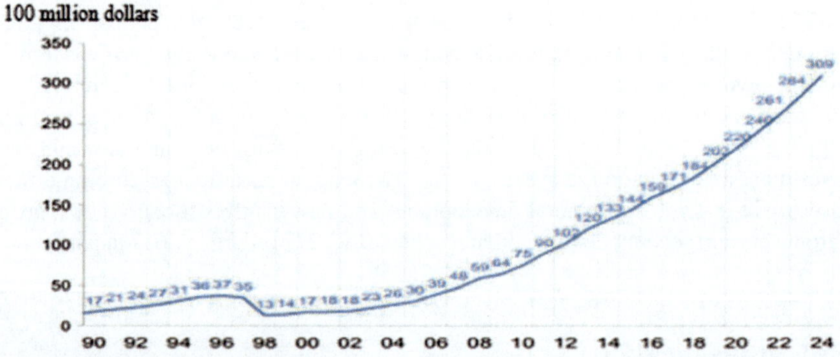


Source: IMF-WEO, Tusheng technology

**Fig. 30.1** Change of total GDP of Laos

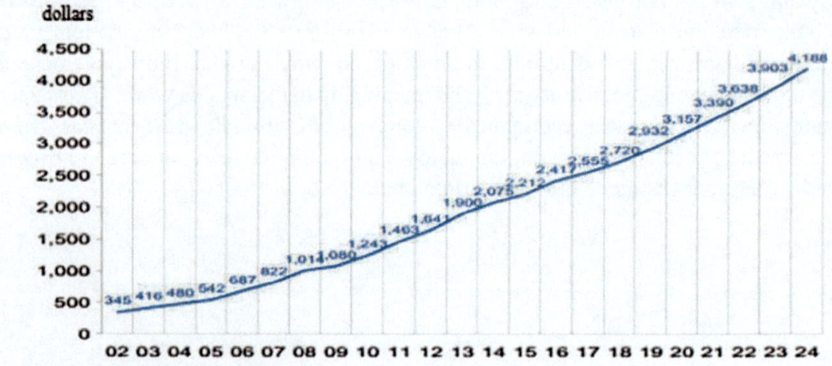


Source: IMF-WEO, Tusheng technology

**Fig. 30.2** Changes of gross per capita GDP in Laos

In 2018, the per capita GNP of Laos was us $2720.3, slightly higher than that of low-income countries. The IMF forecasts a GDP per head of $2932 in 2019. Laos' GDP per head is expected to rise to $4188 by 2024.

Laos is a low- and middle-income developing country with high potential for economic growth. Since the launch of market-oriented reform in 1988, the economy of Laos has maintained a high growth rate, with average annual GDP growth rate reaching 7.1% from 1989 to 2017, and a stable growth range of 7.5–8.0% in most years. From 2013 to 2018, the GDP growth rate of Laos decreased year by year, from 8.5% in 2013 to 6.5% in 2018. Despite the downward trend, Laos's GDP growth rate is still above 6% (see Fig. 30.3). 2017 is the second year after the implementation of the eighth five-year plan and the first year after the implementation of the plan to keep pace with international standards. The GDP growth rate is 6.8%, of which the growth rate of agricultural output is 2.78%, accounting for 16.34% of the GDP. The growth rate of industrial production was 9.53%, accounting for 30% of GDP. The growth rate of service sector output was 6.15%, accounting for 42.08% of GDP. Into □ tariff revenue growth of 6.9%, accounting for 11.53% of GDP. In 2018, GDP growth was 6.5%, but the three sectors of the national economy rose and fell. Year-on-year growth in agriculture was 2.5%, down slightly. The year-on-year industrial growth rate was 7.7%, down significantly. Year-on-year growth in the service sector was 7.5%, up sharply. According to the imf-weo, GDP growth will reach 6.8% in 2024, and the overall economic situation will be stable. The main driving force for Laos' economic growth comes from: firstly, the country is rich in hydraulic resources, and large-scale power station construction has boosted investment and boosted power exports to Vietnam and Thailand. Private consumption is booming, thanks to a booming domestic economy and remittances from Thailand. In recent years, Laos has made great efforts to attract foreign investment from China and other countries through the establishment of special economic zones. The manufacturing sector has developed rapidly, and mineral development and infrastructure have become key areas for foreign investment.

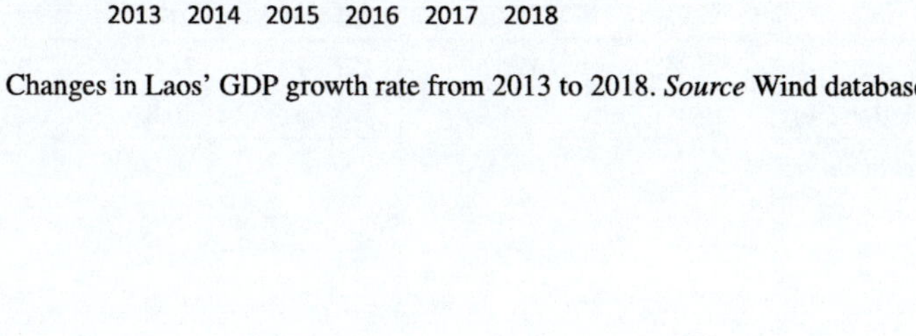


**Fig. 30.3** Changes in Laos' GDP growth rate from 2013 to 2018. *Source* Wind database

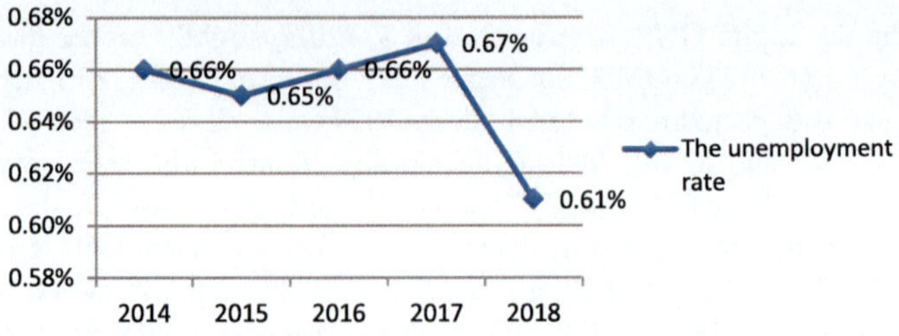


**Fig. 30.4** Changes of unemployment rate in Laos from 2014 to 2018. *Source* Wind database

## 30.3 Unemployment Rate

From 2014 to 2018, the unemployment rate in Laos rose and fell with little fluctuation, and the overall unemployment rate was low, remaining around 0.66%. The highest unemployment rate was 0.67% in 2017; The lowest unemployment rate in 2018 was 0.61% (see Fig. 30.4). Low unemployment provides a guarantee for economic growth.

## 30.4 Price Level

The consumer price index (CPI) increased year by year from 124.22 in 2014 to 131.5 in 2018. According to the CPI, the inflation rate from 2015 to 2018 can be calculated (see Table 30.1). It can be seen that from 2015 to 2017, the inflation rate was lower than 2%. In 2017, it was the lowest, only 0.83%, and rose to 2.04% in 2018. In July 2019, the year-on-year growth rate of the consumer price index in Laos was 3.0%, up from June, while the year-on-year growth rate of the core inflation index was 1.9%, slightly down from June. In the consumer price index basket, food, and non-alcoholic beverage prices rose by 4.3% year-on-year. Clothing and footwear prices continued to fall, with year-on-year growth of 2.1%. Prices for furniture, household equipment and maintenance, and housing and utilities all rose slightly, at 1.8% and 1.0%, respectively, from a year earlier.

**Table 30.1** Inflation rate in Laos from 2014 to 2018

| | 2014 | 2015 | 2016 | 2017 | 2018 |
|---|---|---|---|---|---|
| CPI | 124.22 | 125.80 | 127.81 | 128.87 | 131.50 |
| Rate of inflation (%) | – | 1.27 | 1.60 | 0.83 | 2.04 |

*Source* Wind database

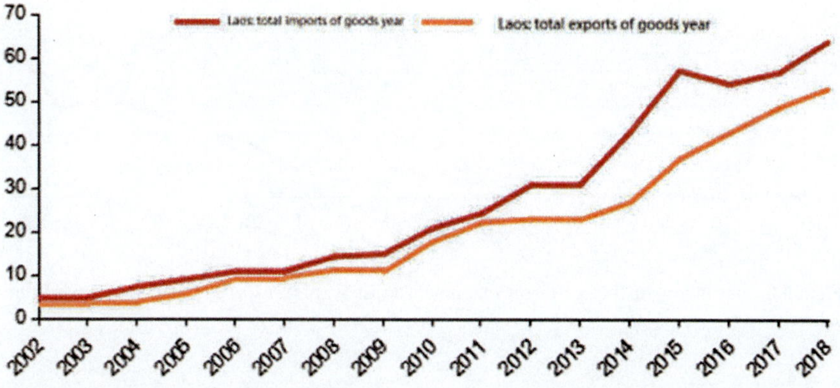


**Fig. 30.5** Total import and export volume of Laos (unit: us $). *Source* Wind database

## 30.5 Import and Export Trade

Since 2002, Laos has been importing more goods than exporting, running a long-term trade deficit (see Fig. 30.5). Currently, Thailand and China are Laos' top two trading partners, among which China is Laos' largest export market. In 2018, Laos' import and export volume reached us $11.2 billion, up 16% year on year, of which us $5.4 billion was exported, up 10.9% year on year, and us $5.8 billion was imported, up 21% year on year.

In May 2019, the monthly export growth rate of Laos was 28% year on year, which was significantly increased, while the monthly import shrank by 1% year on year, which was slightly expanded, and the monthly trade deficit was 93 million us dollars, which was slightly enlarged. In July 2019, monthly export growth to China turned from negative to positive at 16%. On a three-month moving average basis, monthly export growth picked up slightly from the previous month. Monthly imports from China rose slightly to 37% year-on-year. On a three-month moving average basis, monthly import growth from China picked up slightly from a year earlier.

## 30.6 Foreign Direct Investment

To boost its economy, Laos welcomes foreign investment. Foreign direct investment in Laos is mainly concentrated in vientiane and the area along the Mekong River, among which the investment from China plays an important role in the foreign direct investment in Laos, and bucking the trend of the slow recovery of the world economy, it increases year by year [2]. By 2018, Laos had become the second largest country for Chinese outbound investment in Asia. In September 2019, the eu and the rok, respectively, proposed follow-up assistance and investment to Laos [3].

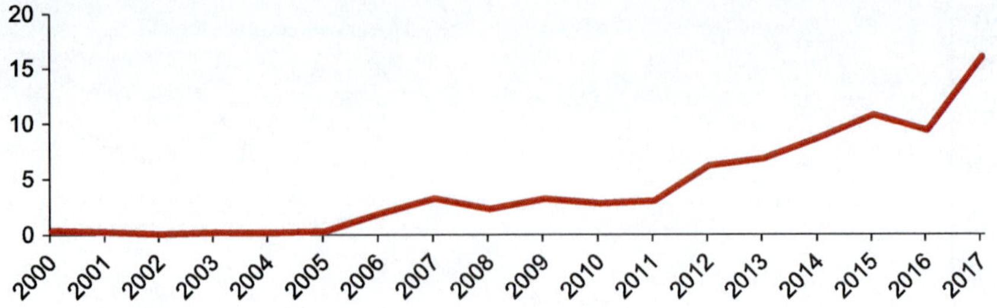


**Fig. 30.6** Net inflow of foreign direct investment in Laos (unit: us $100 million). *Source* Wind database

From the perspective of direct investment, the main areas attracting foreign direct investment in Laos are mineral development and power infrastructure construction. In addition, as the main economic sectors in Laos, agriculture, service industry, and manual manufacturing are also important sectors attracting foreign direct investment, accounting for 30% of foreign direct investment. As can be seen from Fig. 30.6, since 2005, the net inflow of foreign direct investment in Laos has shown an upward trend of volatility. In 2017, foreign direct investment in Laos reached us $3.53 billion, mainly in the electricity, gas, communications, agriculture, forestry, and fishery sectors. From 2015 to 2017, the top industries for foreign direct investment in Laos were wholesale, retail, accommodation, and catering. The main source of investment is China, Thailand, Vietnam, and other countries and regions [4].

## 30.7 Balance of Payments

Laos has long run a huge current account deficit, particularly in 2017, of $2.277 billion (see Fig. 30.2). In the future, with the increase of electricity and other exports, the trend of narrowing. Laos' main exports are copper and coffee, and its labor remittances from Thailand are large. However, its import demand for investment products such as hydropower projects accounts for about a third of its total imports, and a large number of industrial products are also dependent on imports. As a result, Laos has been running a huge current account deficit for a long time [5]. From 2000 to 2018, the ratio of current account balance to GDP reached 18.1% annually, which is a very high level in the world. With a batch of large hydropower projects completed and connected to the grid, electricity exports to Vietnam, Thailand, Laos will rise sharply, the international copper recovery can also help a copper products of its main export commodities—increase income, Laos current-account deficit is expected to improve, but for the development of infrastructure and persistent imports of capital goods will stop the decline in imports a significant improvement, Laos current-account deficit and the proportion of GDP is only slightly narrowed. Laos' total international reserves, including gold, are low, at around $1 billion a year, compared

**Table 30.2** Balance of payments of Laos in 2014–2018 (unit: us $100 million)

| | 2014 | 2015 | 2016 | 2017 | 2018 |
|---|---|---|---|---|---|
| Current account balance | – | −11.93 | −12.34 | −22.77 | −11.78 |
| Total reserves (including gold) | 9.80 | 12.71 | 8.80 | 10.54 | 12.19 |

*Source* Wind database

with $1.219 billion in 2018. As Laos's trade deficit persists, imports will continue to eat into its foreign exchange reserves. Laos' mineral exports are falling due to the decrease in international market demand, and its timber exports have been suspended by the Lao government in recent years. The weakness of these major foreign exchange generating projects makes it difficult for Laos to rebound and increase its foreign exchange reserves for a period of time (Table 30.2).

## 30.8 Conclusion

Overall, growth in Laos has slowed in recent years, with industry and agriculture contributing less, while the service sector remains the main driver of growth. The consumer price index has risen, but prices are generally under control. Exports and related manufacturing have improved. The investment of Chinese private enterprises, the export of electricity and the construction of China-Laos railway have played a certain role in supporting the economic growth of Laos.

## Reference

1. Wei, Y.: Laos: review of 2018 and outlook of 2019. China and Southeast Asia **2019**(01), 36–42 (2019)
2. Huang, C.: Analysis on trade and investment in Laos. China and Southeast Asia **2017**(01), 67–72 (2017)
3. Fang, W.: Development achievements and prospects of Laos since the reform and opening up. J. Tongling Univ. **2015**(04), 67–70 (2015)
4. Qi, P., Vilaiphorn, C.: Influence of foreign direct investment on economic development of Laos. Cooperative Econ. Technol. **2019**(24), 68–70 (2019)
5. Cao, B., Sun, S.: Analysis on the status quo and countermeasures of China's investment in Laos under the background of "One Belt And One Road". Econ. Res. Guide **2018**(18), 177–180

# Chapter 31
# Design of Evaluation System for Competitive Cooperation Capabilities of Cross-Border E-commerce Comprehensive Pilot Area

**Lei Chen, Tingting Pan, and Baojin Zhao**

**Abstract** The establishment of China Cross-border E-commerce Comprehensive Pilot Area indicates that the practice of China's cross-border e-commerce is moving forward at a more open pace. In a way, the qualitative and quantitative research on regional economy can promote the practice of regional development. Therefore, the evaluation of the collaborative development capabilities of Cross-border E-commerce Comprehensive Pilot Area in the new era has become an attractive issue to meet the needs of practice. This paper focuses on the competitive cooperation capabilities of Cross-border E-commerce Comprehensive Pilot Area. Based on the comprehensive evaluation method, it introduces the relevant concepts of Cross-border E-commerce Comprehensive Pilot Area and competitive cooperation capability. In the process of exploring the applicability and scientific nature of the indicator system, it draws the comprehensive evaluation system. With the analytic hierarchy process, it constructs the weight system, hoping to be some path references for the collaborative development practice and evaluation theory.

## 31.1 Introduction

Under the background of China's new economic normality and consumption upgrading, cross-border e-commerce has become a new driving force for China's economic development [1], which is driving China's foreign trade growth and promoting foreign trade transformation and upgrading. According to the "2018–2019 China Cross-border E-Commerce Market Research Report" released by IIMEDIA, China's cross-border e-commerce transaction scale reached 19.1 billion yuan in annual growth rate of 19.74% in 2018, and the user scale exceeded 100 million [2]. The proportion of cross-border online shopping consumption above RMB 1000

L. Chen · T. Pan (✉)
International Business Faculty, Beijing Normal University, Zhuhai 519087, China

B. Zhao
School of Sports Leisure, Beijing Normal University, Zhuhai 519087, China



V. Chang et al. (eds.), *Modern Industrial IoT, Big Data and Supply Chain*,
Smart Innovation, Systems and Technologies 218,
https://doi.org/10.1007/978-981-33-6141-6_31

is 48%. It can be seen that the domestic market demand of cross-border e-commerce industry is huge, and the purchasing power of cross-border e-commerce consumers continues to upgrade. Consumer demand and international exchange under the overall opening up pattern have created a good internal and external environment for the further development of China's cross-border e-commerce industry [3].

The practice of cross-border e-commerce development in China calls for a new era issue. The establishment of a cross-border e-commerce comprehensive pilot area is a new exploration of the development path of cross-border e-commerce in China under the comprehensive open pattern, just as the newborn baby needs breastfeeding. At the time of the advancement of the practice of piloting the pilot in the cross-border e-commerce comprehensive pilot area in China, a number of new issues are needed. From the establishment of special economic areas in the early stage of reform and opening up to the establishment of cross-border e-commerce comprehensive pilot areas in the new era [4], China is exploring the unique path of coordinated regional economic development step by step. Under the two wings of "digital economy" and "Belt and Road" [5], the way to develop the ability of regional coordinated development, with the unique Chinese elements in the "Chinese business card" presented to the world, has attracted the attention of some scholars.

In order to understand the collaborative development practice of cross-border e-commerce comprehensive pilot area, and further study the measurement, analysis, and evaluation of the competitive cooperation ability of China's cross-border e-commerce comprehensive pilot area, this paper intends to select and apply the synergy theory and comprehensive evaluation under operational research, the comprehensive index construction method and the analytic hierarchy method. In the process of exploring the applicability and scientific nature of the indicator system, it draws the comprehensive evaluation system. With the analytic hierarchy process, it constructs the weight system, hoping to be some path references for the collaborative development practice and evaluation theory.

## 31.2 Construction of Evaluation Index System for Cross-Border E-commerce Comprehensive Pilot Area

To measure the competing ability of the cross-border e-commerce comprehensive test area, according to the theory of synergistic competition, the target-level measurement index of competing ability can be divided into the competitiveness index and the cooperation index, and the literature related to the comprehensive evaluation index can be consulted. A variety of indicators, such as evaluation index reports and visiting experts related to regional economic development, obtained a preliminary indicator category, and combined with preliminary design changes, constructed a cross-border e-commerce comprehensive test area competition and competency evaluation index system.

Cross-border e-commerce comprehensive pilot area as a new type of international trade development under the "digital economy", there is still some controversy in the academic community's understanding of its connotation. For the cross-border e-commerce comprehensive test area competition and competition evaluation index design, we must first understand the construction and development essentials of the cross-border e-commerce comprehensive test area. The construction and development of the cross-border e-commerce comprehensive test lie in the creation and maintenance of the e-commerce ecological chain and industrial chain [6]. This requires that the construction and development of cross-border e-commerce comprehensive testing should not only pay attention to the main body of domestic and foreign transactions but also pay attention to the diversification of the participating entities at home and abroad throughout the ecological chain, the improvement of the supporting facilities in each link, and the promotion of comprehensive testing.

This evaluation system comprehensively considers the elements of "cross-border e-commerce comprehensive test area" related to "people, finance, material, and environment" and establishes an evaluation index model for all levels of competing ability in cross-border e-commerce comprehensive test areas. The first is the static basic competitiveness including cross-border e-commerce enterprises, consumers, and venues; the second is the dynamic extension of innovation, brand, management, and culture, which is inspired by the elements of urban core competitiveness. Competitiveness; the third is to consider the internal synergy and complementarity of various roles in the internal ecological chain of the comprehensive test area, including logistics companies, government finance and payment platforms; fourth, the source of regional opening up and cooperation, including funds, population, External coordination and cooperation capabilities, including goods and economic and cultural exchanges. Based on the idea of measuring both "competitiveness" and "cooperative force", this paper has set up various levels of indicators as shown in Table 31.1. Each of these indicators has its own original design intention. This paper intends to explain the ways in which each indicator is constructed.

**Table 31.1** Evaluation indicators of the competing ability of cross-border e-commerce comprehensive test area

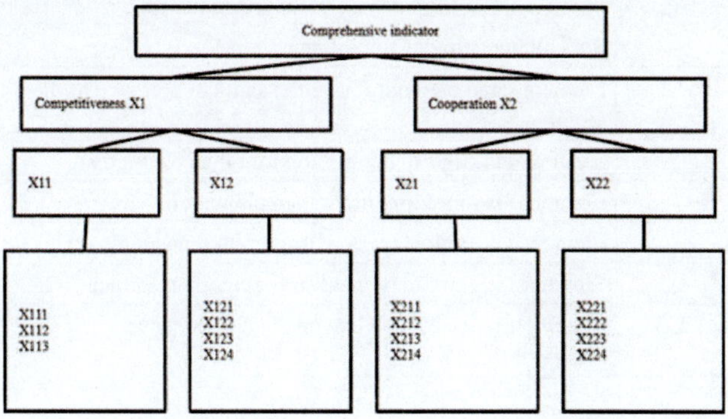

## 31.3 Design of Re-engineering and Index Weight System Based on AHP

The design of index weighting system relies on the analytic hierarchy process: after the establishment of each level of indicators, the two-two discriminant matrix is established. On the basis of verifying the adaptability and rationality of the initial indicators, the weights of each index and their combined weights are calculated and carried out. After the overall ranking of the indicators, the cross-border e-commerce comprehensive test area competition and competition indicator weight system is proposed to be built.

### *31.3.1 Scoring Data Sources and Preprocessing*

Using Delphi (expert) method, this thesis designed three experts to collect opinions, and collected opinions through online mailbox and offline discussion. In order to ensure the reliability of the score, six people were invited from the e-commerce field.

### *31.3.2 Establishment of Two-Two Judgment Matrix*

According to the 9-point scoring matrix method of AHP model in Table 31.2, experts in the field of regional economics, operations research, and supply chain management, and three experts from the statistical bureau and the commercial bureau respectively collect and summarize the opinions of the above experts, eliminate the invalid scores, and replace the missing values with the mean method to obtain the data. The summary score after preprocessing. In order to ensure the correctness of the design of

**Table 31.2** 9-scale assignment method for pairwise comparison matrix

| Score value | Specific meaning of importance |
|---|---|
| $aij = 1$ | Element *i* and element *j* have the same importance to the previous level factor |
| $aij = 3$ | Element *i* is slightly more important than element *j* |
| $aij = 5$ | Element *i* is more important than element *j* |
| $aij = 7$ | Element *i* is obviously more important than element *j* |
| $aij = 9$ | Element *i* is extremely more important than element *j* |
| $aij = 2n, n = 1, 2, 3, 4$ | The importance of the elements *i* and *j* is between $aij = 2n - 1$ and $aij = 2n + 1$ |

indicators at all levels, two professionals were also invited to conduct proofreading of the initial indicators to achieve the most scientific and accurate results (Table 31.2).

### 31.3.3 Establishment of the Weighting System for the Evaluation Index of Competition and Cooperation Ability in Cross-Border E-commerce Comprehensive Pilot Area

Establish a comprehensive evaluation model for the competing ability of the cross-border e-commerce comprehensive test area, and obtain a weighting system for the comprehensive evaluation index of the competing ability of the cross-border e-commerce comprehensive test area. See Table 31.3, multiply the index values $X_{ijz}$ by the combined weights. The product summation of $W_{ijz}$ is obtained by the comprehensive evaluation index of the competing ability of the cross-border e-commerce comprehensive test area, which can be expressed by $F$. The comprehensive evaluation index is: $F =$ cross-border network business index $\times$ 0.0536 + cross-border online shopping index $\times$ 0.0536 + Infrastructure equipment index $\times$ 0.0179 + innovation development potential $\times$ 0.2160 + non-homogeneous brand power $\times$ 0.0834 + district management validity $\times$ 0.0469 + district cultural image strength $\times$ 0.0286 + smart logistics service level $\times$ 0.1786 + payment and financial service level $\times$ 0.0571 + Policy service support $\times$ 0.0960 + information sharing power $\times$ 0.0434 + capital flow level $\times$ 0.0219 + customs clearance level $\times$ 0.0663 + economic exchange level $\times$ 0.0368.

It can be seen that in this indicator system, the most important is the innovation development potential and the level of intelligent logistics services; the second is the policy service support, non-homogeneous brand power, customs clearance level, etc.; because it is comprehensive consideration and passed the test, so other indicators There will be no major disparity between the two, and the evaluation system can be considered scientific.

The statistical indicator system is also adaptable. The ability of the comprehensive test area in the four major sectors of basic strength, competition development, internal coordination, and external cooperation is included, and the framework of the large-scale framework and the indicator system in each sector are grasped. It can be changed according to the development practice of cross-border e-commerce comprehensive pilot area. For example, the significance of information flow has been paid more and more attention, and its role has been increasingly emphasized by cross-border network operators. In the future, it can be re-established according to actual conditions. The new weight of information sharing power (X214).

Based on the above exploration, this paper believes that in order to promote the improvement of the competing ability of cross-border e-commerce comprehensive pilot areas, it is necessary to establish the idea of comprehensive statistical analysis of multiple indicators, and to exert the synergy of multiple actors in the region and to

**Table 31.3** Combination weights and ranking of evaluation indicators for competing ability in cross-border e-commerce comprehensive test areas

| Primary indicator | Secondary indicator | Tertiary indicator | Combination weight | Descending sort |
|---|---|---|---|---|
| Competitiveness (0.5) | Basic competitiveness (0.1250) | Cross-border network business index (0.4286) | 0.0536 | 7 |
| | | Cross-border online shopping index (0.4286) | 0.0536 | 8 |
| | | Infrastructure allocation index (0.1429) | 0.0179 | 14 |
| | Extended competitiveness (0.3750) | Innovation potential (0.5761) | 0.2160 | 1 |
| | | Non-homogeneous brand power (0.2224) | 0.0834 | 4 |
| | | District management validity (0.1251) | 0.0469 | 9 |
| | | District cultural image (0.0764) | 0.0286 | 12 |
| Cooperation (0.5) | Internal Complementation (0.3750) | Smart logistics service level (0.4763) | 0.1786 | 2 |
| | | Payment and financial service level (0.1522) | 0.0571 | 6 |
| | | Policy service support (0.2559) | 0.0960 | 3 |
| | | Information sharing (0.1156) | 0.0434 | 10 |
| | External cooperation (0.1250) | Fund flow level (0.1752) | 0.0219 | 13 |
| | | Customs clearance level (0.5303) | 0.0663 | 5 |
| | | Economic exchange level (0.2945) | 0.0368 | 11 |

create a benign mechanism of competition and collaboration. For the development of cross-border e-commerce in China, we will put in the wings of trial and error to help the sustainable and healthy development of cross-border e-commerce in China. In this regard, based on the above design analysis, the following five points are proposed for the promotion strategy of the cross-border e-commerce comprehensive pilot area in China: First, grasp the core advantages and enhance the core competitiveness of the comprehensive pilot area; Second, improve the service elements, enhance the complementary cooperation of the comprehensive pilot areas; Third, stimulate innovation and enhance the first-come competitiveness of the comprehensive pilot area; Fourth, promote internal and external linkages and enhance the cooperation between the internal and external areas; Fifth, Convergence of all parties to improve the overall pilot area competitive strength.

## 31.4 Significance

Based on the design of the comprehensive evaluation index system, the study of the competing ability problem of the cross-border e-commerce comprehensive pilot area has the following theoretical and practical significance.

In the first 12 years of this century, China's cross-border e-commerce is still in its infancy, so domestic scholars pay little attention to cross-border e-commerce [7]. At present, China's cross-border e-commerce development is in a critical period of multi-party efforts [8]. However, domestic research on cross-border e-commerce has only begun to increase in 2013 [9], and research on cross-border e-commerce comprehensive pilot areas is still in its infancy. It is to be sustained. Based on the design of comprehensive evaluation index system, the problem of competing ability in cross-border e-commerce comprehensive pilot area can be studied, which can enrich the research on the statistical index system of cross-border e-commerce comprehensive pilot area, refine the research perspective, expand research ideas, and deepen can further solve research problems. Based on the principle of applicability and scientificity, this paper innovatively selects the ability to compete and compete as a research point, measuring both "competitive ability" and "cooperative ability", and hopes to comprehensively competing for cross-border e-commerce comprehensive pilot area. The evaluation index system is discussed as completely as possible, so as to make up to some extent the relevant research only measures the lack of "competitive ability", enriches the domestic research system for the cross-border e-commerce comprehensive pilot area, reflects the timeliness, and deepens the cross-border electricity Reflections and prospects of research in the comprehensive pilot area.

This thesis further enriches the research on the statistical index system of the cross-border e-commerce comprehensive pilot area, makes up for the relevant research, expand research ideas, and deepen and solve research problems. At present, China's cross-border e-commerce development is in a critical period of multi-party efforts. However, research on cross-border e-commerce comprehensive pilot areas is still in its infancy and needs to be sustained. Based on the principle of applicability and scientificity, this thesis innovatively selects the cohesive ability as the research point, which not only measures the "competitive ability" but also the "cooperative ability", and hopes to comprehensively competing for the cross-border e-commerce comprehensive pilot area. The evaluation index system should be discussed as completely as possible to enrich the domestic research system for the cross-border e-commerce comprehensive pilot area, reflecting the timeliness, and further deepen the research reflection and prospect of the cross-border e-commerce comprehensive pilot area.

This paper hopes to construct a scientific and valuable comprehensive evaluation index system for cross-border e-commerce pilot areas. In the implementation plan of the cross-border e-commerce comprehensive pilot area, it is proposed to explore statistical standards. This paper explores the design of the comprehensive evaluation index system for cross-border e-commerce pilot areas, and hopes to establish a certain cross-border e-commerce statistical monitoring system. for your reference. Secondly, it will help to explore new paths for linkage optimization, promote balanced strategy development and transformation and upgrading of cross-border e-commerce enterprises, promote brand enterprises to implement "Internet+" strategy, foster local enterprise brands, and simultaneously conduct cross-border e-commerce comprehensive tests. Participants in the district provide action guides. Furthermore, through the study of the factor integration capabilities of e-commerce entities, logistics entities, service organizations and regulatory agencies within the cross-border e-commerce pilot area, it is helpful to explore ways and means to better enhance the competitiveness of the competition. Exploring the ways of "hands-on" economic cooperation with other countries will help the construction of smart supply chain systems for cross-border e-commerce experiments to a certain extent. Finally, to a certain extent, I hope to contribute to the building of cross-border e-commerce reform, innovation, coordinated development and open cooperation. As a big manufacturing country, China must move toward a manufacturing power and cultivate the synergistic development capability of cross-border e-commerce pilot areas. This has become an important part. The improvement of the competing ability of cross-border e-commerce pilot areas will help to consolidate China's international integration. The basis of competition, under the open and mutually beneficial and inclusive open pattern, assists the balanced development practice of cross-border e-commerce comprehensive pilot areas, and releases the wings of China's "Internet + foreign trade" to take off under the blue sky of "digital economy."

## 31.5 Reflection

This paper introduces the related concepts of competing ability in the cross-border e-commerce comprehensive pilot area, and innovatively proposes the application of the synergy theory to the comprehensive evaluation of the competing ability of the cross-border e-commerce comprehensive pilot area to realize the integration of the two forces. Using the comprehensive evaluation analysis method, based on the combination of scientific and pertinent, the combination of adaptability and feasibility, a comprehensive evaluation index system for measuring the competing ability of cross-border e-commerce comprehensive pilot areas was established reasonably; The analytic hierarchy process further strengthens the scientificity and applicability of statistical analysis; calculates the comprehensive measurement model and further discovers the problems, thus effectively conducting a comprehensive assessment of the cross-border e-commerce comprehensive pilot area, and also providing for the promotion of future co-opetition capabilities. There are some forward-looking issues in this paper that need to be further developed and discussed later. First, the method of comprehensive evaluation and analysis is always in the process of development. The comprehensive statistical analysis method includes not only the analytic hierarchy process, but also the factor analysis and other methods worth exploring and using. In the later research, I hope to further apply it. Comprehensive evaluation and analysis methods are conducted in-depth research in related fields. Second, the cross-border e-commerce comprehensive pilot area is also constantly evolving. The sustainable and healthy development of the pilot area still needs internal and external linkage and multi-party efforts. When more theoretical research continues to advance, cross-border e-commerce,the development practice of the comprehensive pilot area still needs to be carried forward. The comprehensive evaluation system of competition and cooperation capability of the cross-border e-commerce comprehensive pilot area designed in this paper can be further tested if it can be applied to the evaluation of the competing ability of each comprehensive pilot area.

## References

1. Xu, R.: Research on the Transformation of Traditional Domestic Trade e-Commerce Cross-Border e-Commerce. Jilin University, Changchun (2018)
2. Liu, X.: China's cross-border e-commerce into the "new format" of import and export. China's Foreign Trade **04**, 62–63 (2018)
3. Zhu, R.T.: The interactive development of China's retail industry "going global" strategy in the context of the Belt & Road initiative. Contemp. Soc. Sci. **02**, 64–73 (2016)
4. Rui, H.: The five key words of "Electric Business Law" reshape the e-commerce industry. Times Financ. **04**, 15 (2019)
5. Hu, X., Du, D.: Study on the niche evaluation of urban science and technology competitiveness—taking 11 cities in Zhejiang Province as an Example. Sci. Technol. Prog. Policy **29**(12), 36–40 (2012)
6. Chen, Y.: Research on the Innovation of Spatial Form of Emerging Industries From the Perspective of Niche. Shanghai Jiaotong University, Shanghai (2014)
7. Liu, J.: Research on Co-evolution of Cross-Border e-Commerce Ecosystem. Changchun University of Technology, Changchun (2018)
8. Hao, B.: Research on Industrial Competitiveness Evaluation of Cross-border E-Commerce Comprehensive Experimental Zone. South China University of Technology, Guangzhou (2018)
9. Ni, P.: Analysis Paradigm and Conceptual Framework of China's Urban Competitiveness. J. Econ. **06**, 14–18 (2001)

# Chapter 32
# How to Market New Energy Vehicles

**Fan Yang**

**Abstract** Along with the rapid development of the national economy, the living standards of the people have been greatly improved, and the private possession of automobiles has shown a spurt development trend. This progress and innovation in the automotive industry have had an impact. On the one hand, the demand for automobiles has increased greatly, and on the other hand, the major damage caused by automobile pollution has become increasingly prominent. In order to comply with the national green environmental protection appeal, new energy vehicles have become an emerging hot spot. However, compared with traditional vehicles, new energy vehicles still have disadvantages such as high price, low market penetration rate, and low public acceptance. In this context, we will focus on the marketing strategy of new energy vehicles, which is particularly important, and it is of great significance to the development of the automotive industry.

## 32.1 Introduction

Any product that wants to occupy a large market share and is accepted by the broad masses of consumers needs to rely on certain appropriate marketing. Therefore, in addition to the continuous improvement and improvement of its own technology, the progress of new energy vehicles must closely follow the changes in the market [1]. Marketing, changing the traditional car consumption concept of the people, and formulating a scientific and rational marketing strategy. It can be foreseen that the new energy vehicles meet the technological development prospects and green environmental protection requirements, and are new things that are in line with the future development direction. The market development prospects are extremely broad [2].

F. Yang (✉)
Guangdong University of Science & Technology, Dongguan 523083, China

 
V. Chang et al. (eds.), *Modern Industrial IoT, Big Data and Supply Chain*,
Smart Innovation, Systems and Technologies 218,
https://doi.org/10.1007/978-981-33-6141-6_32

## 32.2 Marketing Situation of New Energy Vehicles

In general, new energy vehicles have a larger policy dividend in the country. In particular, in recent years, the state has continuously increased the government-related policy subsidies for the purchase of new energy vehicles, which has also greatly encouraged the research and development of automobile manufacturers effect. So new energy vehicles have continued to develop in general and have achieved certain results in sales, but the challenges and difficulties are still very serious.

The main difficulties encountered when selling new energy vehicles are as follows:

1. The public's awareness of environmental protection is lacking and the concept of consumption is outdated.

Most of the public lacks sufficient understanding of new energy vehicles and has an inherent impression of cars, which makes them inclined to buy traditional fuel vehicles. This situation has a great relationship with the lack of environmental education. It is difficult for people to receive scientific and rational environmental education, which brings great difficulties for the sales of new energy vehicles. It is difficult for consumers to fully recognize new energy vehicles psychologically.

2. The market competition continues to expand.

Under the support of the state's policies and the broad market prospects of new energy vehicles, the number of companies participating in the research and development of new energy vehicles is increasing, and the market supply and output are also increasing. In the ever-expanding market, how to rely on the brand influence and the characteristics of the marketing market to stand out, it is increasingly difficult to obtain the favor of the market. In order to win commercial competition, major auto manufacturers have adopted a large number of price reduction promotions and new models, but at present, the number of new models and car prices is not the most important factors affecting car marketing.

3. There is a shortage of professional marketing talents.

Sales is a person's science. In addition to theoretical knowledge, the personal ability of sales people often plays a more important role. The subjective initiative of marketing talents has a huge impact on product sales. However, since new energy vehicles are still new, their own development is not mature enough, and it is difficult for relevant marketing talents to fully grasp product knowledge and lack relevant sales experience.

4. The marketing mechanism needs to be optimized.

The so-called marketing mechanism refers to the long-term mechanism of marketing, that is, the logistics support system. The development foundation of the new energy automobile industry is still not solid, and the overall system is not mature enough. Therefore, there are certain problems in marketing channels, marketing personnel literacy, product logistics support, etc., which will directly affect the marketing effect of new energy vehicles [3]. By observing the current situation, most auto manufacturers have not established a long-term mechanism for new energy vehicles. The logistics support of products is prone to problems, which directly hits consumers' enthusiasm for purchasing new energy vehicles.

5. Insufficient level of enterprise managers.

Business managers lack sufficient professional knowledge, cannot be sensitive to the market, and make full use of customer information, which makes the company lose more potential customers, cannot further expand market share, and even inadvertent management will lead to existing customers lose. At the same time, managers lack systematic and systematic marketing strategies, and often tend to distract their energy, fail to correctly estimate market returns and risks, and fail to make necessary adjustments in time according to market conditions [4].

## 32.3 Analysis of New Energy Vehicle Sales Methods

After analyzing the marketing difficulties of new energy vehicles, this paper attempts to propose some strategies to promote the marketing of new energy vehicles, in order to provide a reference for its future development.

1. Increase publicity and environmental education for new energy vehicles.

Therefore, in the marketing process of new energy vehicles, it is necessary to highlight the unique selling points of new energy vehicles to ensure that they are clearly distinguished from traditional vehicle positioning. In addition, it is important to pay full attention to the propaganda work. After all, the propaganda is crucial in the success of the sales work. Only when the product promotion is in place, its presence in the consumer group will be enhanced, and it will be further enhanced. The effect of new energy vehicle sales will be the ultimate in sales. At the same time, the national environmental protection department should strengthen cooperation with automobile manufacturers, increase education on public environmental awareness, popularize the concept of environmental protection passed down by new energy vehicles to the public, and create a good social environment and atmosphere for the sales of new energy vehicles. Once the public's awareness of environmental protection is improved and the concept of environmentally friendly consumption is developed, the sales of new energy vehicles will become easier.

2. Expand marketing channels and multi-angle marketing.

The new era has not been able to meet the needs of the times with the help of traditional TV and offline marketing. Nowadays, online marketing has gradually occupied the main position. In the marketing of new energy vehicles, we strive to further expand the marketing channels based on the existing sales channels. For example, in addition to grasping the traditional offline marketing methods such as offline stores and mainstream TV media, we should also make full use of the network. From the media, multimedia, and all media to fragment marketing and build a rich, comprehensive, three-dimensional marketing network to maximize the public's understanding of new energy vehicles [5].

3. Cultivate professional new energy vehicle marketing talents.

Automobile manufacturers should strengthen the targeted cooperation with relevant marketing companies and jointly cultivate a professional new energy vehicle marketing team. On the one hand, the introduction of high-level trainers, allowing them to teach marketers the necessary knowledge of new energy vehicles, and improve the level of training teachers. On the other hand, urge marketers to continuously improve their basic knowledge and professional level, and enhance their understanding of the concept of new energy vehicles, so as to ensure that marketers can perfectly communicate the relevant advantages and product concepts of new energy vehicles to consumers and further enhance marketing effect.

4. Improve the information platform

In addition to the above-mentioned measures, we should also actively improve the information platform. In the process of building this information platform, the support of Internet technology is particularly important. The build platform is a huge systematic project covering a wide range of topics, including the establishment of distribution work and the issuance of multimedia electronic publications. Each of these tasks requires a lot of effort and money, and the complexity is strong. For the marketing of new energy vehicles, in order to achieve the smooth marketing of new energy vehicles, it is necessary to make full use of the power of the Internet, and at the same time, to put relevant information of products on the Internet, so that users can be well documented. Build a complete product information network, enhance users' trust in new energy vehicles, and cultivate users' consumption habits and brand loyalty.

5. Handle the problem of capital introduction

The promotion of any new product requires a large amount of upfront funding to capture a certain initial share of the market. At present, although the state has certain policy subsidies and dividends, the healthy development of an industry must not always rely on government support, but should have its own ability to make blood. In order to further open up the market, new energy vehicles must attract investors to invest as soon as possible, and provide necessary reserve funds for the development of sales channels and sales. In the specific process of attracting investment, the focus should be on the city, because the economic development of the big city is relatively stable and the economic situation is good, the available funds should be more adequate, and the probability of successful investment is greater. As far as the level of urban development is concerned, the best strategy should be to start in the first-tier cities, first to expand the sales share in the automobile market in the first-tier cities, and then vigorously promote the publicity, and spread the advanced experience to other second- and third-tier cities and even more developed towns and villages [6].

## 32.4 Conclusion

In summary, this article studies the sales of new energy vehicles, sales bottlenecks, and corresponding sales skills. It can be seen that new energy vehicles are the key industries supported by the state. Although there are many difficulties in promotion at this stage, with the progress and development of society, new energy vehicles will surely win wider and wider due to their advantages in energy conservation and environmental protection. The strategy given in this paper is only a part. The marketing of new energy vehicles is still a hot topic with high discussion and rich content. I hope this article can provide a reference for the progress of new energy vehicles.

## References

1. Ying, Q.: Discussion on marketing strategy of new energy vehicles. Times Auto **14**, 17–18 (2017)
2. Energy; Studies from Guangxi University Describe New Findings in Energy (Enabling strategies of electric vehicles for under frequency load shedding), Energy Weekly News, 2018
3. Rui, F.: Analysis of China's automobile marketing problems and countermeasures. Knowl. Econ. **17**, 87–88 (2016)
4. Rui, W.: Analysis of the current situation and prospects of new energy vehicle marketing. China Manage. Inf. **21**(23), 99–100 (2018)
5. Zhang, X.: Exploration and practice of local economic transformation of automobile marketing professional service from the perspective of new energy automobile industry development. In: Proceedings of 2018 International Conference on Economics, Finance, Business, and Development, 2018
6. Science—Applied Sciences; University of Alabama Researcher Provides Details of New Studies and Findings in the Area of Applied Sciences (Small-Signal Modeling and Analysis for a Wirelessly Distributed and Enabled Battery Energy Storage System of Electric Vehicles), Science Letter, 2019

# Part IV
# Advanced Supply Chain and IIOT-BD Integration

# Chapter 33
# A Two-Sided Matching Method Based on Multi-stage Dynamic Reference Points

**Hong Yu, Mengnan Cui, Guoyin Wang, and Xiaofang Chen**

**Abstract** The problem of two-sided (bilateral) matching is a process of matching demanders and suppliers reasonably in supply chain. The vast majority of existing researches are based on a single-stage static evaluation information. It is necessary to consider the dynamic evaluation information during multiple stages since the performances of both matching parties have some certain dynamic characteristics. To address the problem, this paper proposes a two-sided matching method based on multi-stage dynamic reference points, because the decision-makers often focus on the level of gain or loss of the evaluation object relative to a reference point in the view of prospect theory.

## 33.1 Introduction

With the continuous developing of Industry 4.0, the latest smart technology has become the important way to achieve transformation of traditional manufacturing and industrial practices. In the problem of supplier selection, the decision-makers (DMs) must weigh the tangible and intangible evaluation criteria in order to select the best suppliers for the enterprise [6]. So far, the most widely accepted and used method is based on the multiple criteria decision-making (MCDM) method with unilateral selection.

H. Yu (✉) · M. Cui · G. Wang
Chongqing Key Laboratory of Computational Intelligence, Chongqing University of Posts and Telecommunications, Chongqing, China
e-mail: yuhong@cqupt.edu.cn

G. Wang
e-mail: wanggy@cqupt.edu.cn

X. Chen
School of Information Science and Engineering, Central South University, Changsha, China
e-mail: xiaofangchen@csu.edu.cn

 
V. Chang et al. (eds.), *Modern Industrial IoT, Big Data and Supply Chain*,
Smart Innovation, Systems and Technologies 218,
https://doi.org/10.1007/978-981-33-6141-6_33

However, some practical problems are the two-way selection process, which needs to consider the preferences of both sides. For example, the model for large passenger aircraft [3] is a two-way selection and collaboration process.

The theory of bilateral matching decision problem was originally proposed by mathematicians D. Gale and LS Shapley [2]. The existing studies have made important contributions to solving various bilateral problems, but most of these studies only considered single-stage static evaluation information rather than multi-stage dynamic evaluation.

The purpose of this paper is to gather multi-stage evaluation information, and consider the stage development speed of evaluation information to reflect the level of evaluation information from both horizontal and vertical aspects. Set dynamic reference points according to the prospect theory, construct a dynamic prospect evaluation model, and make scientific and reasonable matching decisions based on the prospect value of the matching subject.

## 33.2 Preliminary Work

### *33.2.1 Prospect Theory*

Prospect theory was originally proposed by Tversky and Kahneman [4], and it is a theoretical model widely used to analyze and describe the decision-making process under risk. We can evaluate the integrated data through the value function in prospect theory. Tversky and Kahneman [5] gave a value function with a two-part power function of the form:

$$v(\Delta x) = \begin{cases} (\Delta x)^a, \ \Delta x \geq 0 \\ -\lambda(\Delta x)^a, \ \Delta x < 0 \end{cases} \tag{1}$$

In Eq. 1, $\Delta x$ represents the degree to which the scheme $x$ deviates from the reference point $x_0$. $\alpha$ and $\beta$ represent the coefficients of decision makers' attitudes towards gains and losses. $\lambda$ represents the risk aversion coefficient. Tversky and Kahneman [4] found that the value function value is more consistent with the actual empirical data when $\alpha = \beta = 0.88$, $\lambda = 2.25$.

### *33.2.2 Problem Formulation*

In order to define our work, suppose that there are m demand individuals and n supply individuals that need to be matched. Let $A = \{a_1, \ldots, a_i, \ldots, a_m\}$ represent the set of demand-side individuals, where $a_i$ represents the $i$th demand entity and $m \geq 2$;

and $B = \{b_1, \ldots, b_j, \ldots, b_n\}$ represents a collection of supplier entities, where $b_j$ represents the $j$th supplier object and $n \geq 2$.

We use the evaluation criterion set $C = \{c_1, \ldots, c_k, \ldots, c_p\}$ to evaluate the demand individuals $A$, that is to say, the supply side evaluates the demand individuals through the criterion set; similarly, the criterion set $D = \{d_1, \ldots, d_h, \ldots, d_q\}$ is used to evaluate the supply individuals $B$, that is, the demand side evaluates the supply individuals through the criterion set. In this matching problem, the satisfaction of the two sides with each other is formed by the evaluation information of multiple stages, and we use the set $\pi = \{1, \ldots, t, \ldots, T\}$ to represent all evaluation stages. The evaluation values expressed in the form of interval numbers. Among them, the evaluation value $c^t_{ij(k)}$ means when the supply individual $b_j$ evaluates the demand individual ai at the stage $t$ through the criterion $c_k$, and we can understand $d^t_{ij(h)}$ in the same way.

The problem to be discussed in this paper is how to obtain a more reasonable matching decision result based on the multi-stage mutual evaluation information of the matching individuals of both sides.

## 33.3 Proposed Method

### 33.3.1 The Dynamic Reference Point

In the multi-stage evaluation information aggregation problem, the setting of the reference point needs to consider the both evaluation objects and evaluation information changes in different stages.

For a series of evaluation value, the average development speed represents the average degree of the development and change trend of the sequence in the overall stages, and it is a standard for judging the chain development speed of each stage. The definitions are given below.

**Definition 1** Stage development speed: indicates the development trend of the current stage of the evaluation criterion, that is, the ratio of the current stage evaluation value to the previous stage evaluation value.

For individuals of supply side, the stage development speed of the evaluation criterion $d_h$ at stage $t$ is $V^t_{(h)} = \overline{d^t_{(h)}}/\overline{d^t_{(h)}}$, where $\overline{d^t_{(h)}}$ represents the expected value of criterion $d_h$ for all suppliers at stage $t$.

**Definition 2** Average development speed: indicates the average level of the development speed of each stage of the time series, calculated by the moving average of the expected value of the evaluation criterion.

In the multi-stage decision-making process, we use the following steps to establish reference points of evaluation criteria at different stages.

**Step 1**. Determine the initial reference point with the expected value.

**Step 2**. Set the dynamic reference points by comparing the development speed of each stage with the average development speed of whole stages. The formula is as follows:

$$rb^{t}_{(h)} = \overline{d^{t}_{(h)}}\{P(V^{t}_{(h)} \geq \overline{V_{(h)}}) - P(\overline{V_{(h)}} \geq V^{t}_{(h)})\} \tag{2}$$

where $P(.)$ represents the possibility relationship between the two interval numbers.

### 33.3.2 Prospect Values Acquisition

Since the reference points and evaluation values of each criterion are both interval numbers, we can obtain six relationships between the two interval numbers [1].

**Definition 3** For interval number $[Z^{-}, Z^{+}]$, let $x$ denote any value on interval $[Z^{-}, Z^{+}]$. Then, the probability density function of $x$ is

$$f(x) = \left\{\frac{1}{Z^{+} - Z^{-}}, Z^{-} \leq x \leq Z^{+}\right. \tag{3}$$

where $\int_{Z^{-}}^{Z^{+}} f(x)$ and there is $f(x) \geq 0$ for any $x$.

We can calculate the gain $\text{GoB}^{t}_{ij(h)}$ and loss $\text{LoB}^{t}_{ij(h)}$ of the supply side to the demand side at each stage. Then, the gain–loss matrices $\text{GLM}_{A}$ of the demand set $A$ are established,

$$\text{GLM}_{A} = \left(\Delta x^{t}_{ij(h)}\right)_{(m*n)*(q*T)} \tag{4}$$

where $\Delta x^{t}_{ij(h)} = \text{GoA}^{t}_{ij(h)} + \text{LoA}^{t}_{ij(h)}$.

Through Eq. 1, we transform the above two matrices into the prospect value matrix of demand side $PM_{A} = v^{t}_{ij(h)_{(\mathrm{m}\times\mathrm{n})\times(\mathrm{q}\times\mathrm{T})}}$. Where $v^{t}_{ij(h)}$ represents the prospect value of demand individual ai to supply individual $b_{j}$ regarding its evaluation criterion dh at stage $t$.

### 33.3.3 Satisfaction Calculation

Because different evaluation criteria often have different physical meanings, the calculated prospect values are often disproportionate. Therefore, the two prospect value matrices need to be dimensionless to eliminate inconsistencies.

We can convert it from $PM_{A}$ into $\overline{PM_{A}} = \bar{v}^{t}_{ij(h)_{(\mathrm{m}\times\mathrm{n})\times(\mathrm{q}\times\mathrm{T})}}$.

Let $W_D^t = \left(\omega_{D(1)}^t, \ldots, \omega_{D(h)}^t, \ldots, \omega_{D(q)}^t\right)$ denote the weight vector of all evaluation criteria at stage $t$, then, the satisfaction degree $O_{ij}$ can be obtained from below.

$$O_{ij} = \sum_{t=1}^{T} \sum_{h=1}^{q} \overline{v_{ij(h)}^t \omega_{D(h)}^t} \tag{5}$$

### 33.3.4 *Construction of the Two-Sided Matching Optimization Model*

Let $x_{ij}$ denote the 0–1 variable. When $x_{ij} = 1$, it indicates that the demand individual ai and the supply individual $b_j$ form a matching relationship; when $x_{ij} = 0$, it indicates that the two have not formed a matching relationship. In the matching of supply and demand, both the demand and the supply individuals hope to cooperate with the one from the other side with the highest satisfaction. Therefore, based on the principle of maximizing the satisfaction of both sides, the model is established to select the optimal matching combinations.

## 33.4 Illustrative Examples

In order to prove the feasibility and effectiveness of the proposed method, this section sets up two experiments to verify. The first experiment uses a simple numerical example to verify the feasibility of the proposed dynamic reference point setting method. The second experiment constructs a simulation numerical example based on the supply and demand relationship between the government projects and the investment companies.

### 33.4.1 *Simulation Numerical Example*

Due to the lack of a suitable data set, in order to verify the effectiveness of the proposed method, we constructed a numerical simulation example as an experimental data. There are multiple government projects in a certain area that require additional investment from enterprises, so the government commissions a third-party intermediary to provide consulting and bidding services. Let $A$ denote the demand set, which is the project side; let $B$ denote the supply set, which is the investment enterprise side. Considering the pressure of project management, one project can only accept investment from one company. In order to reduce the cooperation risk between the project and the investment company, and improve the profitability and matching efficiency of both sides, the intermediary conducted a survey on the actual operation status of five government projects ($a_1, a_2, a_3, a_4, a_5$) and four pre-invested companies ($b_1, b_2, b_3, b_4$) *n* the past three years. In addition, the intermediary invites experts from all he project sides and investment companies to score the performance of multiple valuation criteria in recent three years, so as to help both sides form reasonable matching results. Here, to make full use of the expert preference information, the value range of the expert's score is specified as any number of intervals between [0, 100], and the higher the value, the more satisfied the expert is with the criteria. For the government project side, they evaluate the investment companies according to the four benefit evaluation criteria of technical ability ($d_1$), professional qualification ($d_2$), financial strength ($d_3$), and management experience ($d_4$). We show the obtained interval number evaluation matrix results in Tables 4.3–4.5. For investment companies, they evaluate government projects based on five benefit evaluation criteria: return rate ($c_1$), project prospect ($c_2$), project size ($c_3$), reputation ($c_4$), and capital growth rate ($c_5$).

We explain the specific steps of the method in detail by using examples of project investment. According to 3.1, we can get the initial reference points and dynamic reference points of all the evaluation criteria of the demand side to the supply side, as shown in Tables 33.1 and 33.2.

Table 33.1 shows the initial reference points we set. By comparing Tables 33.1 and 33.2, we can see that at some stages, as the stage development speed of the evaluation criteria is higher or lower than the average development speed, its reference point has increased or decreased to some extent. And according to Eq. 2, its reference point becomes smaller, so that when $t = 2$, the prospect value of the demand side has therefore get promoted. When the evaluation criteria $d1$ is in the third stage, the development speed of the stage is lower than the average development speed, so the satisfaction of the decision-makers with respect to the development of this stage decreases, the corresponding reference point becomes larger, and the prospect

**Table 33.1** Initial reference points for demand side evaluation of supply side

| Stage | $d_1$ | $d_2$ | $d_3$ | $d_4$ |
|---|---|---|---|---|
| $t = 1$ | [67.40, 72.15] | [62.20, 67.80] | [76.40, 82.10] | [78.15, 84.15] |
| $t = 2$ | [78.35, 84.75] | [80.20, 84.05] | [82.70, 87.75] | [81.05, 87.15] |
| $t = 3$ | [87.75, 93.30] | [87.40, 93.15] | [87.20, 92.35] | [78.95, 84.85] |

**Table 33.2** Dynamic reference points for demand side evaluation of supply side

| Stage | $d_1$ | $d_2$ | $d_3$ | $d_4$ |
|---|---|---|---|---|
| $t = 1$ | [67.40, 72.15] | [62.20, 67.80] | [76.40, 82.10] | [78.15, 84.15] |
| $t = 2$ | [63.57, 68.76] | [31.91, 33.45] | [75.31, 79.91] | [63.70, 68.49] |
| $t = 3$ | [102.64, 109.14] | [136.31, 145.28] | [94.91, 100.51] | [95.89, 103.05] |

**Table 33.3** Initial reference points for supply side evaluation of demand side

| Stage | $c_1$ | $c_2$ | $c_3$ | $c_4$ | $c_5$ |
|---|---|---|---|---|---|
| $t = 1$ | [68.10, 75.15] | [60.45, 65.55] | [89.05, 94.60] | [76.05, 80.45] | [75.55, 81.15] |
| $t = 2$ | [75.10, 81.75] | [64.20, 68.65] | [75.90, 82.50] | [80.75, 86.85] | [82.65, 87.30] |
| $t = 3$ | [87.10, 92.85] | [66.05, 71.95] | [67.50, 72.90] | [79.45, 84.65] | [87.65, 94.10] |

**Table 33.4** Dynamic reference points for supply side evaluation of demand side

| Stage | $c_1$ | $c_2$ | $c_3$ | $c_4$ | $c_5$ |
|---|---|---|---|---|---|
| $t = 1$ | [68.10,75.15] | [60.45, 65.55] | [89.05,94.60] | [76.05, 80.45] | [75.55, 81.15] |
| $t = 2$ | [85.45, 93.02] | [60.82, 65.03] | [82.72,89.91] | [52.97, 56.97] | [77.69, 82.06] |
| $t = 3$ | [75.19, 80.15] | [69.33, 75.52] | [60.90,65.78] | [105.44,112.34] | [92.54, 99.35] |

value of this stage also decreases. It should be noted that since the first stage is the base stage for calculating the stage development speed and the average development speed, its reference point cannot and should not be dynamically adjusted by the above method.

Correspondingly, we can obtain the initial reference point and the updated dynamic reference point of each evaluation criterion of the supply side to the demand side, as shown in Tables 33.3 and 33.4. By comparing them, we can see the development and changes of various evaluation criteria at different stages.

Then calculate the overall satisfaction between the two sides. We believe that the weights of evaluation criteria at different stages will not change. Among them, the weight vector of each evaluation criterion of a government project is $W_D = (0.36, 0.15, 0.35, 0.14)$, and the weight vector of each evaluation criterion of an investment company is $W_C = (0.42, 0.16, 0.14, 0.17, 0.11)$. Through Eqs. 3–5, we can obtain the satisfaction of each demand individual to the supply individual and the satisfaction of each supply individual to the demand individual, as shown in Tables 33.5 and 33.6.

Therefore, based on the overall satisfaction between the sides in Tables 33.5 and 33.6, we establish the following multi-objective model.

You can get the results:$x_{12} = 1$, $x_{24} = 1$, $x_{33} = 1$, $x_{51} = 1$, and the rest are 0. In other words, the final matching results in this simulation example are $a_1 \leftrightarrow b_2$

**Table 33.5** Demand side's satisfaction degree on supply side

| – | $b_1$ | $b_2$ | $b_3$ | $b_4$ |
|---|---|---|---|---|
| $a_1$ | −0.582 | 0.132 | −0.221 | 0.776 |
| $a_2$ | −0.422 | 0.003 | −1.051 | −0.296 |
| $a_3$ | −0.142 | −0.928 | 0.435 | −0.276 |
| $a_4$ | −0.047 | −1.074 | −0.950 | −0.745 |
| $a_5$ | −0.223 | −0.317 | −0.172 | −0.470 |

**Table 33.6** Supply side's satisfaction degree on demand side

| – | $a_1$ | $a_2$ | $a_3$ | $a_4$ | $a_5$ |
|---|---|---|---|---|---|
| $b_1$ | −0.564 | 0.050 | −0.104 | −0.281 | −0.033 |
| $b_2$ | −0.202 | −1.043 | −0.300 | −0.408 | −0.493 |
| $b_3$ | −0.467 | −0.346 | 0.506 | 0.004 | −0.845 |
| $b_4$ | −0.800 | 0.283 | −0.582 | −1.397 | −0.532 |

$a_2 \leftrightarrow b_4, a_3 \leftrightarrow b_3, a_5 \leftrightarrow b_1$, the rest of the combinations did not meet the satisfaction needs of both sides and did not form a matching combination. It is worth noting that due to the one-to-one matching relationship between the two sides, the government project $a_4$ failed to match any investment company in this tender. In order to give the project and investment companies more choices, we introduce two variables, $\xi$ and $\eta$, and convert the above multi-objective model into a single-objective model:

$$\max Z_3 = \xi Z_1 + \eta Z_2 = \xi \sum_{i=1}^{5}\sum_{j=1}^{4} O_{ij}x_{ij} + \eta \sum_{i=1}^{5}\sum_{j=1}^{4} O'_{ij}x_{ij}$$

where the variable $\xi \in [0, 1]$ indicates the importance of the project side in the matching relationship. If the intermediary pays more attention to the interests of the project side, then $\xi > \eta$; The variable $\eta \in [0, 1]$ indicates the importance of the investing company in the matching relationship. If the intermediary pays more attention to the interests of the investing company, then $\eta > \xi$. And we stipulate $\xi + \eta = 1$, and if $\xi = \eta = 0.5$, it means that both sides have the same importance.

From Table 33.7, we can see the effect of different values of $\xi$ and $\eta$ variables on the matching results. When matching decision-makers (intermediaries) need to

**Table 33.7** Sensitivity analysis

| $\xi$ | $\eta$ | Matching results | Recommended matching pairs |
|---|---|---|---|
| 0 | 1.0 | $x_{12} = 1, x_{24} = 1, x_{33} = 1, x_{51} = 1$ | $a_1 \leftrightarrow b_2, a_2 \leftrightarrow b_4, a_3 \leftrightarrow b_3, a_5 \leftrightarrow b_1$ |
| 0.1 | 0.9 | $x_{12} = 1, x_{24} = 1, x_{33} = 1, x_{51} = 1$ | $a_1 \leftrightarrow b_2, a_2 \leftrightarrow b_4, a_3 \leftrightarrow b_3, a_5 \leftrightarrow b_1$ |
| 0.2 | 0.8 | $x_{12} = 1, x_{24} = 1, x_{33} = 1, x_{51} = 1$ | $a_1 \leftrightarrow b_2, a_2 \leftrightarrow b_4, a_3 \leftrightarrow b_3, a_5 \leftrightarrow b_1$ |
| 0.3 | 0.7 | $x_{12} = 1, x_{24} = 1, x_{33} = 1, x_{51} = 1$ | $a_1 \leftrightarrow b_2, a_2 \leftrightarrow b_4, a_3 \leftrightarrow b_3, a_5 \leftrightarrow b_1$ |
| 0.4 | 0.6 | $x_{12} = 1, x_{24} = 1, x_{33} = 1, x_{51} = 1$ | $a_1 \leftrightarrow b_2, a_2 \leftrightarrow b_4, a_3 \leftrightarrow b_3, a_5 \leftrightarrow b_1$ |
| 0.5 | 0.5 | $x_{12} = 1, x_{24} = 1, x_{33} = 1, x_{51} = 1$ | $a_1 \leftrightarrow b_2, a_2 \leftrightarrow b_4, a_3 \leftrightarrow b_3, a_5 \leftrightarrow b_1$ |
| 0.6 | 0.4 | $x_{12} = 1, x_{24} = 1, x_{33} = 1, x_{41} = 1$ | $a_1 \leftrightarrow b_2, a_2 \leftrightarrow b_4, a_3 \leftrightarrow b_3, a_4 \leftrightarrow b_1$ |
| 0.7 | 0.3 | $x_{14} = 1, x_{22} = 1, x_{33} = 1, x_{41} = 1$ | $a_1 \leftrightarrow b_4, a_2 \leftrightarrow b_2, a_3 \leftrightarrow b_3, a_4 \leftrightarrow b_1$ |
| 0.8 | 0.2 | $x_{14} = 1, x_{22} = 1, x_{33} = 1, x_{41} = 1$ | $a_1 \leftrightarrow b_4, a_2 \leftrightarrow b_2, a_3 \leftrightarrow b_3, a_4 \leftrightarrow b_1$ |
| 0.9 | 0.1 | $x_{14} = 1, x_{22} = 1, x_{33} = 1, x_{41} = 1$ | $a_1 \leftrightarrow b_4, a_2 \leftrightarrow b_2, a_3 \leftrightarrow b_3, a_4 \leftrightarrow b_1$ |
| 1.0 | 0 | $x_{14} = 1, x_{22} = 1, x_{33} = 1, x_{41} = 1$ | $a_1 \leftrightarrow b_4, a_2 \leftrightarrow b_2, a_3 \leftrightarrow b_3, a_4 \leftrightarrow b_1$ |

match according to their different focuses, we can provide them with a variety of matching results through the setting of such variables and sensitivity analysis.

## 33.5 Conclusions

The most of existing two-sided (bilateral) matching methods are based on the single-stage static evaluation information without multi-stage dynamic characteristic. This paper addressed the problem of two-sided (bilateral) matching based on multi-stage dynamic reference points in the view of prospect theory. Using multi-stage evaluation information to approximate the characteristics of time series, the reference points of different stages are dynamically set by comparing the average development speed with the speed of the stage. And, calculating the prospect value of the evaluation criterion in different periods. Then, by establishing a multi-objective model to solve the optimal stable matching relationship between the two sides. Finally, the practicality and validity of the proposed method are verified by simulation examples.

## References

1. Fan, Z.P., Zhang, X., Chen, F.D., Liu, Y.: Multiple attribute decision making considering aspiration-levels: a method based on prospect theory. Comput. Ind. Eng. **65**(2), 341–350 (2013)
2. Gale, D., Shapley, L.S.: College admissions and the stability of marriage. Am. Math. Monthly **69**(1), 9–15 (1962)
3. Kai-kai, Y., Jian-jun, Z., Ming, Z., He-hua, W.: Analysis on the main manufacture-supplier cooperative strategies of airliner based on dynamic games of incomplete information. Chin. J. Manage. Sci. **025**(5), 125–134 (2017)
4. Tversky, A., Kahneman, D.: Prospect theory: an analysis of decision under risk. Econometrica **47**(2), 263–291 (1979)
5. Tversky, A., Kahneman, D.: Advances in prospect theory: cumulative representation of uncertainty. J. Risk Uncertain. **5**(4), 297–323 (1992)
6. Wetzstein, A., Hartmann, E., Benton Jr., W., Hohenstein, N.O.: A systematic assessment of supplier selection literature–state-of-the-art and future scope. Int. J. Prod. Econ. **182**, 304–323 (2016)

# Chapter 34
# A Food Safety Traceability System Based on Blockchain

**Yao Juan, Tian Fang, and Zhang Cheng**

**Abstract** Food safety traceability is an important method used to guarantee food safety, but the traditional data center-based food safety traceability system is not of strong data security. A blockchain-based food safety traceability system was constructed in this study and divided into four layers using a hyperledger architecture. It is verified through a test that the traceability system with blockchain technology can effectively prevent nodes from tampering data, so the data becomes safer and more credible.

## 34.1 Background and Significance

### *34.1.1 Background*

With the continuous improvement of living standard, more and more extensive attention has been paid to food safety. Scientific and technological progresses have considerably promoted the development of the food industry, and food becomes more diversified and food processing becomes more complicated. Food safety is the foundation underlying all efforts to build a country. However, food safety accidents have frequently appeared in the recent decades, and food fraud has become one of global difficulties. Even those famous food production and processing enterprises are found with extreme safety problems, for example, 2008 Sanlu "melamine milk" event, 2011 German Dioxin "toxic fodder" event and 2013 event of Clostridium botulinum detected out of concentrated whey protein powder in New Zealand Fonterra Co-operative Group, etc. So far, food safety problems have become very severe in China. The public are full of worries when selecting food. Raw food materials mainly come from agricultural and sideline products which need to go through various steps including plantation, storage, processing, packaging, transportation and sales, where

Y. Juan · T. Fang · Z. Cheng (✉)
College of Informatics, Huazhong Agricultural University, Wuhan, Hubei, China
e-mail: sameboy@mail.hzau.edu.cn

 
V. Chang et al. (eds.), *Modern Industrial IoT, Big Data and Supply Chain*,
Smart Innovation, Systems and Technologies 218,
https://doi.org/10.1007/978-981-33-6141-6_34

problems may exist in any link. Constructing a food traceability system is one of important measures of improving food safety. Taking food certification and supply chain tracking as measures to guarantee food safety can rapidly search and discover food pollution sources.

With the emergence and high-speed development of Bitcoin in recent years, blockchain starts being brought to the forefront. Considered as the cradle of blockchain, Bitcoin is a distributed P2P network 4, which is maintained by miners through mining, and new blocks are generated for trade information ledger. A blockchain system can be divided into data layer, network layer, consensus layer, incentive layer, contract layer, and application layer 5. The birth of blockchain technology has broken through cognitive constraints for traditional trade, while the emergence of Bitcoin has changed the traditional mode of digital currency transaction and exerted a far-reaching effect on the development of distributed ledger technology. The important characteristic—smart contract—is introduced into Ethernet program, and virtual machine which can complete scripting language via Turing is developed so that the functions of digital currency transaction become more flexible and complicated. Meanwhile, thanks to the birth of smart contract, blockchain is applicable to other categories besides currency transaction, for example, financial service, ownership management of credit investigation, resource sharing, trade management, IoT, etc. In addition, blockchain technology has attracted high attention from governmental organizations and capital market.

The domestic food safety quality has encountered a severe trust crisis. Reconstructing a food safety governance pattern so as to further realize trustworthy wide-range monitoring is the only solution to the current trust crisis. With the high-speed development of blockchain technology, blockchain has attracted high attention from the business circles and academic circles. Because of real-time data sharing and tamper-proof characteristics of blockchain technology, applying blockchain to the food field, taking it as an underlying technology to develop a food safety traceability system, and constructing new-type traceability logic using distributed ledger combined with smart contract will greatly improve trustworthiness, safety and effectiveness of food traceability.

### *34.1.2 Research Significance*

Traceability technology has been brought forward for a long time since 2003. Government organs have exert all their energies to advance the construction of food traceability system, but the final result is not ideal. As a result, problems concerned with food safety have occurred frequently all the time, principally because the traditional food traceability data are stored in a centralized way and are easily influenced by human factors. The person in charge of each process in the traceability process stores the data and information in a database, and no one can access to the detailed information except himself and the system provider, thus the unreliability of personal stored data and the risk of being tampered with will tremendously increase. Hence, it is not quite easy for consumers and mass users to accept the traceability system, and thus they will not recognize and use it. Nevertheless, the decentralized (also called multi-centralized) distributed accounting method of blockchain technology can effectively address this problem. If the food safety traceability system uses blockchain technology, it can enable people who doubt each other to build trust and address the trust crisis with high efficiency and at low cost.

Since 2017, the Internet industry and traditional food enterprises have been aiming at the development potential of food traceability system in the blockchain, such as TACChain, Vechain, and Wo'er food traceability system. Traceability chain is a common chain, which is designed to address the problem of reliable traceability of products in production, processing, sales and other processes, thus addressing the trust problem of consumers. Apart from that, on December 14, 2017, Wal-Mart, IBM, JD.com and Tsinghua jointly set up China's first "Alliance for Traceability of Safe Food Block Chains".

Despite that the above traceability systems all regard blockchain as the basic technology, it can be learned from the published report that they are still under research and development, and the others are also different to some extent in application scenarios, underlying implementation technologies, research and development, and operation modes. In the first place, the traceability chain TAC perfectly combines the superiority of Ethernet and Bitcoin, VeChain uses Ethernet technology, while Wal-Mart's food traceability system adopts Hyperledger technology. Blockchain technology is recognized as the core technology that can push forward human development after steam engine, power system and Internet. In the future, mankind will initiate an era in which artificial intelligence, Internet of things, and blockchain are dominant technologies. If the three are combined, enormous value will be created as a result. The blockchain can be regarded as an object to build trust, thus changing the way people interact with information. The data stored in the block is transparent while the user's identity is encrypted, which can only be used after authorization, thus ensuring the security of information and the privacy of users. Currently, the market demand in anti-counterfeiting traceability has become more urgent with the increase of anti-counterfeiting efforts by the state and consumers' unbelief in food. Blockchain technology boasts strong superiority in anti-counterfeiting traceability system, like decentralization, data tamper-proof and unique timestamp, etc., which will be more effective than other technologies. To sum up, if the blockchain is combined with traceability, it will give rise to great impetus to the food field.

## 34.2 Principle of Food Safety Traceability System Based on Blockchain

### *34.2.1 Food Safety Traceability*

Currently, there is no uniform definition of food traceability on a global scale, and most of the definitions only impose common constraints on the requirements of food traceability system. For instance, the food traceability system should complete the quality inspection work of the health management department in an efficient manner, which renders it impossible to enhance the self-control capability of food production, efficiently analyze various factors affecting the product quality and then play a good monitoring role on food. The system needs to contain public information of the entire supply chain from production to sales. Apart from that, through the food traceability system, enterprises can tremendously reduce the cost of recalling food and effectively save losses and build consumers' trust. The European Commission defines food traceability as the tracking process of substances that may be used as food ingredients in the processes of planting, processing, and sales. The International Organization for Standardization defines it as the capability to effectively trace back and track products. Traceback refers to the process of identifying the detailed information of one or a batch of food from the product forward in sequence, and tracking refers to the process of recording and collecting product information from the seed backward. Food traceability system can identify the source of food, provide all the information from planting to sales, and identify the product identity in each stage of the supply chain. It is a whole composed of a series of traceability processes such as product information collection, storage and quality inspection.

To make the traceability system work, the priority work is how to code and mark food. Moreover, the quality and safety monitoring of products and origin inspection are vital links in food production and sales, and are also the issues that the Food and Agriculture Organization should attach importance to first. Every process of food from production to distribution needs to receive the coding information for connecting the previous process, generate detailed information of this process, and then add relevant transaction information to be stored in the database. Building a food safety traceability system is beneficial to finding out the problem efficiently and recalling the food in question in a prompt manner, thus effectively controlling the food quality and furnishing a good food safety environment for consumers. Meanwhile, it also offers a platform for food regulatory agencies and food manufacturing enterprises to share information, which has strengthened the transparency of food information, ensured the reliability of transactions, protecting consumers' rights to know, rendering vital strategic guiding significance for establishing market credit.

### *34.2.2 Introduction to Blockchain Technology*

**Introduction to Blockchain**: Blockchain technology employs a consensus mechanism to address the double-payment problem without a centralized organization, making a major breakthrough in digital cash. When it comes to blockchain, bitcoin often comes our minds first. The core of blockchain technology appears in bitcoin, but blockchain is not equal to bitcoin. Blockchain is a structure in which blocks are connected in series based on time sequence. Transactions (i.e., operations on account books) are stored in blocks for a period of time and encrypted by cryptography technology, thus ensuring user privacy and data security, and besides, distributed account books can only be added but not deleted. Nevertheless, the blockchain is not just a database. Even though the blockchain is essentially a point-to-point distributed account book, the core problem it addresses enables participants to trust each other, principally by using a consensus mechanism to ensure the data consistency of each node.

The first block in the blockchain is called Chuangshi block, and all subsequent blocks need to contain the Hash value of the previous block header, which is a digital fingerprint gained from its secondary Hash by SHA256 algorithm. This Hash value ensures the security of transaction information. If the block information is modified, its value will also change, and the subsequent block Hash value needs to be recalculated. This work can be completed only with its extremely strong calculation force, so this workload ensures the safety of the block. Apart from that, besides Hash value, identifying blocks can also be subject to the block height. The block height of the first block is 0, and then the height of the block generated later is one position higher than that of the previous block.

**Consensus Mechanism**: Consensus mechanism can be called to be the soul of blockchain technology. The core problem to be addressed in distributed systems is to ensure the consistency of information of each node. As there will be delays in P2P networks, the order in which each node observes transactions may be different to some extent. Thus, consensus mechanism can be said to be a kind of algorithms for reaching a consensus on the order of multiple transactions. Currently, there exist numerous consensus algorithms, but they are not perfect yet. All algorithms have their advantages and disadvantages. The algorithms are improving step by step, and some consensus algorithms are only born to address some specific problems. Consensus algorithms can be generally classified into two categories, namely Crash Fault Tolerance (CFT) and Non-Crash (Byzantine) Fault Tolerance (BFT). The former refers to the fact that the failure of a node may be caused by network delay or machine shutdown (downtime), and will not falsify information, including Paxos, RAFT, etc. It is relatively easy to solve this situation, while the latter is used to address the problem of malicious nodes, etc. Currently, several common consensus algorithms are roughly as below:

*PoW*: Namely the proof mechanism of workload; the typical application of PoW is bitcoin. Just like the literal meaning, the greater the workload, the greater the revenue.

*PoS*: Namely the proof mechanism of rights and interests; the typical application of PoS is peercoin, which is similar to the share certificate and voting system. The probability of acquiring the right to book accounts is determined by calculating the percentage of currency held and the holding time. It can also be called the share certificate algorithm.

*PBFT*: Namely practical Byzantine fault-tolerant algorithm; the Byzantine Generals Problem is to model the real problem and deal with the situation where a few nodes do evil. The algorithm idea is to copy the duplicate of the state machine, which is based on calculation and has no token reward. It furnishes $(n - 1)/3$ fault tolerance while ensuring safety.

### 34.2.3 Hyperledger Fabric

Fabric, as a representative early project of Hyperledger community, furnishes a distributed account book platform for alliance category for the first time, and realizes the complete function of authority management. Hence, the most obvious difference between Fabric and other projects lies in private ownership and license. Unlike allowing participants with unknown identities to join the network without license in the common chain, Fabric register all members using Membership Service Provider (MSP), avoiding the need to ensure the validity of transactions through the proof mechanism of workload, thus shortening the calculation cycle and realizing effective expansion to meet all types of application requirements in the industry. MSP furnishes a function of identity registration, allowing users to apply for a transaction certificate from the transaction authentication center after receiving the certificate. Only in this way can transactions be carried out and the information of transaction certificate is permanently stored on the blockchain, thus ensuring the security of the network.

The core components in the Fabric include the following modules: Ledger, Block, Node, Channel, Chain-code, Consensus Mechanism, Authority Management, CA (Certificate Authority, Certification Center and Digital Certificate Issuing System), etc. Various resources can be configured and managed via the Fabric network. Therein, the FabricCA project furnishes certificate service for the network based on PKI system, adopting the typical CS architecture. The CA server furnishes a complete function of certificate management, which can serve other CAs and can also be used to issue user certificates, including a single RootCert and an ECert (Enrollment Certificate, owned for a long term and for identity authentication), as well as a large number of TCerts (Transaction Certificate, generated during each transaction, used for signature of the transaction), and TLSCert used for TLS transmission. CA clients can interact with the server with fabric-ca-client commands, and can log-in to access to ECert, register users, issue certificates, revoke certificates, etc.

## 34.3 Food Traceability System Design

### *34.3.1 Hierarchical Structure of Food Safety Traceability System*

The food traceability system architecture (Fig. 34.1) is mainly divided into four layers: Data layer, consensus layer, contract layer, and application from bottom up. The concrete functions of each layer are as follows:

**Data layer**: This layer provides maintenance for distributed ledger and World State, and blockchain functions like product life-cycle management for smart contract, and it can realize data tamper-proof and Chain-code business logic. In addition, it provides member registration and cancellation via fabric-ca. It implements distributed storage of data acquired by digital means and data encryption.

**Consensus layer**: This layer encapsulates various consensus mechanisms between network nodes and it is a guarantee for normal operation of distributed network. The Practical Byzantine Fault Tolerance (PBFT) is used at consensus layer to ensure that data will not be tampered by malicious nodes during network communication and storage process and that a consensus can be still reached even under unknown malicious nodes in the network. For a protocol technology, it can become mature without relying on digital currency, thus reducing the waste of resources caused by mining or in reward mechanism.

**Contract layer**: Smart contract at contract layer is implemented via Chain-code, including Chain-code invoking functions like data release, data signing/receiving,

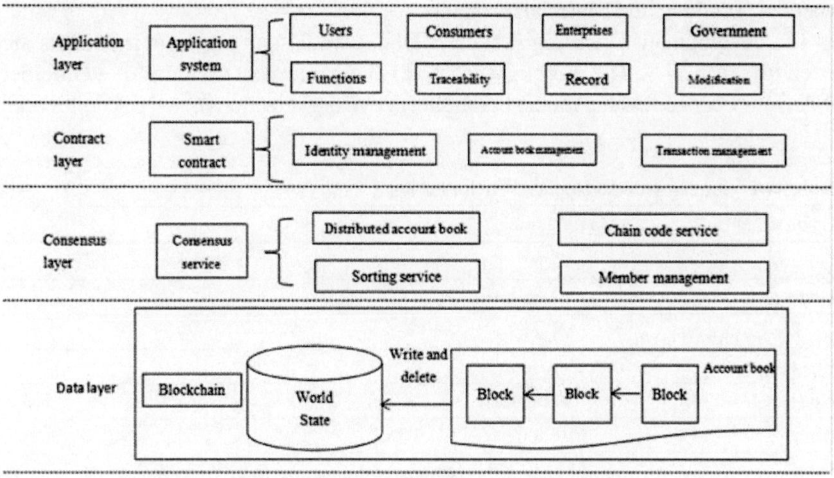


**Fig. 34.1** Hierarchical structure of blockchain-based food safety traceability system

and data transfer. Chain-code inquiry includes inquiring relevant personnel information of agricultural products, and information of the agricultural products, etc.

**Application layer**: Web application is established using express frame. It provides interface operation of user interaction, including user operation and business operation functions. Common users can perform registering and log-in operations, but they can only inquire food information and realize authenticity traceability (forward traceability) and safety tracking and safety traceability (reverse traceability).

### *34.3.2 Implementation of Food Safety Traceability System*

**Development environment**: Since channels are instances created by sorting services to isolate transactions on different chains, nodes added to different channels will store different data. As far as this system is concerned, all data do not need privacy, so only one channel is needed. Nevertheless, it can be learned from Sect. 34.3.1 that the food traceability process involves multiple links. The alliance chain environment of Fabric is tested by two organizations (Org 1 and Org 2), and each organization is responsible for different links. Thus, three Chain-codes are designed, namely myccfoodall, myccfoodone, and myccfoodtwo. Therein, myccfoodall is installed on all participating nodes for operating User and Person; myccfoodone is installed only on nodes in Org 1 and is used to operate Drug, Seed, Seedling, Seedsoakdrug, Seedlingspraydrug, Plant, Plantusedrug, Seed2seedling and Seedling2plant; myccfoodtwo is installed only on nodes in Org2 and is used to operate Input, Warehouse, Warehouse2Feed, Plant2input, Input2warehouse, Feed2product, Feed, Material, Product, and Material2product.

The development operating system is Ubuntu16.04 and other environments are listed in Table 34.1. The established Fabric1.0 environment includes one Orderder node, four Peer nodes (separately belonging to two organizations), two CA nodes (one

**Table 34.1** Summary of development environments

| Environment name | Version |
|---|---|
| Go | V2.0 |
| Docker | V18.09.03 |
| Docker-Compose | V1.24.1 |
| Python-pip | V9.0.3 |
| Node | V10.15.3 |
| Npm | V6.8.0 |
| Fabric | https://github.com/hyperledger/fabric/tree/release-1.4 |
| Fabric CA | https://github.com/hyperledger/fabric-ca/tree/release-1.4 |

in each organization), and one client-side node. The needed Fabric Docker images include fabric-peer, fabric-orderer, fabric-ca, fabric-tools and fabric-couchdb, all of which can be downloaded using download-dockerimages.sh in source code. Libtool and libltdl-dev libraries need to be installed.

**Sorting Service Based on Kafka**: The default consensus of Fabric is Solo, that is, single-node consensus. In actual production environment, Kafka cluster needs to be built for multi-node consensus. Kafka can orderly manage information and enable participating copies to maintain data consistency, and kafka cluster needs to be managed by zookeeper. To satisfy the fault tolerance of crash, at least 4 sets of kafka clusters need to be built, and one set can be fault-tolerant at this time. The number of zookeeper clusters needs to be an odd number exceeding 1 in order to prevent split-brain and single-node failures. Clusters are started based on dependency relation, i.e. zookeeper, kafka and Orderer clusters in sequence. In this environment, Client can send the transaction to any Orderer node to complete the global sorting. In the stand-alone test environment, there are 3 zookeeper nodes, 4 kafka nodes, 3 orderer nodes, 4 peer nodes, and 1 cli node, and the Docker container runs after completion.

**Traceability implementation based on block chain**: Languages used to compile Chain-code mainly include Go and Java. As Fabric is developed by Go itself, Chain-code is developed using Go under most circumstances. The Chain-code compiled by Go language needs to contain a structural body named Simple Chain-code as well as three basic functions: main, Init, and Invoke, where main is Chain-code entry function, and shim. Chain-code StubInterface parameter in other functions contain many interface methods, mainly including Chain-code invoked parameter parsing, trade information parsing, operation status data, invoking of other Chain-codes, event handling and auxiliary operation, etc.

All Chain-codes in this system can invoke other Chain-codes via InvokeChaincode (chaincodeName string, args [][]byte, channel string), which can realize modularization programming. []string needs to be converted into [][]byte for the sake of invoking. If the invoked Chain-code and invoking Chain-code are not on the same chain, a new trade simulator will be generated, Response executed by invoking will be returned to the Chain-code, but no read/write set will be added into the trade. In other words, Chain-codes which invoke different chains can only inquire status data but cannot be written in.

Chain structure and Channel are one-to-one correspondence. Channel is a vital feature of Fabric, which is quite similar to the concept of topic in message queue. The information can only be received by the Peer that joins it. If a Peer joins several channels, then the Peer node maintains the information of several chains. Parameters required for API access are Peer number and Channel name.

There is an initLedger branch method in all of the three Chain-codes of this system, and this method is used to initialize ledgers and save some data. As non-relational database (NoSQL) is used by the ledgers, data inquiry becomes troublesome and compound key needs compiling.

## 34.4 Conclusion

The features of blockchain technology such as decentralization, collective maintenance, and information transparency have provided a new idea for solving problems existing in the agricultural products traceability system, including low information reliability and poor credibility. Each organization in the traceability system constructed in this study takes charge of different food links and maintain their different blockchain nodes, which jointly constitute a blockchain network. Fabric is used as underlying technology to verify type-in data integrity and the consistency of ledgers is guaranteed using the consensus mechanism in the network. Any data tampering phenomenon by nodes can be rapidly discovered through inquiry and verification. Furthermore, this system can be matched with other supervision and management means so that organizations participating in the traceability system cannot tamper data. The deployment of the traceability system is transformed from traditional centralized management mode into multi-center co-management, which contributes to improvement of authenticity and safety of food information.

**Acknowledgements** Project 2662018YJ016 supported by the Fundamental Research Funds for the Central Universities.

## References

1. Jin, S., Zhang, Y., Xu, Y.: Amount of information and the willingness of consumers to pay for food traceability in China. Food Control **77**, 163–170 (2017)
2. Kim, Y.G., Woo, E.: Consumer acceptance of a quick response (QR) code for the food traceability system: application of an extended technology acceptance model (TAM). Food Res. Int. **85**, 266–272 (2016)
3. Wang, J., Yue, H., Zhou, Z.: An improved traceability system for food quality assurance and evaluation based on fuzzy classification and neural network. Food Control **79**, 363–370 (2017)
4. Huckle, S., Bhattacharva, R., White, M., et al.: Internet of things, blockchain and shared economy applications. Procedia Comput. Sci. **98**, 461–466 (2016)
5. Ken, A.: Digital blockchain networks appear to be following Metcalfe's Law. Electron. Commer. Res. Appl. **24**, 23–29 (2017)

# Chapter 35
# Application of Blockchain Technology in Management Accounting: Based on the Perspective of Frontier Literature Review

**Yongyi Cai and Xuefen Xiao**

**Abstract** With the rise of technologies such as big data, cloud computing, and blockchain, financial accounting is in the stage of lean management. The application and transformation of management accounting tools and methods are undoubtedly the only way to achieve lean management. In this context, combing and reviewing relevant research literature are of great significance. Taking the core technology and infrastructure of the blockchain as the entry point, based on the cutting-edge hot literature in recent years, the main results and progress of the research on the application of blockchain technology in management accounting are summarized. It is hoped that it can provide important theoretical guidance for future research points the way.

## 35.1 Introduction

In recent years, blockchain technology is in continuous development and evolution, various new applications emerge in endlessly, and the maturity is gradually improved. At the same time, inspired by the application of blockchain, various fields try to reshape the operation logic of traditional industries from the perspective of blockchain, so as to let the development of the industry find new momentum. It is generally believed that financial accounting will be replaced by artificial intelligence, and accountants will be transformed into talents who can provide financial information for effective operation and optimal decision making, and can help middle and grass-roots managers to make decisions, plan, and control their daily business activities. Management accounting needs to use a large number of forecast financial data to describe the future business situation, and blockchain technology can provide real-time data and clearer cash flow diagram, which greatly promote the decision making of management, the formulation of company strategy, and operation.

Y. Cai · X. Xiao (✉)
Neusoft Institute Guangdong, Guangdong, China
e-mail: xiaoxuefen@nuit.edu.cn



V. Chang et al. (eds.), *Modern Industrial IoT, Big Data and Supply Chain*,
Smart Innovation, Systems and Technologies 218,
https://doi.org/10.1007/978-981-33-6141-6_35

## 35.2 Entry Point of Blockchain Core Technology in Management Accounting Application

### *35.2.1 Distributed Ledger*

Blockchain, also known as distributed ledger technology, provides a consistent verification mechanism through a computer network to facilitate peer-to-peer transactions, without the need for intermediaries or centralized authorization to update and maintain information generated by transactions. Distributed ledgers can be shared and synchronized across multiple sites, institutions or geographic locations, and across networks.

#### 35.2.1.1 Build a Cross-Organization Trust System and Transform Management Accounting Thinking

The traditional management accounting information system is usually limited to the enterprise itself, and the data is often stored in different systems, which is not conducive to mutual integration and communication. Therefore, it cannot provide managers with sufficient decision-making information about whether to form cross-organizational cooperative relationship with other enterprises. The advantages of blockchain, such as decentralization, tamper proof and distributed data storage, are conducive to the formation of a trust system across organizations and departments, the sharing and common governance of economic activities, the strengthening of internal management of enterprises, and the improvement of economic benefits [1]. In the cross-organizational cooperation environment, the management needs not only the data of the enterprise but also the reference data of other enterprises in order to identify and analyze the business efficiency of the enterprise. In the blockchain network, the multi-party participants of a business are closely related, and the activities and resources of all participants should be taken into account. In addition, in the cross-organizational management accounting network embedded in blockchain, horizontal value chain management is very necessary to compare and analyze the production of similar products between the enterprise and competitors in the same industry. Based on the open and transparent network environment brought by blockchain technology, enterprises have the opportunity to obtain the production information of different competitors through the chain, so as to overcome the production constraints and determine the optimal production scheme. Enterprises are no longer limited to rely on their own financial data to judge their operating conditions, but more comprehensively consider and compare the information of other blocks in the network, so as to realize the evaluation and control of their own business model, core competitiveness, and sustainable innovation ability [2].

With the emergence of a buyer's market characterized by individualized consumer needs, competition between markets has become increasingly fierce. Enterprises have to change the original "vertical integration" organizational model and establish a "horizontal integration" organizational model, namely more and more companies have formed strategic alliances based on supply chains. In order to adapt to this change in organizational mode, management accounting has also transformed from single-dimensional management to multi-dimensional management. Chen Suyun (2020) pointed out that distributed accounting technology may promote the "value creation theory" to become the mainstream management accounting goal. Each node of the blockchain system inputs, transmits, and stores transaction information at the same time, which means that the system independently provides decision-making financial data to the management. And management information, "decision theory" and "information theory" may no longer be suitable as the goal of management accounting. In the blockchain technology environment, the concept of resources may not be limited to practical assets, intangible assets, etc., and may need to be redefined. The "benefit theory" view needs to be improved. Under the blockchain technology, the generation and transmission of information are synchronized, and the division of execution, decision making, and strategic stages has also become blurred, which blurs the existence basis of the "stage multi-objective theory" [3].

#### 35.2.1.2 Improve Decision-Making Efficiency and Speed up Capital Turnover

In order to maximize the economic value of an enterprise, management accounting needs to reflect relevant information in a timely manner and make targeted decisions. The task-based cost management in the future accounting industry is a new type of centralized cost management that can improve the economic efficiency of enterprises and enhance the value of customers. This method can be directly collected according to the cost, through the dynamic reflection of time, cost, etc., through the operation motivation, and then make a cost chain analysis based on these motivations and operation costs, and provide the most effective suggestions and decision-making information for the managers of the enterprise [4]. The application value of management accounting has been improved due to blockchain technology, which is embodied in three aspects: activity cost management, value chain management, and responsibility accounting. In terms of operating cost management, the blockchain can reflect all the operating activities of the enterprise and dynamically reflect it, providing the traceability and positioning function of operating activity data. Any operating activity can be traced to the source through a chain structure. Activity-based cost management can track and dynamically reflect all business activities of the enterprise. Blockchain also has this function. Each block is connected in turn. When a block is transferred, it is a process of transferring one job to another. At the same time, it is also a process of value transfer, and finally, a blockchain is formed, which records all the data of the operating activity and can provide the traceability and positioning function of the operation activity data of the blockchain. This method can directly collect according to expenses, dynamically reflect the time, cost, etc. through operation drivers, and then make cost chain analysis based on these drivers and operation costs, and provide enterprise managers with the most effective suggestions and decision-making information [6, 7]. In addition, each node in the blockchain contains not only financial information, but also non-financial information. Users can obtain all data in the blockchain in time, thereby eliminating the risk of information asymmetry. Therefore, management accounting can use this information to redefine the sales channels of the highest value chain and the lowest value chain of the enterprise, in order to reduce the total cost and expenditure level of the enterprise, and accelerate the formation of the capital turnover currency and the core competitiveness of the enterprise [5]. In terms of responsibility accounting, based on the integrity of the blockchain's link trajectory and the characteristics of distributed accounting, the completion of each sub-point in each region can be tracked in real time, so work performance can be evaluated objectively and fairly, and costs and profits can be increased [6].

### *35.2.2 Asymmetric Encryption*

The "non-tamperable" feature of the blockchain comes from asymmetric encryption algorithms. The realization of asymmetric encryption relies on a pair of public and private keys. Each user has two keys: a public key and a private key. The two keys match each other. The public key is disclosed to the outside world, and the private key is only held by the user. When communicating, the sender uses the receivers public key to encrypt. Once the letter reaches the receiver, the public key becomes invalid and only the private key can decrypt it. This fulfills the requirement that only the sender and receiver can view the data. Therefore, asymmetric encryption technology can guarantee the privacy of information and prevent the transaction information stored on the system from being tampered with and destroyed.

#### 35.2.2.1 Contain Fraud and Reduce Supervision Costs

In the current accounting information environment, information asymmetry and corporate governance defects make motivated levels to manipulate accounting information risks within their own responsibilities, creating favorable conditions for fraudulent behavior. The "non-tamperable" of blockchain technology can fundamentally eliminate the possibility of fraud and fraud, and improve the stability and reliability of financial data. The "voluntary principle" of blockchain design creates a decentralized distributed accounting system in which everyone can participate in the entry of information. Participants in the entire network record transactions together, which can decentralize the accounting responsibilities of accountants and corporate leaders. The distributed accounting responsibility feature of the blockchain makes all participating employees of the entire network share the responsibility of writing legal, true, and complete accounting information. All employees of the company jointly implement accounting supervision to ensure the compliance and legality of economic activities [7]. Li [8] pointed out that the significance of this mechanism is that lower-level employees can supervise senior management, and lower-level companies can also supervise higher-level companies, changing the previous single supervision mode and forming a two-way, all-round supervision corporate governance model. The premise of accounting fraud is to create false transactions and build a false transaction environment. The traceability of blockchain technology makes false transactions no longer exist, ensuring the authenticity and reliability of accounting information.

#### 35.2.2.2 Optimize Audit Process and Improve Audit Efficiency

Zhang Fengyuan (2018) pointed out that the optimization of the audit process by the blockchain can exist in many aspects. For example, when the auditor performs an audit and obtains the signature and vouchers of the audited entity, there are often problems such as communication difficulties and long waiting times. And the "untamperable modification" of the blockchain can directly use the data exported by the enterprise, reducing the communication and waiting process with the financial personnel of the audited unit, compressing the audit steps, and ensuring the authenticity of the information [9]. The blockchain verifies the integrity of electronic files by using a hash function. The transaction enters the digital fingerprint into the blockchain, and the digital fingerprint can be time stamped indefinitely. Subsequently, the integrity of the file can be proved by regenerating the same fingerprint and comparing it with the fingerprint stored in the blockchain. If the fingerprints are the same, it means that the document has remained unchanged and has not been tampered with since the first submission, which ensures the authenticity of audit data and reduces the misreporting of audited entities, deliberate provision of false accounting data, and improper audits issued by external auditors. The possibility of reporting greatly improves the quality and efficiency of audit practice.

### *35.2.3 Consensus Mechanism*

When one participant wants to send information to another, other nodes in the blockchain communicate with each other using a predetermined mechanism to check whether the newly added transaction is valid. This is known as the "consensus mechanism". Different from the traditional ledger, no node has the special authority to edit or delete the transaction, so there is no transaction intermediary in the purchasing process. Each node has distributed bookkeeping, synchronous settlement between departments and timely reconciliation, saving the cost of reconciliation and consolidation. The transaction information stored in the blockchain cannot be changed or revoked unless more than 51% of the participants in the network agree to change the transaction information in subsequent transactions. Based on the trust of blockchain technology, the trust crisis and transaction cost can be reduced to the maximum extent, and the management value of accounting can be improved.

#### 35.2.3.1 Optimize Performance Evaluation Methods

The problems of management accounting are as follows: (1) The internal management system of enterprises is not perfect, lack of sound rules and regulations and incentive and restraint mechanism, thus affecting the development of management accounting work. (2) The quality level of enterprise management personnel and accounting personnel is not high, the decision-makers do not understand the importance of accounting, and pay less attention to the management accounting work, which leads to the outdated concept of enterprise accounting personnel, shallow management accounting knowledge and low application level. The current performance evaluation may be unable to record truthfully and abuse power for personal gain, resulting in the authenticity of the record in doubt. The false high or low performance appraisal affects the normal operation of the enterprise, resulting in the "talent mismatch" of managers or the inability to objectively identify the talents with high ability, which increases the management and supervision costs. The introduction of blockchain technology provides feasible ideas for the establishment and implementation of scientific performance evaluation mechanism. Decentralization and consensus mechanism make performance evaluation not only arbitrary by personnel with special authority, but also subject to the supervision of each node in the network, which greatly enhances its fairness. With the endorsement of technology, the performance evaluation of employees has higher credibility, which can effectively encourage employees to contribute to the development of the company. Based on the report issued by the system, the salary and level of managers can be determined. In the talent market, the ability of job seekers can also be effectively recognized by the company to avoid the waste of resources and promote the social labor force toward a more suitable era. The direction of demand development.

#### 35.2.3.2 Causes Short-Termism Problems

Xu Ying (2019) believes that the blockchain crosses the boundaries of organizational departments, breaks the closeness of a single value system, penetrates all value systems, and integrates and creates an unprecedented, huge value chain that can reflect the long-term economic interests of enterprises. Therefore, the introduction of blockchain can prevent managers from limiting their vision to short-term performance goals while ignoring the long-term economic interests of enterprises. However, Potekhina [10] believes that the real-time nature of blockchain will lead to automatic audit and highly transparent financial statements, that is, it will aggravate the problem of short-term doctrine, and management tries to meet the profit expectations of stakeholders on stock prices and rejects long-term projects that may be profitable.

The real-time nature of the blockchain may bring about the negative effects of short-termism in the following ways [11]: In the course of operations, companies should always pay attention to the fluctuations of financial information, and conduct comprehensive analysis, inspection, and control of the factors that cause fluctuations.

Strengthen risk prevention and pass on the causes of fluctuations to stakeholders; establish a credit risk sharing mechanism, transfer risks rationally, enhance credit risk management capabilities, and timely adopt appropriate credit risk management tools to assess, prevent and eliminate potential risk factors; Management can provide stakeholders with reports that reflect the overall long-term goals and strategies of the company, and provide communication about the strategy and long-term vision to help investors obtain the information needed to evaluate the value of the company; build the compensation and long-term management of the company Remuneration system consistent with goals and strategies to achieve long-term strategic goals and value creation.

### *35.2.4 Intelligent Contract*

Intelligent contract is a kind of computer protocol which aims to spread, verify, or execute the contract in an information-based way. If the set conditions are met, the program can be executed accurately and automatically, and the efficient peer-to-peer transaction and value transfer can be realized. It plays an important role in the establishment of inter-organizational trust mechanism. Xu Jinye (2020) believes that the intelligent contract is a contract signed for the future business, which must be signed by the future-oriented management accounting as the basis for signing the contract. The distributed account book in the blockchain is financial accounting, which can only perform the function of "fiduciary responsibility", but cannot realize the function of "decision relevance", and cannot be used as the basis for the conclusion and performance of intelligent contracts. Management accounting is the basis for the conclusion and performance of intelligent contracts [12].

### *35.2.5 Improve Contract Performance Efficiency and Save Execution Costs*

The traditional contract consumes a lot of human resources and time, and needs legal intervention to ensure its implementation, while the constraint of the intelligent contract depends on consensus, the working principle is simple, and the treaty can be executed automatically only by the code. Improve the efficiency of contract performance, and ensure that the data is safe, true, and accurate. Intelligent contract eliminate human errors and omissions, and both parties do not have to worry about the risk of counterparty breaking the contract. Jie [13] pointed out that because the data of blockchain technology is open and transparent, both sides of the transaction often check the solvency of the trading object in advance and choose enterprises with strong ability to trade. Thus, the risk of turning accounts receivable into bad debts can be reduced, and the management mode of accounts receivable will also change.

The modification of the traditional contract after signing will cause the deviation between the actual business and the economic business of the accounting records, and the accountants have to adjust the wrong accounts, so as to increase the intensity of work. Intelligent contract can avoid such situations, reduce the burden on accountants, and reduce costs and time costs. Therefore, the function of the accountant will no longer be the traditional bookkeeping, but will be transformed into an accountant who integrates new technology and management accounting thinking to realize the creation of management value.

## 35.3 The Application Conception of Blockchain Infrastructure in Management Accounting

The infrastructure of blockchain is divided into six layers, including data layer, network layer, consensus layer, incentive layer, contract layer, and application layer. Each layer completes a core function, and each layer cooperates with each other, thus realizing a decentralized trust mechanism. Hu [14] starts with the application of blockchain technology in management accounting from five aspects: data layer, network layer, consensus layer, incentive layer, and contract layer: The technical elements of the data layer play an important role in management accounting, for example, chain structure can trace the origin of data, and its application is valuable in supply chain and value chain. Timestamp provides convenience for process optimization, cost management, etc. Asymmetric encryption improves information security through double guarantee and improves the distinction between powers and responsibilities. The network layer realizes the autonomy of management accounting and improves fairness and science through networking, message dissemination protocol and data verification mechanism, while the consensus layer can be combined with management accounting performance evaluation and incentive mechanism. Establish a comprehensive evaluation system. The incentive layer can avoid mistakes caused by subjective judgment, improve transaction efficiency and decision-making quality, and intelligently program performance evaluation indicators, incentive mechanism, value chain analysis, supply chain management and other aspects to optimize benefits.

For the "application layer" that Hu Ruixue (2018) did not analyze, Du [15] pointed out that under the support of blockchain technology, the management accounting reporting system has great flexibility, which is manifested in the fact that it does not have a fixed report format and reporting time. Decision makers can generate management meetings in different fields according to their own needs and time requirements in combination with smart contract technology Report and make business decisions accordingly. This way of generating management accounting report greatly improves the relevance of management accounting information, and then improves the decision support and value creation function of management accounting information. Taking the infrastructure of blockchain technology as the main line, Du Yong (2020) rebuilt the management accounting reporting system, forming a multi-dimensional reporting system of management accounting at the strategic level, the management level and the operation level, so that all data of the enterprise can be shared among various departments, and the problems such as poor relevance of management accounting information, untimely information transmission and imperfect traceability system have been changed It improves the efficiency of decision making and control, and strengthens the value creation and decision support function of management accounting report.

Li [16], according to the functional advantages brought by blockchain infrastructure, the "four in one" accounting model architecture based on blockchain technology was constructed. Therefore, it can realize the bottom-up extension from manual business accounting function to strategic accounting function and intelligent accounting function. Finally, the transformation and upgrading of enterprise accounting from "accounting type" to "intelligence type" will be helpful for enterprise value creation, value transmission and value management. Chen and Xiang [17] proposed to build an intelligent cockpit of management accounting tools based on blockchain technology, and build different dashboards according to each generated management data block and different business units and performance objectives, including strategic management dashboards, budget management dashboards, cost management dashboards, operation management dashboards, investment and financing management dashboards, risk management dashboards, performance management dashboards, etc. By using blockchain technology to engrave the key indicators determined by various management accounting tools into the dashboard, the dynamic data and information of the enterprise in the operation process can be displayed in each dashboard in time, and the comparison of various data in operation with the scales of each dashboard is helpful for the management to make decisions and formulate the company's strategic measures to deal with and prevent risks.

Liu [18] proposed "intelligent management accounting" based on the perspective of blockchain. Through the use of advanced financial management methods and theories, with the help of man–machine intelligent system, and through the automatic accounting and processing of financial relations by machines, it provides forward-looking basis for enterprise decision-making, and realizes the management mode of intelligent financial management independent operation. Intelligent management accounting covers three levels of intelligence: the foundation of intelligent management accounting is the integration of industry and finance, the core is intelligent management accounting platform based on business intelligence, and the development trend is intelligent financial platform based on artificial intelligence.

## 35.4 Conclusion

Based on the basic functional characteristics of the blockchain, in recent years, the academic community has generally recognized that blockchain solutions can bring many benefits to management accounting, and are optimistic about the development of blockchain, but these concepts have not been applied and In practice, few scholars have comprehensively considered the technical bottlenecks of blockchain technology, combined with its shortcomings to conduct in-depth exploration of its feasibility, and future research can start from here to enrich relevant research results.

**Acknowledgements** I would like to express my sincere thanks and high respect to all the scholars who have contributed to the academic circle.

## References

1. Ying, Xu.: Research on the application of blockchain technology in management accounting reconstruction. Big Data Age **12**, 25–29 (2019)
2. Wu, Q.: Cross-organizational management accounting innovation embedded in blockchain. Manage. Accounting Res. **2**(04):76–84+88 (2019)
3. Suyun, C., Fangfang, Z.: The influence of blockchain technology on the future development of management accounting. Bus. Accounting **06**, 78–81 (2020)
4. Cheng, H.: Blockchain adds technology "weapon" to management accounting. China Accounting News, 09–06(008) (2019)
5. Juan, Lv.: Application of blockchain technology in management accounting. Cooperative Econ. Technol. **03**, 149–151 (2018)
6. Yue, S., Qi, Z.: Analysis of the application of blockchain technology in accounting. Times Finance **32**, 102–103 (2019)
7. Jinye, Xu., Yangyang, Z.: Research on the impact of blockchain information technology on accounting supervision. Friends Accounting **01**, 156–160 (2018)
8. Xiangyu, Li.: Research on accounting fraud governance based on blockchain technology. Market Weekly **03**, 88–89 (2019)
9. Zhang Fengyuan, Wu., Shuqi, Y.C.: Opportunities and challenges of auditing under blockchain technology. Friends of Accounting **03**, 153–155 (2018)
10. Potekhina, A., Riumkin, I.: Blockchain–A New Accounting Paradigm: Implications for Credit Risk Management (2017)
11. Yongyi, C., Xuefen, X.: Research on enterprise credit risk management for blockchain accounting. Finance Finance **03**, 24–30 (2020)
12. Jinye, Xu.: Intelligent management accounting: conclusion and performance of intelligent contracts. Friends Accounting **10**, 156–160 (2020)
13. Jie, W., Haixia, Z., Peipei, Z.: Research on the characteristics and accounting application of smart contracts based on blockchain technology. China Bus. Rev. **11**, 192–193 (2019)
14. Ruixue, Hu.: Blockchain: technical weapon of management accounting-based on information economy. Econ. Trade Pract. **05**, 307–308 (2018)
15. Yong, Du., Biao, X., Qin, Li.: Construction of management accounting reporting system based on blockchain. Friends Accounting **08**, 153–159 (2020)
16. Li, K.: Research on accounting model and application based on blockchain technology. Accounting Monthly (2020)
17. Hui, C., Xianhua, X.: Research on the application of management accounting tools based on blockchain technology. Friends Accounting **13**, 157–161 (2020)
18. Liu, G., Gan, S., Jiang, Q.: Application logic of smart management accounting from the perspective of blockchain. Accounting Monthly (2020)

# Chapter 36
# Blockchain Certificate Deposit, Electronic Signing, Strengthening Notarization—A New Model of Dealing with Creditor's Rights Disputes

**Jiejing Yao and Peng Hui**

**Abstract** Under the background that the state is promoting Puhui finance, we can data fix, store, and scenario application the process of strengthening notarization handled by traditional methods, and realize the distributed and differential storage of big data at the loan end and between financial, notarial and judicial institutions with the help of blockchain technology. Through the structured atlas big data acquisition and processing, the architecture of the whole technology system and sub-system is constructed, and the legal demonstration of the data technology on the chain is carried out in combination with the court's intelligent execution system. The integration of financial institutions is to solve the difficulties and pain of online lending.

## 36.1 Introduction

The new mode of financial innovation under the background of Puhui finance is realized by promoting the target market cooperation mode of Internet, industry, and finance integration [1]. Its goal is to enable financial institutions to fully release the online comprehensive financial service capabilities, and jointly serve the C-end users and b-end users in the cooperation platform, so as to achieve a win–win situation.

J. Yao
Shanghai University of Political Science and Law, Shanghai, China

P. Hui (✉)
Shanghai Adcademcy of Social Sciences, Shanghai, China
e-mail: penghui@sass.org.cn

 
V. Chang et al. (eds.), *Modern Industrial IoT, Big Data and Supply Chain*,
Smart Innovation, Systems and Technologies 218,
https://doi.org/10.1007/978-981-33-6141-6_36

### 36.1.1 Demand and Supply

In order to change the problem that Internet finance risks rise sharply due to the diversity of subjects, the state policy guides and encourages financial institutions such as state-owned banks to enter and develop this field. According to document No. 76 issued by the Insurance Regulatory Commission of the Bank of China in 2018, it is encouraged to develop self-operated consumption loans with scenario support and designated purpose. Subsequently, on August 27, 2019, in the "opinions on accelerating the development of circulation and promoting commercial consumption" issued by the general office of the State Council, financial institutions were encouraged to innovate consumer credit products and services. Taking bank of Shanghai as an example, by the end of 2019, the total number of online personal customers of Bank of Shanghai had exceeded 30 million, the balance of Internet consumer loans had reached 107.6 billion yuan, the cumulative number of Internet transactions was 318 million, and the amount of Internet business transactions had reached 2.65 trillion yuan. As a licensed institution and a formal institution in the financial field, banks actively respond to the call of regulatory regulation of the market and participate in and provide a growing number of standardized online loan products.

### 36.1.2 Shortboard of Relief

Undoubtedly, with the rapid development of Internet online finance, a series of problems emerge, such as the high cost of loan collection and the high risk of bad debts. Especially in the online loan transaction, there is a great difficulty in the proof, because this transaction is conducted online, and all the borrower's operations are completed online rather than to the bank, so there will be a lack of paper-based text evidence [2]. At the same time, the basic verification operation is to verify the authenticity of the borrower's operation through face recognition and other basic means. Important files such as loan agreement have neither paper certificates nor electronic storage. Electronic receipts and electronic contracts are designed and completed by the bank unilaterally. Therefore, when the number of online loans is rising rapidly, the negative problems in customer complaints, denial of transactions, bad debt rates, and other aspects are also prominent.

There are two main deficiencies in the relief response. First, the default cost of online loans is low. Financial institutions have limited collection means for online loans. For example, the main collection means of banks depend on SMS and telephone. At the same time, financial institutions cannot be restricted by way of restricting high consumption through the list of dishonest people, and the pressure on customers mainly comes from the so-called credit records. Still, the warning function of credit records is relatively limited, so the extent of restricting transactions in practice is obviously insufficient. Second, the lack of effective means of asset preservation. The traditional way of asset preservation and recovery are to realize the real right of security or to put it into arbitration and court action. However, the legal ways of these remedies are long cycle, high cost of error correction, difficult to deal with in batch, difficult to put on the judicial record, and difficult to provide evidence. Therefore, the process and efficiency of the above relief methods are difficult to meet the large-scale demand for online loans [3].

## 36.2 The Chain Structure of Strengthening Notarization

On September 30, 2018, the Supreme People's Court issued the "provisions on Several Issues concerning the enforcement of notarized creditor's rights documents," in which it clarified the standardization of the notarization and enforcement of creditor's rights documents with enforcement effect, which has great significance for the construction of the intelligent system for the enforcement of creditor's rights documents. Objectively speaking, the strengthening notarization blockchain is to build an intelligent credit document execution system through the use of big data, cloud computing, artificial intelligence, and other technologies, which specifically solves the pain points of Puhui finance, such as high collection cost, poor efficiency, long cycle, and so on.

### *36.2.1 Financial Online Electronic Certificate*

In order to solve the Puhui finance problem, the key is to enable financial institutions and notaries to establish a blockchain data distributed storage and mutual trust system and obtain legal documents with enforcement force by means of online notarization, so as to promote the whole process management before and after online loans. In the process of the loan application, this model can alert customers and block the risk of potential fraudulent customers. In the post-loan collection process, the notarial institution verifies and urges the collection, thus reducing the collection workload of relevant departments of financial institutions. At the same time, in the current situation of strict collection supervision, it can also solve the problems of traditional collection methods, such as high collection cost, low efficiency, long cycle, and non-compliance of collection methods [4]. Once there is a non-performing loan, with the electronic execution documents issued online, it can directly enter the stage of court enforcement, which greatly improves the efficiency of court enforcement. At the same time, after the court fails to recover, the dishonest person can be timely included in the list of dishonest persons to be executed for punishment.

Among them, the online loan of financial institutions takes the notarization of empowerment as the entry point, that is, in embedding the loan application into the notarization process, the introduction of notarization institutions can directly identify the borrower, clearly inform the customers of the process of notarization of empowerment, and give continuous legal consequences prompt and early warning before the loan. Then, the two sides of the loan agreement are signed online, and the customer's transaction data of each payment has an online deposit certificate, so as to build a complete loan evidence chain. From the perspective of financial institutions, all the customer information obtained by mandatory notarization before the loan is based on the customer authorization meeting the legal requirements of notarization and transmitted and stored through an encryption algorithm. At the same time, the loan transaction data is also transmitted in the form of ciphertext. With the help of such a scheme, the big data on the blockchain can be stored safely and reliably on the client side and the financial institution side. It is the credibility of this distributed storage that enables financial institutions to complete all online pre-loan process applications, post loan reminders of non-performing customers from notarial institutions, and localized application courts to enforce and connect the list of dishonest executees.

### *36.2.2 Notarized Hash Value Differential Storage*

Supporting and supporting the electronic financial certificate is the "cloud notarization and empowerment platform." This platform includes two functions: on the one hand, through the effective docking with the online product technology platform of financial institutions, it completes the notarization procedure of the whole process of the loan contract signed by financial institutions and customers online. Finally, it issues the notarization certificate with "enforcement effect." In this platform, the identity of the borrower is verified by using three elements of identity verification, biological living identification, OCR, third-party human image ratio, and other technologies to ensure the authenticity of the borrower. Then, the financial institutions connect with the "cloud notarization and strengthening platform" of the notary office, and the "seal and sign" behavior of signing the contract with the borrower is completed on the cloud platform, and data is stored. By using the authentication function of the cloud platform digital certificate, it ensures that the behavior of seal and signature cannot be tampered with and ensures the authenticity and objectivity of the whole signing behaviour [5]. Then, through the standardized and modular design of the relevant steps, and the use of electronic signature of notary institutions and notaries, so that the system can directly generate a notarial certificate with enforcement effect, and complete the "empowering" notarization.

On the other hand, through the data exchange in the loan, record the performance of the financial institution, and the borrower in the loan. Once the borrower breaches the contract, the financial institution can submit the corresponding application online, the notary office issues the execution certificate, and then through the data push after the docking of Cloud platform and the court's execution filing system, complete the online filing and subsequent measures of judicial limitation. Financial institutions push the behavior data of their lending and borrower's repayment on schedule to the cloud platform to generate a hash value for differential storage. Once the borrower breaches the contract, the financial institution initiates an application to the notary office to issue an execution certificate, and the cloud platform can confirm the fact of breach of contract through hash value comparison and directly send an online reminder verification letter to urge the borrower to repay. If the borrower fails to perform the contract after a certain period of time, the cloud platform can push the execution certificate containing the execution elements to the execution department of the court through the docking with the execution filing system of the court, so as to complete the judicial measures such as online batch filing, high upper limit measures for the executee, online seizure and freezing of the property of the executee, and finally help the financial institutions realize the creditor's rights.

## 36.3 Court Batch Data Docking

Enforcement notarization is generally governed by the court in the place where the property is located or where the defendant is located. After entering the implementation phase, the court will not return to the data stream, and due to the pioneering nature of the network-based notarization, it still lacks the deep care of technology and system in the exploration stage. Therefore, in a short period of time, the courts, which are positioned in the steady role of national governance, are unable to realize the synchronous data layout of blockchain with information retained by customers, electronic financial certificates, and notarized hash values. Instead, they will take the blockchain data acquisition method of accepting batch data connection of filing based on the filing registration system.

One of the key problems is that when the court where the property is located or where the defendant is located refuses to accept a large number of network empowerment cases, such disputes will not form the final closed-loop in the whole chain. Therefore, all courts in the country can consider referring to the cases of cross-administrative district courts, intellectual property courts, financial courts, and Internet courts follow the example of administrative litigation cases, food and drug cases, bankruptcy and liquidation cases, resource and environment cases and large traffic cases, and adopt the case centralized jurisdiction mechanism of special courts or special courts in the face of network empowerment cases. Before this mechanism is determined, financial institutions can also make clients promise to open a class II account in the financial branch of the place where they intend to accept the loan, and actually pay the loan into the class II account at one time, so as to ensure that the court will accept the case with the property in the place where the dispute occurs. Of course, the client can easily transfer the amount deposited into the class II account into the actual use online. The purpose of mandatory settlement allowance is to ensure the consistency and coverage of the place where the loan is applied for and the jurisdiction.

## 36.4 Data Knowledge Map of Blockchain

The design of data analysis and database has the judicial characteristics of the intelligent system for the execution of creditor's rights documents, which requires careful analysis of the whole process of data generation, processing, and storage. The system structure chart is a process of data flow that gives the enforcement power of creditor's rights documents to individuals and financial institutions after they have a loan relationship. Among them, the more important is the data distribution structure of the blockchain, which mainly includes the following basic information input data, the label data of the lending financial institutions, the e-commerce label data of the lender's image, the anti-deception data, and the verification data.

- Input data. The powerful notarization cloud platform has a large number of input data for each loan, and the authenticity and accuracy of the data are guaranteed through the verification of notarization. A large number of data will form a big data pool with a legal evidence effect. On the basis of input data, a large number of wide table data will be gradually formed [6].
- Financial label data. For this type of data, it ensures the "double verification" of financial compliance before the loan and the notarization of the cloud platform [7]. Its data acquisition scope actually extends to the stage of loan or even court enforcement, which has the effect of factual evidence for financial institutions, lenders, notaries, and courts.
- Based on the data acquisition of distribution, social relationship network, psychological behavior characteristics, online consumption records, and dynamic characteristics of other asset transactions, through deep mining of the social relationship and transaction behavior characteristics of the lender, a preliminary decision can be made to judge its credit status.
- Anti-fraud data. Through internal and third-party channels to obtain high-quality multi-source data, so as to further mine the sequential data and the spatial structure of the data on the cross section and put forward a more refined data analysis framework of time and space tight coupling lenders, in order to make a conclusion on the multi-modal information of lenders.
- Verification type data. In fact, by integrating the internal data of financial institutions, the above-mentioned e-commerce labels, and anti-fraud data, a multi-level Lender Credit knowledge map database and time–space change model are constructed. For the verification data, it is the knowledge map containing multiple relationships to carry out accurate error correction, completion, and reasoning.

**Acknowledgements** This paper obtained the national key R & D plan "Research on Fine Execution Technology and Equipment of Full Process Management" (2018YFC0830403)

## References


1. Vallee, B., Zeng, Y.: Marketplace lending: a new banking paradigm? Rev. Financial Stud. **32**(5), 1939–1982 (2019)
2. Riggins, F.J., Weber, D.M.: Information asymmetries and identification bias in P2P social microlending. Inf. Technol. Dev. **23**(1), 107–126 (2017)
3. Lenz, R.: Peer-to-peer lending: opportunities and risks. Eur. J. Risk Regulation **7**(4), 688–700 (2016)
4. De Beer, M., Van der Merwe, M., Ball, L., Fourie, I.: Legal deposit of electronic books —a review of challenges faced by national libraries. Library Hi Tech **34**(1), 87–103 (2016)
5. Aysan, A., Disli, M., Fethi, M.D., Ozturk, H.: Islamic banks, deposit insurance reform, and market discipline: evidence from a natural framework. J. Financial Serv. Res. **51**(2), 1–26 (2016)
6. Esposito, C., De Santis, A., Tortora, G., Chang, H., Choo, K.K.R.: Blockchain: a panacea for healthcare cloud-based data security and privacy? IEEE Cloud Comput **5**(1), 31–37 (2018)
7. Meng, W., Tischhauser, E., Wang, Q., Wang, Y., Han, J.: When intrusion detection meets blockchain technology: a review. IEEE Access **6**(1), 10179–10188 (2018)

# Chapter 37
# New Operators Based on Pythagorean Fuzzy Information and Their Application in Reliability Estimation of Military Sealift Transportation

**Yang He, Lei Wang, Sheng Hu, Changhua Yao, and Xiang Zheng**

**Abstract** Pythagorean fuzzy sets (PFSs) have been widely used in practical decision-making problems, and research on Pythagorean fuzzy multi-attribute decision-making methods has emerged in an endless stream. One new Pythagorean fuzzy operation law is proposed, and some averaging operators are presented and aggregated information. Then, an example of reliability estimation of military sealift transportation is shown that this operator is feasible and useful.

## 37.1 Introduction

Risk and reliability of transportation system are important issues, which are related to the safety and benefit of the whole navigation process [1]. The problems of risk and reliability estimation are of great significance to maritime traffic safety and navigation efficiency [2]. The complexity and diversity of the ship navigation environment and constraints makes route planning one of the most challenging topics in mission planning. The current research focuses on the optimization based on the shortest route and the choice of route based on meteorological elements [3]. Usually, risk and reliability estimation is a MADM problem, and we solve this issue by using the method of PF information.

The rest of the paper is organized as follows: In Sect. 2, some basic concepts related to Pythagorean fuzzy set (PFS) and basic operation laws will be discussed. In Sect. 3, some new operations will be introduced, and the properties will be proved. We will propose a new method and prove the rationality of these operators in Sect. 4.

Y. He · S. Hu
Naval Command College, Nanjing, China

L. Wang (✉) · X. Zheng
Army Engineering University of PLA, Nanjing, China
e-mail: iponly@126.com

C. Yao
Nanjing University of Information Science and Technology, Nanjing, China

 
V. Chang et al. (eds.), *Modern Industrial IoT, Big Data and Supply Chain*,
Smart Innovation, Systems and Technologies 218,
https://doi.org/10.1007/978-981-33-6141-6_37

## 37.2 Preliminaries

As we know, since Aanassov [4] proposed intuitionistic fuzzy sets (IFSs), it has received broad attentions by many researchers all over the world. Then, Pythagorean fuzzy sets were discussed by Yager [5], which hold more space and more flexible than IFS. Some MCDM problems have been developed under PF environment. Ren [6] used PF information and solved the MCDM problems. Yager [7] investigated operations of PFSs that are closed under numbers and considered the problem of MCDM problems.

### *37.2.1 Pythagorean Fuzzy Set*

**Definition 1** [6] Let X be a fixed set. A Pythagorean fuzzy set Γ on X can be defined as follows:

$$\Gamma = \{\langle x, \mu_\Gamma(x), \nu_\Gamma(x)\rangle | x \in X\} \tag{37.1}$$

The membership function called $\mu_\Gamma(x) : X \to [0, 1]$, and the non-membership function is $\nu_\Gamma(x) : X \to [0, 1]$.

For convenience, $\Gamma = \langle \mu_\Gamma, \nu_\Gamma \rangle$ is called a PF number (PFN) and $(\mu_\Gamma(x))^2 + (\nu_\Gamma(x))^2 \leq 1$.

**Definition 2** Let $\Gamma = \langle \mu_L, \nu_L \rangle$ be a PFN, the score function [8] is defined as:

$$S(\Gamma) = (\mu_\Gamma)^2 - (\nu_\Gamma)^2 \tag{37.2}$$

The accuracy function [16] is represented as:

$$H(\Gamma) = (\mu_\Gamma)^2 + (\nu_\Gamma)^2 \tag{37.3}$$

In which $S_\Gamma \in [0, 1]$ and $H_\Gamma \in [0, 1]$. The largest the value of $S(\Gamma)$ is the best choice. To the contrary, the biggest value of $H(\Gamma)$ is the highest accuracy.

**Definition 3** [9] For any two PFNs $\Gamma_1, \Gamma_2$, let $S(\Gamma_i)$ be the score value of $\Gamma_i$, then.

(1) If $S(\Gamma_1) < S(\Gamma_2)$, then $\Gamma_1 \prec \Gamma_2$.
(2) If $S(\Gamma_1) > S(\Gamma_2)$, then $\Gamma_1 \succ \Gamma_2$.
(3) If $S(\Gamma_1) = S(\Gamma_2)$, then $\Gamma_1 \sim \Gamma_2$.
where $S(\Gamma) \in [-1, 1]$.

### 37.2.2 Operation Laws

**Definition 4** [10] Let $\gamma=\langle\mu,\nu\rangle$, $\gamma_1=\langle\mu_1,\nu_1\rangle$ and $\gamma_2=\langle\mu_2,\nu_2\rangle$ be PFNs, the commonly used operations are as follows:

(1) $\gamma_1\oplus\gamma_2=\left\langle\sqrt{\mu_1^2+\mu_2^2-\mu_1^2\mu_2^2},\nu_1\nu_2\right\rangle$

(2) $\gamma_1\otimes\gamma_2=\left\langle\mu_1\mu_2,\sqrt{\nu_1^2+\nu_2^2-\nu_1^2\nu_2^2}\right\rangle$

(3) $\lambda\cdot\gamma=\left\langle\sqrt{1-(1-\mu^2)^\lambda},\nu^\lambda\right\rangle$

(4) $\alpha^\lambda=\left\langle\mu^\lambda,\sqrt{1-(1-\nu^2)^\lambda}\right\rangle$.

**Definition 5** [11] The Pythagorean fuzzy weighted geometric averaging (PFWA) operator can be defined as:

$$PFWA_\omega(\alpha_1,\alpha_2,\ldots,\alpha_n)=\omega_1\alpha_1\oplus\omega_2\alpha_2\oplus\cdots\oplus\omega_n\alpha_n$$

$$=\left\langle\sqrt{1-\prod_{j=1}^n(1-\mu_j^2)^{\omega_j}},\prod_{j=1}^n\nu_j^{\omega_j}\right\rangle \tag{37.4}$$

where $\omega_j\in[0,1]$ and $\sum_{j=1}^n\omega_j=1$.

## 37.3 New Pythagorean Operators

### 37.3.1 New Operational Rules on Pythagorean Fuzzy Environment

Pythagorean fuzzy information is mainly aggregated by these aggregation operators. We introduce some new PF operation laws under algebraic environment.

**Definition 6** Let $a=(\mu_1,\nu_1)$, $b=(\mu_2,\nu_2)$ two PFVs, the new Pythagorean operation laws can be defined as:

(1) $a\oplus b=\left(\sqrt{\frac{\mu_1^2+\mu_2^2-2\mu_1^2\mu_2^2}{1-\mu_1^2\mu_2^2}},\sqrt{\frac{\nu_1^2\nu_2^2}{\nu_1^2+\nu_2^2-\nu_1^2\nu_2^2}}\right)$

(2) $a\otimes b=\left(\sqrt{\frac{\mu_1^2\mu_2^2}{\mu_1^2+\mu_2^2-\mu_1^2\mu_2^2}},\sqrt{\frac{\nu_1^2+\nu_2^2-2\nu_1^2\nu_2^2}{1-\nu_1^2\nu_2^2}}\right)$

**Definition 7** For PFV $a=(\mu_a,\nu_a)$, the following operations can be defined:

(1) $\lambda a = \left(\sqrt{\frac{\lambda\mu_a^2}{1+(\lambda-1)\mu_a^2}}, \sqrt{\frac{\nu_a^2}{\lambda-(\lambda-1)\nu_a^2}}\right)$

(2) $a^\lambda = \left(\sqrt{\frac{\mu_a^2}{\lambda-(\lambda-1)\mu_a^2}}, \sqrt{\frac{\lambda\nu_a^2}{1+(1-\lambda)\nu_a^2}}\right)$

### *37.3.2 New Pythagorean Fuzzy Weighted Averaging Operators*

We introduce a new PF weighted arithmetic averaging operator.

**Definition 8** Let $a_i = (\mu_{a_i}, \nu_{a_i})$ be a set of PFVs, the mapping PF weighted arithmetic averaging operator (PFWAA): $PFV^n \to PFV$ satisfies:

$$PFWAA(a_1, a_2, ..., a_n) = w_1 a_1 \oplus w_2 a_2 \oplus ... \oplus w_n a_n$$

$$= \left( \sqrt{1 - \frac{\prod_{i=1}^{n}(1-\mu_{a_i}^2)}{\sum_{j=1}^{n}\left(w_j \prod_{\substack{i=1\\ i\neq j}}^{n}(1-\mu_{a_i}^2)^2\right) - \left(\sum_{j=1}^{n} w_j - 1\right)\prod_{i=1}^{n}(1-\mu_{a_i}^2)}}, \sqrt{\frac{\prod_{i=1}^{k}\nu_{a_i}^2}{\sum_{j=1}^{k}\left(w_j \prod_{\substack{i=1\\ i\neq j}}^{k}\nu_{a_i}^2\right) - \left(\sum_{j=1}^{k} w_j - 1\right)\prod_{i=1}^{k}\nu_{a_i}^2}} \right) \tag{37.5}$$

Then, PFWAA is called a PF weighted arithmetic averaging operator.

## 37.4 The Application in Reliability Estimation of Military Sealift Transportation

In order to strengthen maritime safety, relevant departments of all countries have formulated a series of policies and regulations, in which risk management approaches are one of the important measures [3] (Table 37.1).

Criteria to be considered in the selection of sea routes are determined by the expert team. At the end of this pre-study, it turned out to be that these four alternative routes do not have different characteristics with regard to all twelve criteria. All four routes seemed to satisfy the requirements of the eight criteria presented; remaining four criteria seemed to differentiate for the alternative rifles. Four alternatives have

**Table 37.1** Various decision variables

| Stage | Decision variable | Scale V Variable notation | |
|---|---|---|---|
| I | Experience of the navigation crew | 0–30 years | $x_1$ |
| | Experience of the maintenance | 0–30 years | $x_2$ |
| | Effectiveness of the maintenance | 0–10 | $x_3$ |
| | Overall past operational history of the vessel | 0–10 | $x_4$ |
| | Unforeseen events | 0–10 | $x_5$ |
| II | Congestion at the source harbor | 0–1 (congestion factor) | $x_6$ |
| | Congestion at the sea | 0–1 (traffic intensity factor) | $x_7$ |
| | Congestion at the destination harbor | 0–1 (congestion factor) | $x_8$ |
| III | Weather or environmental conditions | 0–10 | $x_9$ |
| | Delivery date of vessel | Year 1970–2009 | $x_{10}$ |
| | Technological up-gradation of the vessel | 0–10 | $x_{11}$ |
| | Region, place or yard where the vessel was built | 0–10 | $x_{12}$ |

**Table 37.2** Pythagorean fuzzy decision matrice $R$

| | $G_1$ | $G_2$ | $G_3$ | $G_4$ |
|---|---|---|---|---|
| $A_1$ | {0.54, 0.4} | {0.67,0.42} | {0.81,0.24} | {0.71,0.3} |
| $A_2$ | {0.87, 0.22} | {0.75,0.31} | {0.72, 0.29} | {0.65, 0.38} |
| $A_3$ | {0.45, 0.67} | {0.63,0.48} | {0.67, 0.45} | {0.9,0.15} |
| $A_4$ | {0.42,0.7} | {0.85,0.24} | {0.66, 0.4} | {0.73,0.35} |

remained in the candidate list. There are four attributes that need to be considered and the weight vector of the attributes. The following decision matrices in Table 37.2.

Then, method steps can be outlined as follows:

Step i. First, convert the actual data into PF numbers according to the trichotomy method.

Step ii. According to the weight parameters ($\lambda$) and Pythagorean fuzzy numbers above, we can aggregate these data based on definition 8:

$$PFWAA = \left( \sqrt{1 - \frac{\prod_{i=1}^{n}(1-\mu_{a_i}^2)}{\sum_{j=1}^{n}\left(w_j \prod_{\substack{i=1\\ i\neq j}}^{n}(1-\mu_{a_i}^2)^2\right) - \left(\sum_{j=1}^{n} w_j - 1\right)\prod_{i=1}^{n}(1-\mu_{a_i}^2)}}, \sqrt{\frac{\prod_{i=1}^{k}\nu_{a_i}^2}{\sum_{j=1}^{k}\left(w_j \prod_{\substack{i=1\\ i\neq j}}^{k}\nu_{a_i}^2\right) - \left(\sum_{j=1}^{k} w_j - 1\right)\prod_{i=1}^{k}\nu_{a_i}^2}} \right)$$

Step iii. Calculate the formula:$S(L) = (\mu_L)^2 - (\nu_L)^2$.
We use PFWAD operator as a contrast:

$$\begin{aligned} & PFWAD(\alpha_1, \alpha_2, \ldots, \alpha_n) \\ & = \omega_1(D_{\kappa_{\alpha_1}}(\alpha_1)) \oplus \omega_2(D_{\kappa_{\alpha_2}}(\alpha_2)) \oplus \ldots \oplus \omega_n(D_{\kappa_{\alpha_n}}(\alpha_n)) \end{aligned}$$

$$= \left\langle \sqrt{1 - \prod_{j=1}^{n}(1-(\mu_{\alpha_j}^2 + \kappa_{\alpha_j}\pi_{\alpha_j}^2)^2)^{\omega_j}}, \prod_{j=1}^{n}(\nu_{\alpha_j}^2 + (1-\kappa_{\alpha_j})\pi_{\alpha_j}^2)^{\omega_j} \right\rangle$$

The score value is shown in Fig. 37.1:

From Fig. 37.1, we can find that when $\lambda \in (0, 0.3692]$, the ranking is $A_4 \prec A_2 \prec A_3 \prec A_1$, and we choose $A_1$. When $\lambda \in (0.3692, 1]$, the best choice is $A_1$.

## 37.5 Concluding

This paper introduces some new operational law for PF values, studies the attributes, and gives some important conclusions. Also, we develop a series of averaging operators to aggregate PF information. The developed method can be used to solve most Pythagorean fuzzy MADM problems.

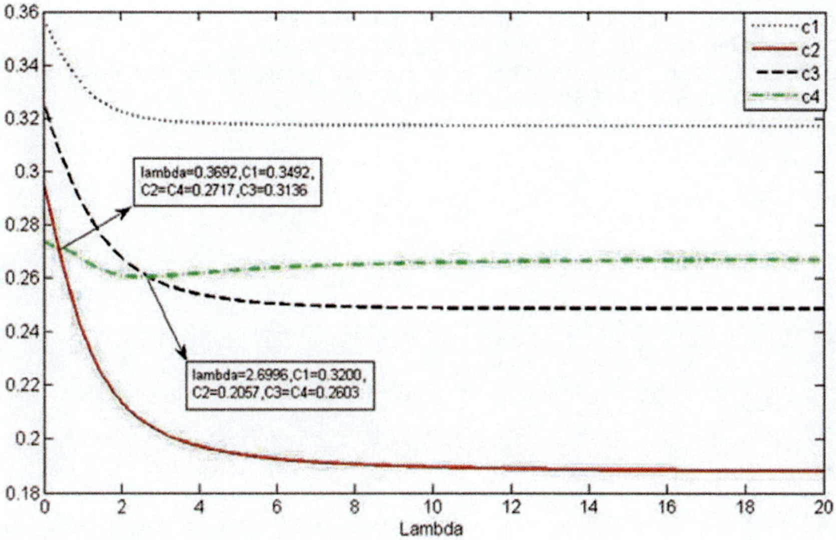


**Fig. 37.1** Score value of PFWAD operator

**Acknowledgements** This work has been partly supported by the National Natural Science Foundation of China (grant numbers 61702543, 61971439), the Natural Science Foundation of Jiangsu Province of China (grant numbers BK20191329), and the China Postdoctoral Science Foundation Project (grant numbers 2019T120987).

## References

1. Sumalee, A., Kurauchi, F.: Guest editorial: reliability and emergency issues in transportation network analysis. Netw. Spat. Econ. **6**(3–4), 169–172 (2006)
2. https://en.wikipedia.org/wiki/Military_Sealift_Command
3. Rajesh, S., Gaonkar, P., Xie, M., Fu, X.: Reliability estimation of maritime transportation: a study of two fuzzy reliability models. Ocean Eng. 721–710 (2013)
4. Atanassov, K.: Intuitionistic fuzzy sets. Fuzzy Sets Syst. **20**, 87–96 (1986)
5. Yager, R.R.: Pythagorean membership grades in mul-ticriteria decision making. IEEE Trans. Fuzzy Syst. **22**, 958–965 (2014)
6. Ren, P.: Pythagorean fuzzy TODIM approach to multi-criteria decision making. Appl. Soft Comput. **42**, 246–259 (2016)
7. Yager, R.R.: Pythagorean membership grades, complex numbers, and decision making. Int. J. Intelligent Syst. **28**, 436–452 (2013)
8. Gary, H.: A novel accuracy function under interval-valued Pythagorean fuzzy environment for solving multi-criteria decision making problem. J. Intelligent Fuzzy Syst. **31**, 529–540 (2016)
9. Yager, R.R.: Pythagorean fuzzy subsets. In: Proceeding Joint IFSA World Congress and NAFIPS Annual Meeting, Edmonton, Canada, pp. 57–61 (2013)

10. Zhang, X.L., Xu, Z.S.: Extension of TOPSIS to multiple criteria decision making with Pythagorean fuzzy sets. Int. J. Intell. Syst. **29**, 1061–1078 (2014)
11. Xia, M., Zeshui, Xu.: Generalized point operators for aggregating intuitionistic fuzzy information. Int. J. Intell. Syst. **12**, 1061–1080 (2010)

# Part V
# Emerging Areas: Education and Emerging Applications

# Chapter 38
# Prospects for the Online Education Promoting the Reform of Traditional Higher Education Under the COVID-19 Situation

**Yu Liao**

**Abstract** With the development of educational technology modernization, online education is booming in recent years. E-learning is a method of content dissemination and fast learning through the application of information technology and Internet technology. With the rapid development of information technology, especially from the Internet to the mobile Internet, it has created a way of life, work and learning across time and space, which has changed the way of knowledge acquisition. Teaching and learning can be independent of time, space and location, and access to knowledge has become flexible and diversified. The rapid global spread of COVID-19 at the beginning of 2020 has affected about 190 countries around the world by suspending offline classes and affecting the educational progress of more than 1.5 billion students. Most schools in China try to provide courses through online education, which has a great impact on the traditional education model. Based on the analysis of the origin and development of online education, this paper discusses that online education has promoted the rapid transformation of traditional higher education and looks forward to the trend of the transformation of China's higher education after the COVID-19 disease.

In this year's Spring Festival in Gengzi, the people across the country have encountered a sudden and huge test—Corona Virus Disease 2019—it spreads from Wuhan and gradually spreads to the whole country and the world. For nearly half a year, in view of the impact of the epidemic of pneumonia caused by the new coronavirus infection on the normal opening of schools and classroom teaching, many schools have started to teach online.

Y. Liao (✉)
City College of Wuhan University of Science and Technology, Wuhan 430083, China
e-mail: tomliao-898@163.com



V. Chang et al. (eds.), *Modern Industrial IoT, Big Data and Supply Chain*,
Smart Innovation, Systems and Technologies 218,
https://doi.org/10.1007/978-981-33-6141-6_38

## 38.1 The Origin and Development of Online Education

E-Learning is a method of content dissemination and fast learning through the application of information technology and Internet technology. The USA is the birthplace of E-Learning, and the staff training of 60% of enterprises is through the network. Since 1998, E-Learning has sprung up all over the world, from North America and Europe to Asia.

Since online education began to flourish in 2014 in China, online education has enjoyed rapid development, and tens of thousands of online education enterprises have been born, not only education giants such as Vipkid, at the same time, there are many small and medium-sized online education enterprises [1]. There are four big trends in the future of online education.

First, recording and broadcasting are back again, and the era of King of vertical screen is coming. The rise of the live broadcast a few years ago has created a large number of live classroom. This time, the return of the broadcast is no longer the original boring long lectures, but the short and pithy carefully designed short video courses. The content of the course must be high-quality content, both full of dry goods, but also lively and interesting. Online education courses gain access to customers, no longer focus on WeChat groups, Qq groups and other community portals, Douyin and other short video apps have bucked the trend in recent years and accumulated a large number of active users. Now there are education and training enterprises have won millions of Douyin fans. Because the popular video is more suitable for mobile phone to see the vertical, so the course form will gradually transition from horizontal screen to vertical screen, and vertical screen is the king of the era which is coming.

Second—online and offline integration. As the cost of acquiring customers online remains high, more and more online education companies will switch to offline work. Just as the new retail sector, big profit and Alipay integration, E-Commerce and offline store cooperation to form a new retail, the education sector will appear in a similar form. Offline classes will become standard fare for online education companies as well as live streaming and double-teacher classes. So even if the offline training institutions are doing well, but also have a sense of crisis, it must be careful to be subverted by the Internet—this "Black Swan".

Third—publishing and online education organic integration. The introduction of mobile phones and iPads into the classroom has become an obstacle as education authorities have stepped up efforts to prevent Myopia among young people, prompting a proliferation of paper-and-pencil-based interactive systems. The pen-and-paper system does no harm to a child's eyes and automatically corrects homework and performs big data analysis. The best way for the educational publishing industry to move into online education is through a pen-and-paper interactive system, which combines thousands of years of pen-and-paper technology with modern smart pen technology, to obtain the students' writing track and time, correct and wrong key indicators, so as to provide the possibility of artificial intelligence and big data in education application.

Fourth—the combination of artificial and AI. The rise of artificial intelligence technology in recent years will bring revolutionary changes, artificial and artificial intelligence technology organic combination, will produce a new form of online education products. Ai will end one-on-one teaching with some teachers, and now some online foreign teaching enterprises are already using AI, the effect is not bad. In the future, under the impact of AI, educational enterprises will adjust their direction, integrate more AI technology into teaching, improve teaching effect with AI teachers and significantly reduce the number of online foreign teachers. Not only online foreign teachers, other K12 subjects of teaching will gradually be replaced by AI teachers.

## 38.2 The New Crown Disease Disaster Has Shaken the Normal State of Education

### *38.2.1 Traditional Forms of Higher Education*

China is one of the oldest countries in the world, China as early as the pre-Qin period has formed higher education institutions, with more than 3000 years of Imperial College history, once was the world's most developed education countries. However, the modern higher education in China has fallen far behind the advanced level in the world. Neither the Westernization School nor the university has a substantial relationship with the original institutions of higher education such as Imperial College, imperial college and Academy of classical learning. The institutions of higher education in modern China were mainly the products of learning from the West and drawing lessons from the West [2]. However, most of the traditional classroom teaching still does not get rid of the shackles of traditional education ideas, and there are dogmatic, unified, static, isolation and students from the reality of life and other abuses. For example, in the traditional classroom, teachers teach and students are taught; teachers speak and students listen; teachers establish discipline and students obey it; teachers make choices and impose them on students and students obey them; teachers choose learning content and students adapt to it; the teacher is the subject of the learning process, while the student is merely the object. It is to mold and train students according to the same standard, the unified training goal, the unified teaching program, the unified curriculum, the unified time, the unified teaching progress, the unified examination, the unified answer and the unified evaluation standard. This kind of teaching seriously neglects the main body status of students in classroom teaching, neglects the individual differences among students, which leads to repress students' individuality and strangles their creativity.

### 38.2.2 Under the Epidemic Situation, Online Education Impacts the Traditional Form of Higher Education

High-quality higher education has always been a scarce resource, and colleges and universities are the most important places to obtain limited educational resources. Under the epidemic situation, the national large-scale online education practice has broken down the traditional higher education to the time and space many restrictions, the university "wall" really began to disappear slowly. The emergence of learning spaces that are perceived as informal and informal by traditional education has gradually blurred the boundaries of traditional classroom education. Under the epidemic situation, online education impacts the traditional form of higher education.

First, online education technology expands the pool of knowledge systems. The establishment and storage of large-scale database based on common sense and professional knowledge in network and online education platform greatly enrich the knowledge reserve of intelligent system. In the daily teaching, the study of these memorization knowledge is no longer the first and necessary, educators and educates can through a simple search, in the shortest time to obtain relevant materials and teaching content. The accuracy, timeliness, extensiveness and dynamic updating of these materials cannot be compared with traditional teachers' artificial memory. In a sense, the paths and clues to knowledge are more meaningful than the knowledge itself. The ability to search becomes a necessary skill and basic literacy. At the same time, because of the unlimited replication of data and extensive accessibility, but also greatly expanded the range of educated people.

Second, the network and online education makes the relationship between teachers and students change. Traditional education is teacher-centered, knowledge-oriented and knowledge-imparting through unified learning. In view of the completeness and convenience of the educational network database, the repetitive traditional teaching based on knowledge accumulation is most likely to be gradually replaced by artificial intelligence. At present, artificial intelligence technology has been able to realize the functions of automatic problem-solving, homework marking, learning disability diagnosis and feedback, physical and psychological monitoring, comprehensive quality evaluation and personalized teaching. Learning will be student-led, and students can choose their own learning content, learning time and learning space. This process will also lead to collaboration and joint learning between teachers and students. This is consistent with the trend that many colleges and universities put forward in the educational reform that "education and teaching should be student-centered".

Thirdly, online education promotes the construction of new learning model. The modern university system is based on the class teaching system and standardized teaching materials and is standardized and large-scale learning. Artificial intelligence technology provides students with a game-like virtual learning environment, breaking the traditional learning model. One is to learn the choice of space and time more freedom. Learning is no longer limited to traditional, fixed classroom learning and can even integrate various types of formal learning, informal learning, blended learning and distance learning to enhance learning effect. Second, it is more convenient to "teach students according to their aptitude". Artificial intelligence technology can make students study more pertinently by mining macro-educational data and analyzing the data produced in the process of students' learning behavior [3].

## 38.3 Online Education Promotes the Rapid Transformation of Traditional Higher Education

### *38.3.1 Change in the Way Knowledge is Acquired*

In an era of information explosion, knowledge is expanded by an exponential curve, but the individual's cognitive ability is very limited, it is obviously impossible to store everything in a student's brain. With more and more complex society, more and more information, and our learning time is very limited, human beings only have to change the way of learning and cognition, only to accept the new way of cognition, that is, man–machine combination of thinking system. In the past, no matter where we are in the world, access to a large number of effective information method is almost only one that read books, read newspapers, the best place to obtain knowledge is the library. Compared with today, the popularization of the Internet, the traditional library has been "moved" into the network, massive paper books, global information can be well preserved and shared, coupled with the development of retrieval technology, access to knowledge has been unimaginable. The use of large and small databases, multimedia, paper media has long been far behind.

Now students want to learn, they can go to the Internet, they can access the electronic library, and there are huge teaching resources around the world, such as watching Harvard professors' video lectures online and learning about the open classes of Tsinghua and Peking University Teachers, you can even have one-on-one English learning with a foreign teacher [4].

### *38.3.2 Constructing Learner-Centered Teaching Behavior Based on Technology*

With the advent of education information, a variety of modern teaching methods have come. At first, modern technologies such as slides, projectors, audio recording, broadcasting, audio laboratory and live broadcast were widely used, which not only greatly enriched teaching methods, but also contributed to the emergence of "individualized teaching", a new approach to education. Since 2001, information technology has been widely and deeply applied in the field of education, and the four elements of classroom teaching, namely teachers, students, teaching materials, media and the relationship among them have changed profoundly.

In the Internet age, the research of teaching mode presents the characteristics of diversification and modernization. Information technology and network make knowledge presentation more in line with students' cognitive needs, teachers' knowledge authority status which is gradually broken, and personalized learning becomes possible. Each learning platform brings together high-quality educational resources from all over the world. Students can open computers to choose their own teachers, choose their own time, choose their own courses and have face-to-face communication with online teachers after completing the course. The core of teaching is students, students can make appointments according to their own preferences, their own state, which is the core of the student-centered teaching model, and the student-centered is the true teaching.

### *38.3.3 Leading to the Formation of Online Education Ecological Chain*

Online education chain is an ecological platform for online education based on blockchain technology, which uses blockchain data structure to store data and generates and updates data through distributed data. When the blockchain technology is applied to the online education network structure, the teaching records information of teachers' lectures and students' lectures is left traces in the system throughout the whole process, combined with recording, verification and other mechanisms, and it makes teachers' evaluation, students' feedback and teaching results transparent and credible, realizes the management and control of non-manual operation, assists intelligent decision-making, improves the efficiency of education management, and is beneficial to the scientific planning of after-class management, to deal with parents' concerns about being left unattended after school.

The education eco-chain is based on blockchain technology and makes great efforts to build a brand-new online education business model with blockchain technology, for the platform user transformation has laid a good foundation. At the same time, the education ecological chain is implanted into the mining function of blockchain to increase the stickiness and activity of users for the platform. With the development of blockchain technology in all walks of life, education ecological chain will apply the latest technology of blockchain to realize the "blowout" development of online education industry.

## 38.4 Prospect of Higher Education Reform in China After New Crown Disease

### *38.4.1 Change in Educational Philosophy*

With the continuous expansion of college enrollment for several years, China has reached the USA in less than 10 years to spend 50 years to complete the number of college students. The number of students has increased rapidly, making the original elite university education become popular education. In particular, schools pay great attention to the cultivation of students' skills and ignore the formation of students' correct values, leading to a good job, earn a lot of money and then buy a house, car and other pure material pursuit, has become the pursuit of most college students. This epidemic has inspired many higher education. Ideological and political education should be carried into the mind and heart. We should use fresh examples of epidemic prevention and control, to guide students to better understand the party's centralized and unified leadership, stick to the people-centered principle, "support from all sides when one side is in difficulty," concentrate their efforts to accomplish important tasks and fully understand the Chinese strength and spirit, strengthen the "four self-confidence". Health education, life education, personality education, ecological civilization education and other short board.

### *38.4.2 Subject and Specialty Optimization*

The discipline and specialty of higher education are the basis and carrier of personnel training. The establishment and construction of disciplines and specialties are related to the objectives and specifications of personnel training, to the allocation and coordination of educational and teaching resources, and to the quality and efficiency of higher education, it is also related to the coordination and adaptation of higher education and economic and social development. The cultivation of high-quality professionals to meet the needs of national and regional economic and social development depends to a large extent on the rational structure of disciplines and specialties in universities. Setting up the subject and specialty scientifically is the basic guarantee for optimizing the structure, improving the quality of higher education, meeting the needs of economic and social development and people's all-round development.

The data show that the proportion of seven disciplines (engineering, science, law, agronomy, medicine, history and philosophy) in the enrollment structure of various disciplines in China's general universities has decreased in the past ten years, the percentage of four disciplines (management, literature, economics, education) increased. Among them, management and literature (including art) enrollment scale increased significantly. In 2016, the media survey of 36 top students found that 61.11 percent of the top students tended to apply for economics majors, 33.33 percent for management majors, and none for medicine, agronomy or military science. In 2016, only 1.31 percent of the top students in China's provincial college entrance exams chose to study medicine, and more than 40 percent chose to major in economic management, according to a survey of the top students in China's national college entrance examination. We should pay more attention to the discipline and specialty construction related to people's livelihood and public welfare undertakings. If a country has a serious shortage of doctors and nurses, a shortage of preventive medicine professionals, a shortage of life sciences and high-end medical research personnel, the social status of social workers is worried, and it is hard to imagine winning a melee.

### *38.4.3 Follow up the Development of Educational Technology and Speed up the Reform of Teaching*

During this year's outbreak, students were unable to return to school. The Ministry of Education launched 22 online course platforms and 24,000 courses. Universities are also working hard to develop and use technology to ensure that students are not suspended from classes. Online education and artificial intelligence, big data and cloud computing have given higher education wings, and educational technology cannot afford to be slow or slow to catch up. "How to teach well and how to learn well" is the biggest challenge to teachers and students in colleges and universities. The teaching concept of "learning as the center" is gradually established, and students' autonomous learning ability is gradually improved. After the outbreak, we need to re-evaluate the quality of online courses and keep some of the best ones available.

### *38.4.4 Accelerating the Formation of "New Ecology" of Web-Based Interactive Teaching*

Interactive teaching should realize more three-dimensional, vivid and efficient teaching through multi-dimensional organization of 'online and offline integration, in-class and out-of-class integration". In addition, university teachers should actively establish a "learning community" to promote "group learning, autonomous learning" and strive to change the role from "giver, demander, professor" to "guide, encourager, organizer". Only by adapting to the new environment and assuming a new role, we can further develop and achieve ourselves.

Now the general online education enterprise or platform is divided into service live broadcast, content recorded broadcast, item bank, tools take photos and search questions, home–school interaction, O2O, integrated platform including service, content, tools and more than four categories. The future is likely to favor fast, live and AI. The nature of education is service, and live services, which are closer to the offline training experience, will be the most important industry outlet in the future. AI big data precision teaching: Take "teacher" and "student" as the core, respectively, carry on the accurate portrait to the student based on the big data in the future, provide the personalized service to the student with "the effect and the efficiency" standard, hopefully, we will be able to really teach them to their students.

The rise of online education and the current development of Internet + and 5G technology will gradually change people's definition of the education industry, from traditional face-to-face education and training to the use of the Internet to bring about a broader world view, at the same time, increasing the social synergy, knowledge is also increasingly decentralized. In this process, because the Internet has changed people's cognition, education must be innovative, from teaching ideas, teaching ideas, teaching organizational forms, teaching methods and so on will change.

**Fund Projects**
The 2020 Teaching and Research Project of City College Of Wuhan University of Science and Technology—"online teaching improvement in universities from the perspective of online education under the COVID-19 disease—a case study of human resource management in City College.", Item number: 2020CYYBJY021, Document number: Academic Affairs of City College (2020) 7.

## References

1. Four trends of online education in 2020-the age of the king. https://www.sohu.com/a/345859508_466950
2. "University" Baidu knows. https://zhidao.baidu.com/question/1115523999871232219.html
3. Ai Enables Higher Education Transformation: Space and limits-china Education News. https://www.jyb.cn/zcg/xwy/wzxw/201810/t20181030_1255887.html
4. Ruwei, D.: Social Intelligence Science. Shanghai Jiao Tong University Press, Chongqing (2007)

# Chapter 39
# A Multi-view Learning Method Based on Feature Correlation

**Zhixin Zhao, Feng Hu, Jin Dai, and Hong Yu**

**Abstract** The multi-view learning can improve the generalization performance of the classifier by using features from different sources. However, many multi-view data sets are constructed by unsupervised methods, which may cause uncertain to the classifier. In this paper, a supervised multi-view learning method based on the correlation between features is proposed, which mainly includes two parts: view evaluation and view construction. First of all, canonical correlation analysis and correlation matrix based on mutual information are employed to extract the correlation information between different feature sets. After that, a supervised data view construction algorithm, named data view construction (DVC), is proposed. Compared with traditional view construction methods, DVC can be used as a general view construction method to construct multi-view data set. The experimental results show that DVC achieves higher generalization compared with other view construction algorithms.

## 39.1 Introduction

In most data analytics problems, data is collected from diverse domains or various feature extractors. Each variable source is referred as a particular view. The multiple views for a particular problem may take different forms, such as in the web classification task, and the web page data can be divided into web hyperlinks view and web contents view [1]; in the face recognition tasks, face images from different angles can be considered as different views to improve the accuracy of face recognition [2]; in the video classification tasks, audio data and image data can be used as two views [3].

At present, multi-view learning mainly includes three categories: co-training, subspace learning and multiple kernel learning. The co-training algorithm makes full use of the complementary information between views to achieve higher performance.

Z. Zhao · F. Hu (✉) · J. Dai · H. Yu
Chongqing University of Posts and Telecommunications, Chongqing 400065, China
e-mail: hufeng@cqupt.edu.cn



V. Chang et al. (eds.), *Modern Industrial IoT, Big Data and Supply Chain*,
Smart Innovation, Systems and Technologies 218,
https://doi.org/10.1007/978-981-33-6141-6_39

The co-training algorithm has many applications and derivatives, including deep co-training [4], correlation recognition [5], and semi-supervised co-training [6].

The subspace learning algorithm treats the view as a subspace. The feature dimensions obtained by subspace learning are lower, so it can effectively avoid dimensional disasters. Representative methods of subspace learning include canonical correlation analysis (CCA) [7], metric learning [8], and subspace clustering [9].

The essence of the multiple kernel learning is to replace the single kernel function with a combination of multiple kernel functions. Currently, researches on multiple kernel learning include (1) sparse and non-sparse multiple kernel learning [10]; (2) integrated radius multiple kernel learning [11]; (3) multiple kernel clustering [12].

Although the existing multi-view research has achieved great results, it still faces many challenges (1) lack of efficient methods to construct multi-view data; (2) insufficient use of correlation information between views. To solve the above problems, a data view construction algorithm (DVC) based on feature correlation is developed.

The content of this paper is organized as follows. The DVC algorithm is developed in Sect. 2. The experimental results are showed to express the effectiveness of the proposed method in Sect. 3. This paper is concluded finally.

## 39.2 Proposed Methods

### *39.2.1 View Evaluation Method*

Generally speaking, multi-view data needs to satisfy two conditions: (1) sufficient redundancy. Each view is sufficient for correct classification and the predict result of different views of the same sample should be consistent; (2) conditional independence. Different views are conditionally independent given the label. According to the two conditions, we propose a view evaluation method based on the correlation between features. The sufficient redundancy is described by the mean value of mutual information between features and label. The conditional independence is described by the mean value of the coefficient calculated by canonical correlation analysis. The two view evaluation metrics are described in detail as follows.

Suppose that data set $D = \{V_1, V_2 \dots V_{|V|}\}$ (where $|V|$ represents the number of views), any view $V_i$ in $D$ is expressed as $V_i = \{A_1, A_2, \dots, A_m\}$ (where $m$ is the number of feature in $V_i$) and label set $C = \{C_1, C_2, \dots C_r\}$ (where $r$ is the number of label values).

The mutual information between two random variables $X$ and Y is defined as follows.

$$I(X; Y) = \sum_{x \in X} \sum_{y \in Y} p(x, y) \log \frac{p(x, y)}{p(x)p(y)} \tag{39.1}$$

Then, the sufficient redundancy metric of $V_i$ is:

$$\text{Score}^{V_i}_{\text{intra\_ view}} = \frac{1}{m(m-1)} \sum_{i \neq j}^{m} I(A_i; A_j) + \frac{1}{m} \sum_{k=1}^{m} I(A_k; C) \tag{39.2}$$

Canonical correlation analysis (CCA) is proposed by Hotelling [7], by which a linear relationship is found between two multi-variables. If we consider each feature as a random variable, each view can be considered as a multi-variable. The coefficient between the two views can be obtained using CCA.

Suppose that $k$ represents the number of coefficients of two views calculated by CCA and $\rho^{pq}_n$ represents the $n$-th coefficient calculated by CCA for $p$-th and $q$-th views. The conditional independence of view $V_i$ is defined as follows.

$$\text{Score}^{V_i}_{\text{inter\_ view}} = \frac{1}{|V| - 1} \sum_{p \neq q}^{|V|} \left( \frac{1}{k} \sum_{n=1}^{k} \rho^{pq}_n \right) \tag{39.3}$$

The subtraction between formula (39.2) and formula (39.3) is used as the final evaluation metric of the view. The final evaluation metric of $V_i$ is:

$$\begin{aligned} Score_{V_i} &= Score^{V_i}_{\text{intra\_ view}} - Score^{V_i}_{\text{inter\_ view}} \\ &= \frac{1}{m(m-1)} \sum_{i \neq j}^{m} I(A_i; A_j) + \frac{1}{m} \sum_{k=1}^{m} I(A_k; C) - \frac{1}{|V| - 1} \sum_{p \neq q}^{|V|} \left( \frac{1}{k} \sum_{n=1}^{k} \rho^{pq}_n \right) \end{aligned} \tag{39.4}$$

### 39.2.2 *Data View Construction Algorithm (DVC)*

Data view construction algorithm DVC is a supervised view construction algorithm based on view evaluation. It adds features to the view by maximizing the view evaluation metric. The specific method is as follows.

Firstly, we calculate the mutual information between features and label according to formula (39.1) to obtain the correlation matrix.

$$Corr\_Matrix = \begin{pmatrix} I(A_1; A_1) & \ldots & I(A_1; A_n) & I(A_1; C) \\ \vdots & \ddots & \vdots & \vdots \\ I(A_n; A_1) & \cdots & I(A_n; A_n) & I(A_n; C) \end{pmatrix} \tag{39.5}$$

Secondly, initialize the views. We select |V| features from feature set randomly and add them into |V| views. As a result, there is only one initial feature in each view.

Then, compute the view evaluation metrics. If a feature is added to view, we should calculate the final evaluation metric according to formula (39.4).

**Table 39.1** Data sets after preprocessing

| Data sets | Number of samples | Number of features | Number of labels |
|---|---|---|---|
| Mfeat | 2000 | 649 | 10 |
| Madelon | 2000 | 500 | 2 |
| SECOM | 1567 | 562 | 2 |
| Internet Ad | 3279 | 1558 | 2 |
| Arrhythmia | 452 | 279 | 16 |
| ESR | 11,500 | 178 | 2 |
| HAR | 10,299 | 561 | 6 |
| Musk (version2) | 7074 | 166 | 2 |

Finally, construct the views. We add features to each view in turn and calculate the final evaluation metric. The feature is added to the view with the highest evaluation metric. Repeat the above steps until all features are added to the view.

Through the above steps, the feature set is divided into multiple views by DVC algorithm. The final multi-view data form is $\{V_1 : [A_1, A_2, \ldots], V_2 : [A_3, A_4, \ldots], V_3 : [A_5, A_6, \ldots]\}$.

## 39.3 Experiments

### 39.3.1 Experimental Data Sets

We use eight high-dimensional UCI public data sets in the experiment. The data sets have different number of features to compare the influence of feature dimensions to view construction performance. Before the views are constructed, the data sets are preprocessed by removing the features with missing values more than half and complete the remaining missing values with the mode. Table 39.1 is the detailed information of data sets after preprocessing.

### 39.3.2 Experimental Setting and Comparison Algorithms

In the experiment, we construct three views for all methods. Firstly, we use decision tree as the base learner on all views. Secondly, we use mean fusion for continuous label and relatively majority voting fusion for discrete label. In the experiment, the ratio of the training data set and the test data set is 9:1. In order to ensure the generalization ability of the experimental results, we use the average result of ten-fold cross-validation as the final result.

View construction algorithms include two categories: unsupervised and supervised view construction algorithms. Unsupervised view construction methods, also known as random view construction methods, include random subspace method (RSM) [13] and feature clustering [14]. Supervised view construction methods include information gain ratio feature ranking (GRFR) [15], optimal feature set partitioning algorithm (OFSP) [16], and UM2L algorithm [17]. We also use the single view algorithm as a baseline for comparison.

### *39.3.3 Results and Discussion*

To compare the performance of different view construction algorithms, we use precision and F1-socre as evaluation metrics. Table 39.2 shows the comparison of the precision and F1-score of the eight data sets constructed as multi-view data using different view construction algorithms. In each data set, the former value in the table is precision, and the latter is F1-score. In the last two rows, the precision and F1-score ranking of each algorithm are described as P Ranking and F1 Ranking, respectively.

We can see clearly from Table 39.2 that compared with single view algorithm, all view construction algorithms obtain better performance on precision and F1-score ranking. According to the ranking of precision, we can see that the DVC algorithm proposed in this paper is ranked first on three data sets. According to the ranking of

**Table 39.2** Precision and F1-score of view construction algorithms on different data sets

| Data Sets | Single View | DVC | OFSP | K-means Cluster | RSM | GRFR | UM2L |
|---|---|---|---|---|---|---|---|
| Mfeat | 0.932<br>0.9312 | 0.9465<br>0.9457 | 0.9385<br>0.9535 | 0.9435<br>0.9402 | 0.946<br>0.9434 | 0.9625<br>0.9587 | **0.971**<br>**0.9692** |
| Madelon | 0.6865<br>0.6837 | 0.7612<br>0.7598 | 0.4904<br>0.4953 | 0.6619<br>0.6618 | 0.7602<br>0.7604 | 0.6692<br>0.6692 | **0.795**<br>**0.794** |
| SECOM | 0.8994<br>0.5569 | **0.9414**<br>**0.5932** | 0.928<br>0.5356 | 0.935<br>0.5777 | 0.9223<br>0.4951 | 0.9229<br>0.5884 | 0.9172<br>0.5002 |
| Internet Ad | 0.9625<br>0.9377 | 0.9652<br>0.9342 | **0.9774**<br>**0.9531** | 0.9405<br>0.8514 | 0.9713<br>0.9387 | 0.9716<br>0.9397 | 0.9713<br>0.9403 |
| Arrhythmia | 0.5783<br>0.355 | **0.6413**<br>0.4112 | 0.6391<br>0.3716 | 0.5587<br>0.3478 | 0.6152<br>0.4081 | 0.6087<br>0.3317 | 0.6369<br>**0.4507** |
| ESR | 0.9439<br>0.9103 | **0.9627**<br>**0.9384** | 0.9542<br>0.9265 | 0.9602<br>0.9353 | 0.9543<br>0.9289 | 0.959<br>0.9349 | 0.9611<br>0.9315 |
| HAR | 0.9394<br>0.937 | 0.9626<br>0.9613 | 0.9663<br>0.9666 | 0.8575<br>0.8589 | 0.9616<br>0.9613 | **0.9725**<br>**0.972** | 0.9658<br>0.9693 |
| Musk (version2) | 0.9475<br>0.9123 | 0.9662<br>0.9423 | 0.9718<br>0.9553 | 0.9619<br>0.9418 | 0.9651<br>0.951 | **0.9871**<br>**0.9796** | 0.9614<br>0.938 |
| P Ranking | 6.25 | **2.5** | 3.625 | 5.125 | 4.25 | 3 | 3.25 |
| F1 Ranking | 5.75 | **2.875** | 3.875 | 5.25 | 4.375 | 3 | **2.875** |

F1-score, the DVC algorithm is also ranked first with UM2L algorithm. The UM2L algorithm ranks first on three data sets, while DVC algorithm ranks first on two data sets. Due to the lack of samples and the diversity of label values, DVC obtains the best precision while UM2L achieves the best F1-score on Arrhythmia data set.

Although the UM2L algorithm also achieves great performance, UM2L needs to map the feature set to a new space as a view using the metric matrix generated by the algorithm. So, the time complexity of UM2L algorithm is higher than DVC algorithm. In general, compared with other view construction algorithms, the views constructed by DVC better satisfy the sufficient redundancy and conditional independence conditions. We can obtain the conclusion that views constructed by the DVC algorithm are higher performance compared to other view construction algorithms.

## 39.4 Summary

We propose a supervised view construction algorithm DVC based on the correlation between features. Compared with unsupervised view construction algorithms, DVC algorithm uses correlation information to reduce the uncertainty of unsupervised view construction. Compared with the supervised method, the DVC algorithm fully considers the sufficient redundancy within the view and the conditional independence between views. Thus, it has better generalization performance. Experimental evaluation results on public data sets have proved the effectiveness of the DVC algorithm.

In this paper, the view construction algorithm has been developed based on the view evaluation method. In the future work, a more accurate view evaluation method would be studied based on metric learning to measure the sufficient redundancy within the view and the conditional independence between views.

## References


1. Blum, A., Mitchell, T.: Combining labeled and unlabeled data with co-training. In: Proceedings of the Eleventh Annual Conference on Computational Learning Theory, pp. 92–100. Association for Computing Machinery, New York (1998)
2. Wang, B.J.: Research on Multi-view Feature Fusion Method. Beijing Jiaotong University (2018)
3. Wang, Y.Q.: Multi-view Learning Algorithm and Application Research. National University of Defense Technology (2017)
4. Chadi, H., Shady, E.: Arabic named entity recognition via deep co-learning. Artif. Intell. Rev. **52**(1), 197–215 (2019)
5. Lu, Y.J., Xv, M., et al.: Cross-lingual implicit discourse relation recognition with co-training. Front. Information Technol. Electronic Eng. **19**(5), 651–661 (2018)
6. Wang, Y., Du, L., Dai, H.: Target discrimination method for SAR images based on semisupervised co-training. J. Appl. Remote Sens. **12**(1), 015004–015004 (2018)
7. Hotelling, H.: Relations between two sets of variates. In Breakthroughs in Statistics, pp. 162–190. Springer, New York (1992)

8. Zhai, D.M., Chang, H., Shan, S.G., et al.: Multiview metric learning with global consistency and local smoothness. ACM Trans. Intelligent Syst. Technol. **3**(3), 1–22 (2012)
9. Cao, X., Zhang, C., Fu, H., et al.: Diversity-induced multi-view subspace clustering. In Proceedings of the IEEE Conference on Computer Vision and Pattern Recognition, pp. 586–594. Boston America (2015)
10. Yan, F., Kittler, J., Mikolajczyk, K., et al.: Non-sparse multiple kernel fisher discriminant analysis. J. Mach. Learn. Res. **13**(1), 607–642 (2012)
11. Liu, X., Wang, L., Yin, J., et al.: An efficient approach to integrating radius information into multiple kernel learning. IEEE Trans. Cybern. **43**(2), 557–569 (2013)
12. Nebehay, G., Pflugfelder, R.: Clustering of static-adaptive correspondences for deformable object tracking. In Proceedings of the IEEE Conference on Computer Vision and Pattern Recognition, pp. 2784–2791. Boston America (2015)
13. Ho, T.K.: The random subspace method for constructing decision forests. IEEE Trans. Pattern Anal. Mach. Intell. **20**(8), 832–844 (1998)
14. Di, W., Crawford, M.M.: View generation for multiview maximum disagreement based active learning for hyperspectral image classification. IEEE Trans. Geosci. Remote Sens. **50**(5), 1942–1954 (2011)
15. Kumar, V., Minz, S.: Multi-view ensemble learning: a supervised feature set partitioning for high dimensional data classification. In Proceedings of the Third International Symposium on Women in Computing and Informatics, pp. 31–37. Association for Computing Machinery, Kochi India (2015)
16. Kumar, V., Minz, S.: Multi-view ensemble learning: an optimal feature set partitioning for high-dimensional data classification. Knowl. Inf. Syst. **49**(1), 1–59 (2016)
17. Ye, H.J., Zhan, D.C., Yuan, J., Zhou, Z.H.: What makes objects similar: a unified multi-metric learning approach. IEEE Trans. Pattern Anal. Mach. Intell. **41**(5), 1257–1270 (2019)

# Chapter 40
# Smart Stove Automation Using Wireless Network and Mobile Application

**Arish Ali, Muhammad Faheem Mushtaq, Muhammad Umer, Maqsood Ahmad, Saima Sadiq, Michele Nappi, and P. Vijayakumar**

**Abstract** Stove automation that can remotely work for temperature-based boiling, time-based cooking, and gas leakage control has not been devised yet. So keeping this in mind, this work presents smart stove automation system providing impressive features including Wi-Fi-based remote control of stove, temperature-based liquid boiling control, timer-based cooking control, and gas leakage protection using gas sensor. The proposed system supports both mobile application and keypad. To validate the accuracy, efficiency, and affordability, the proposed system is designed as a laboratory-scale prototype that can be easily incorporated to the conventional gas stove, very easy to use, and can be easily assembled in other hardware circuits. Arduino UNO is used for controlling components and taking decisions, LCD provide user interface, keypad for user input, servo motor to control stove knob, and some other sensor units (LM35 temperature sensor, MQ-5 gas sensor). User needs to enter time or set temperature in case of boiling liquid and set the position level of required flame. Then, the stove starts its working automatically, and there is no need to control the stove. The proposed design cooking model will be very useful for children and working individuals when implemented for real conditions. In near future, stove automation is used at every place where waiting for cooking time is not feasible.

---

A. Ali · M. F. Mushtaq (✉)
Department of Information Technology, KFUEIT, Rahim Yar Khan, Pakistan
e-mail: faheem.mushtaq@kfueit.edu.pk

M. Umer · M. Ahmad · S. Sadiq
Department of Computer Science, KFUEIT, Rahim Yar Khan, Pakistan
e-mail: umersabir1996@gmail.com

M. Nappi
Department of Computer Science, University of Salerno, Fisciano, Italy
e-mail: mnappi@unisa.it

P. Vijayakumar
Department of Computer Science and Engineering, University College of Engineering, Tindivanam, India
e-mail: vijibond2000@gmail.com

 
V. Chang et al. (eds.), *Modern Industrial IoT, Big Data and Supply Chain*, Smart Innovation, Systems and Technologies 218,
https://doi.org/10.1007/978-981-33-6141-6_40

## 40.1 Introduction

Automation is the use of hardware and computer-controlled devices to expect control of processes [1]. The projected impact in the field of automation will lead to the rapid growth and development in the applications of automation which provide routes to the innovation [2, 3]. Different researches have been done on the automation that presents the automation systems which can complete their task without or with little human effort such as home automation system, automated irrigation system, door automation system, robotic-based item selection system, automatic liquefied petroleum gas (LPG) ordering system, stove automation, etc. [4–10] The Internet of Things (IoT) is a extremely dynamic and cosmopolitan network system. It is a system comprising several recognizable elements which will communicate and to move with one another and with end-users or other entities within the network [4, 5]. Stove automation is the smart devices that sense physical occurrences and translate them into a stream of information data, maximizes safety, comfort, convenience, and energy savings [6–12].

Gas stove is used in our daily life for cooking purpose. It is fueled by combustible gas such as natural gas, liquefied petroleum gas (LPG). With the passage of time, every household appliance such as the dryer, refrigerator has been automated, but the gas stove is in use manually from the previous years [13]. Stove automation is one of the classical problems which has been investigated by many research groups in the last decade [12–17].

Various approaches on traditional stove problems are proposed and evaluated. An automatic gas stove provides safety features during gas leakage such as alarm and automatically message sending facility [12]. The automation of conventional stove system automates the whole cooking process [13]. In this system, user needs to enter time and flame level, and after that the process is completed automatically. A programmable gas stove facilitates the user by solving the daily life issues of gas wastage and leakage [15]. The problem of gas wastage is solved by timer-based controlled cooking system, and the issue of gas leakage is solved using the gas sensor. Automatic gas stove is controlled by servo motor and Wi-Fi in which user can control the stove anytime and anywhere [16]. A gas stove with microcontroller-based embedded system is manufactured to control misuse of gas [17]. This system allows the gas to flow through valve of gas stove for cooking purpose. So, there is no chance to use gas for other purposes such as dry clothes or warm room.

There are many problems in daily life regarding the conventional gas stove that still needs to be improved and enhancement in the existing systems. The drawback of this programmable gas stove is that user needs to stand near the stove until liquid completes its boiling [15]. The process of boiling liquid is time-consuming. User needs to stand near the stove until the liquid completes boiling; otherwise, liquid will be wasted. The limitation of this automatic gas stove using servo motor and Wi-Fi is that it does not provide feature of timer-based cooking [16]. Furthermore, cooking of some foods such as beef and egg takes much time. If user forgets to turn off stove knob on time, food will be burnt or wasted. An issue in automatic gas stove using servo motor and Wi-Fi is that it does not provide safety from gas leakage [16]. Gas leakage is hazardous for human health as it causes different diseases or even death. Some of the limitations are partially solved by the existing researchers, but still there is a gap that needs to resolve it.

In contrast of existing research, our work presents a stove automation system which is manufactured to solve daily-life problems of cooking, boiling, and protection from gas leakage. It provides all the features needed for automated stove. The features of this system include Wi-Fi-based remote control of stove, temperature-based liquid boiling control, timer-based cooking control, and the gas leakage protection using gas sensor. The combination of these features makes the stove a complete automated system. Wi-Fi plays a very vital role in the process to controlling and monitoring the stove temperature and flame. Wi-Fi-based wireless sensor network (WSN) has the features of high bandwidth and rate, non-line-transmission ability, and large data collection [18]. The major use of Wi-Fi technology is to control the stove through mobile application.

In this research, IoT-based smart stove automation system as shown in Fig. 40.1 is proposed by using Wi-Fi module that used to make the cooking process much better than the previous efforts. Wi-Fi system embedded with a stove automation system helps to control the stove. There is an android application based on a user-friendly and attractive display connected with the stove automation system through Wi-Fi. This application provides controls to set temperature, flame level, and time to cook. When user sets value of temperature, flame level, and time, all the information are sent through the Wi-Fi network to stove. The main purpose of this research is to facilitate cooking process. The proposed architecture of this study provides the solutions that help the cooks to eliminate the time-consuming process of the cooking. The main contributions of this paper are following:

- Wi-Fi module with a stove automation is proposed that connected with mobile application that helps to cook food in less time
- The temperature sensor module helps to get the temperature of liquid, i.e., milk, water, and send to the microcontroller, to turn stove off
- Gas sensor protects from gas leakage and turns off the stove knob
- Wireless communication provides facility to control the stove system through mobile application.

This paper is organized as follows. Section 40.2 discusses all the components and hardware required for this system. Section 40.3 describes the system diagram and description of the circuit. Section 40.4 presents material and methods of this research. Section 5 explains the experimental design of the proposed method. Section 6 discusses the case studies that show the importance of each feature. Section 7 summarizes the results and gives the outlook on future work.

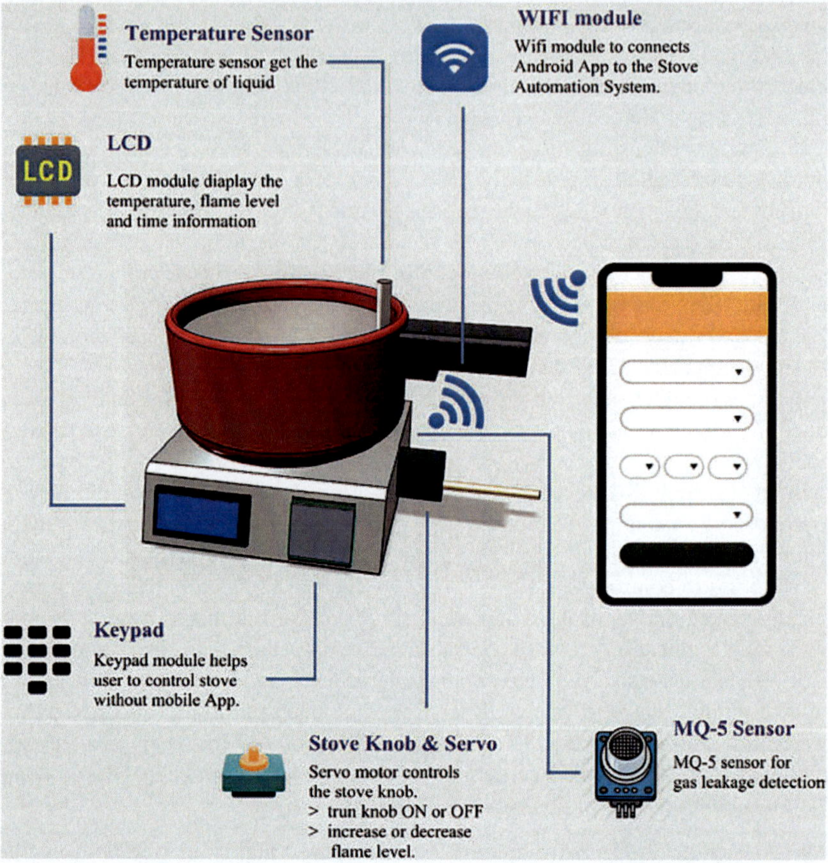


**Fig. 40.1** The proposed design of stove controlling manually and by using smartphones

## 40.2 Materials and Methods

The architecture of the proposed system is divided into six main modules for getting input from the user and provides output to user. The central part of this system, which is an Arduino microcontroller, controls these modules and takes decisions. The input modules such as sensors and keypad gather data from user, temperature from liquid, and send to Arduino microcontroller for taking decisions. The output modules such as LCD and buzzer receive data from Arduino and produce output. Figure 40.2 shows the architecture diagram of the proposed system controlling manually and mobile application.

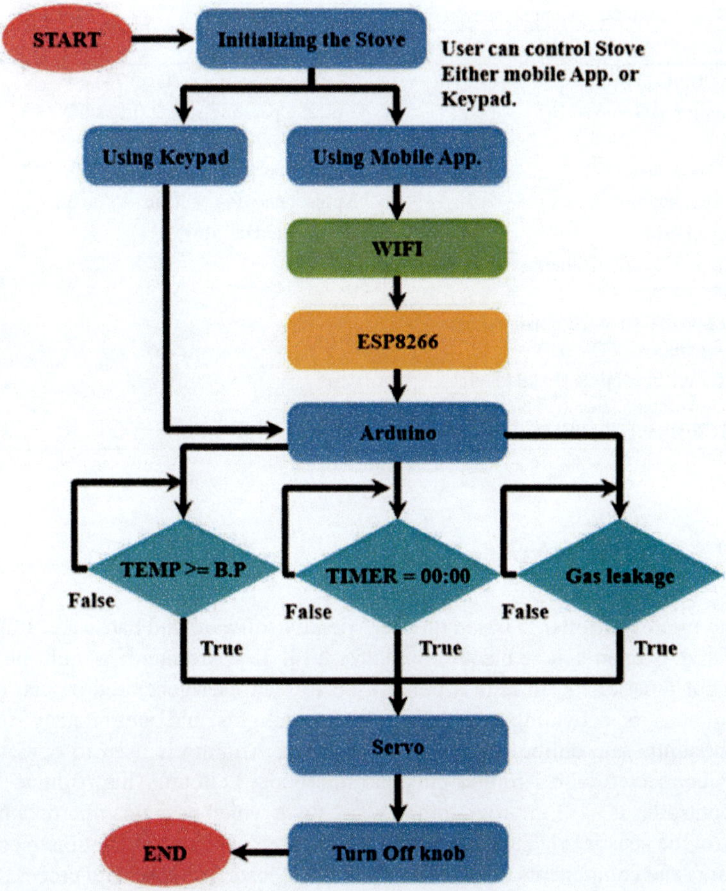


**Fig. 40.2** Flow diagram of using Mobile App. and keypad to control smart stove

### *40.2.1 Electronic Components*

The electronic components and hardware are used to design proposed system. Some of these components are used for input and some for output. Consequently, proposed circuit diagrams also contain those components and hardware that are specified in Table 40.1.

**Table 40.1** Proposed system electronic components specification

| Components | Specifications |
|---|---|
| Arduino UNO [11 , 15] | 28 pins; operating voltage: 7–12 V |
| Servo motor MG-966R [16] | 3 pins; Operating speed: 0.17 s/60° (4.8 V), 0.14 s/60° (6 V) |
| MQ-5 Gas Sensor [17] | 4 pins; operating voltage: 5V ± 0.1 |
| LM35 Temperature Sensor [18] | 3 pins; operating Voltage: 4 V to 30 V |
| Gas stove [19] | Household Gas Stove |
| LCD 16 × 2 With I2C interface [5] and 4 Pins; operating voltage: 7V | |
| 4 × 4 Keypad [20] and 8 pins; maximum operating voltage: 24V | |
| ESP8266 Wi-Fi Module [7] and 8 pins; operating voltage: 3.0–3.6V; Network Protocol: IPv4, TCP/UDP/HTTP/FTP | |

### *40.2.2 Arduino UNO*

Arduino microcontroller is based on user-friendly software and hardware. It is also considered as open-source electronic platform [4, 15]. Arduino is a multi-purpose board, but most of the time it is being used to read user-generated inputs, check status of sensors, activating a servo motor, turn on lights, and continuously streaming/publishing data online. In this research work, Arduino is used to operate the sensors connected with Arduino pins and microboard circuit. This Arduino UNO microcontroller is used in programmable gas stove which uses this microcontroller to control the sensor [5]. This microcontroller is used in stove automation to control all sensors and components. It receives input from sensors, keypad, and process input to take decision and send output result to the LCD.

### *40.2.3 Servo Motor MG-966R*

The MG996R servo motor is a type of servo motor which gives torque of 10 Nm, along with 360° rotation. The weight of this servo motor is 55 g and operates on 4.5V–6V DC voltage. One of the PWM-enabled pins on the Arduino UNO is used to drive the servo motor. Servo motors are used for multiple tasks specially where rotation of 360 is required. It is also used for monitoring and storing the database of registered vehicles. This concept is called RFID-based monitoring system which uses a 13.56 MHz RFID module. This RFID-based design model also sends SMS upon any violation of the vehicle using GSM module. The monitoring is done using ultrasonic sensor HC-SR04 and camera module that detect and capture images of those vehicles which are entering in the prohibited locations [16]. The servo motor is used in this automation system for turning the stove knob ON and OFF. It is fixed with the knob of the stove.

### *40.2.4 MQ-5 Gas Sensor*

For the detection of leakage in the home and industry, the MQ-5 gas sensor is mostly used. The reason for its wide usage is that it detects $H_2$, CO, LPG, $CH_4$, alcohol, and smoke. The other reason for its wide usage is fast response time, high sensitivity, and measurements can be taken and stored within no time. The potentiometer is used to adjust the sensitivity of the MQ-5 sensor. This is used to design a cigarette disposal blower that detects smoke and produces automatic freshener using MQ-5 sensor [17]. This sensor is used in this research for detection of gas leakage and sends level of gas present in the surrounding. If level of gas is higher, then it turns off the gas knob.

### *40.2.5 LM35 Temperature Sensor*

Whenever we want to take temperature readings with an output voltage linearly proportional to the centigrade temperature, LM35 series-integrated circuits are used. Like linear temperature sensors which are calibrated in Kelvin, users are not required to subtract the constant value to obtain readings in centigrade. The main advantage of the LM35 temperature sensor is that trimming or external calibration is not required for getting accuracy of $\pm$ 0.75 °C at room temperature and $\pm$ 0.75°C over a full –55 °C to 150 °C temperature range. This sensor is used to design and implement the microcontroller-based digital thermometer that uses LM35 for temperature measurement [18]. The sensor is used in proposed system to measure the temperature of the liquid and sends the value of temperature to the Arduino.

### *40.2.6 Gas Stove*

A gas stove is an appliance that is fueled by combustible gas such as syn-gas, natural gas, propane, butane, liquefied petroleum gas, or other flammable gas that are used for the cooking purpose. Exploring the pollutant emission such as PM2.5 and CO from natural gas and LPG which effects the cooking [19]. This gas stove is used in this research to make an automated stove by placing all sensors and components.

### 40.2.7 LCD 16 × 2

Whenever we are designing a standalone project, we need LCDs to get output of that project. A regularly used LCDs 16×2 require a large number of pins embedding to operate while the I2c interface-based LCDs require fewer pins. I2c interface just requires four connections SDA, GND, SCL, & VCC. Using I2c scanner, we can easily find I2c sensor address on Arduino. This LCD is used in programmable gas stove that is employed to display the sensor values and results [5]. This 16×2 LCD with I2C module is used to display the output of the user input and Arduino decision.

### 40.2.8 Keypad

A 4 × 4 keypad provides a useful interface to the microcontroller for variety of projects. The adhesive back provides a simple way to place the keypad to different applications. The 4 × 4 keypads have 16 keys which uses a combination of 4 rows and 4 columns to provide button states to microcontroller. Each key is a push button, with one end connected to one row and the other end connected to one column [20]. This keypad is used in development of an Arduino-based automated household utility power monitoring system that proposed an automated household utility power monitoring system and data logging in real time [21]. The 4×4 keypad is used in this system for user input to set time for timer-based cooking or to set temperature value.

### 40.2.9 ESP8266 Wi-Fi Module

This Wi-Fi module has the capability to host any type of application. It can also use to exchange networking functions among application processors. This Wi-Fi module is pre-programmed using AT commands. These commands help to know about the Wi-Fi characteristics. Multiple task scan be done using this Wi-Fi chip like a water flow control board, website communication board, soil moisture and temperature sensor board can be design using it [7]. This Wi-Fi chip is used in this research to control the stove knob and set the boiling point or time through Android application.

## 40.3 Designing Methodology

The description of main system components contains the Arduino UNO which takes input from sensors and user and take decision, MG996R servo motor to turn stove knob on or off, MQ-5 gas sensor module to detect the gas leakage, buzzer to produce

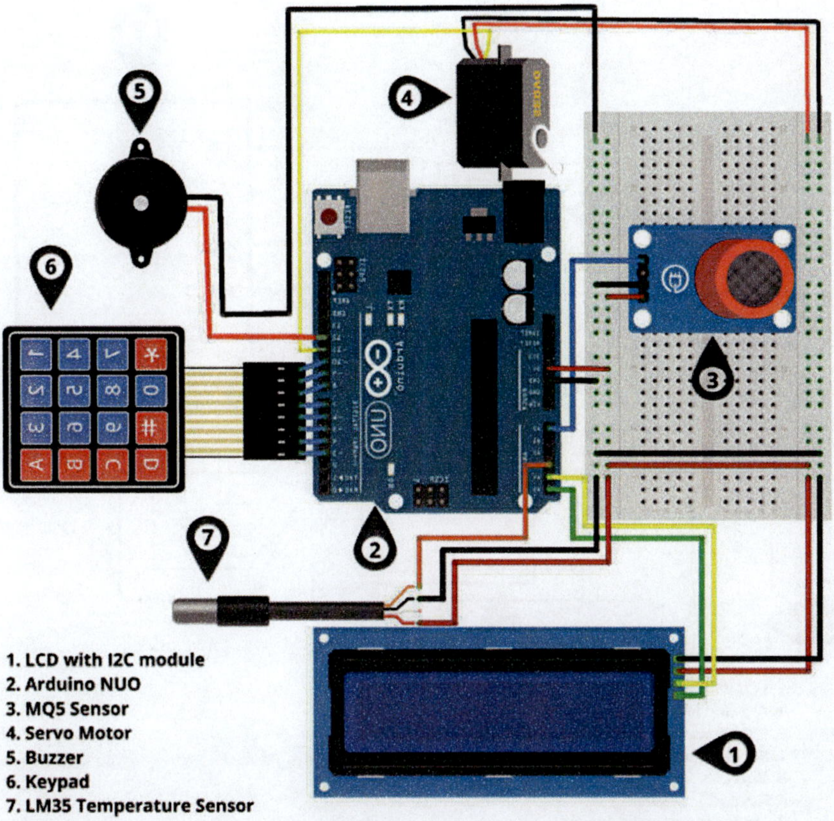


**Fig. 40.3** Description of main system components

sound in case of gas leakage, 16×2 LCD display with I2C interface to display all functions and menus, 4×4 keypad for user input and a temperature sensor LM35 to get liquid temperature are shown in Fig. 40.3. The temperature sensor (LM35) measures the temperature of the liquid and returns to the Arduino which takes decision, on whether turn off the knob of stove or not. Gas sensor (MQ-5) senses the leakage of gas in the air and turns off the knob of the stove.

### *40.3.1 Circuit Design of Electronic Components*

Circuit diagram, in the Fig. 40.4, shows the connections of the different components (such as temperature sensor LM35, 16×2 LCD with I2C interface module, 4×4 Keypad, gas sensor MQ-5, servo motor) and microcontroller (Arduino UNO). All the sensors and components are connected to the Arduino. Arduino receives data

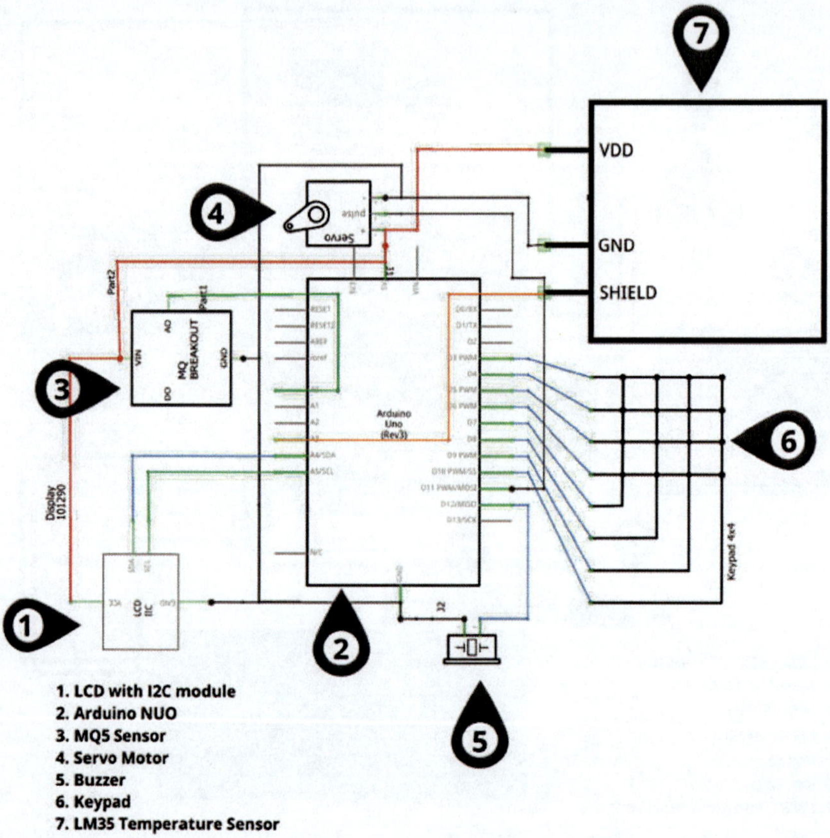


**Fig. 40.4** Circuit diagram of the system

from the sensors and keypad and makes decisions. Arduino processes data and sends to the LCD. The LCD is connected to inter-integrated circuit (I2C) module which allows to communicate the digital pins of LCD with analog pins of the Arduino. I2C module has four pins VCC, GND, SDA, and SCL for communication, and it consumes only two analog pins of the Arduino. Temperature sensor LM35 has three pins VCC, GND, and OUT. OUT pin sends temperature data to the Arduino. Red wire in the circuit shows VCC.

### *40.3.2 Working Phenomena*

The working of proposed system is divided into two parts including stove knob movement control, and input and output devices can be explained as follows:

#### 40.3.2.1 Stove Knob Movement Control

The MG996R servo motor is used in this system to control the stove knob. There are four possible positions of the knob which is used to set the flame level of the stove. These positions show 0° which corresponds to off, the 90° which corresponds to low flame, the 135° which corresponds to medium flame, and 180° which corresponds to high flame. In case, if knob is not at off position or servo motor is not at 0° whenever the system is turned on, servo is brought back to 0° before carrying out any other functionality. In order to operate the servo motor, the library Servo.h is included in the Arduino Integrated Development Environment (IDE) [12]. Figure 40.3 shows the connected servo motor with the stove knob.

#### 40.3.2.2 I/O Devices

A 4×4 keypad is used in this proposed system to set cooking time or liquid temperature. Selection of option from main menu, entering temperature of boiling liquid or set time for timer-based cooking; these functions are performed through keypad. In order to check current status and to take input from the user, a 16×2 LCD display is used. Using this LCD display, users set the values like time to adjust flame level to a particular valve level and the other necessary specifications. When the user set values according to the requirement of cooking material, these values get stored in the Arduino board. Until the completion of timer set initially by the user, the stove keeps the required flame level. The position of knob comes back to zero, as soon as the last set of flame and delay is completed. In order to boil a liquid, such as milk, user needs to select boil option from main menu. When liquid's temperature reaches its boiling point, stove automatically turned off.

### *40.3.3 Results and Discussions*

The LM35 temperature sensor is used to get temperature of liquid which dips in the liquid, and it sends the temperature values to the Arduino. LM35 temperature sensor is fully covered for waterproof use. The current temperature of liquid is displayed on the LCD. The keypad operation and LCD display operations are performed by including their respective header files, Keypad.h [12] and Liquid-Crystal_I2C.h [12], in the Arduino IDE. The code was uploaded to the Arduino UNO using Arduino IDE, an open-source IDE which can be used to upload code in high-level languages to the Arduino boards. The experimental prototype is shown in Fig. 40.5. The proposed system consists of the four main features that include flame level, liquid boiling, timer-based cooking, gas leakage protection. The three main options in LCD display are flame level, boil, and timer shown Fig. 40.6a. Flame-level option sets the level of flame. The boil option is used when boiling a liquid, and timer is used for timer-based cooking.

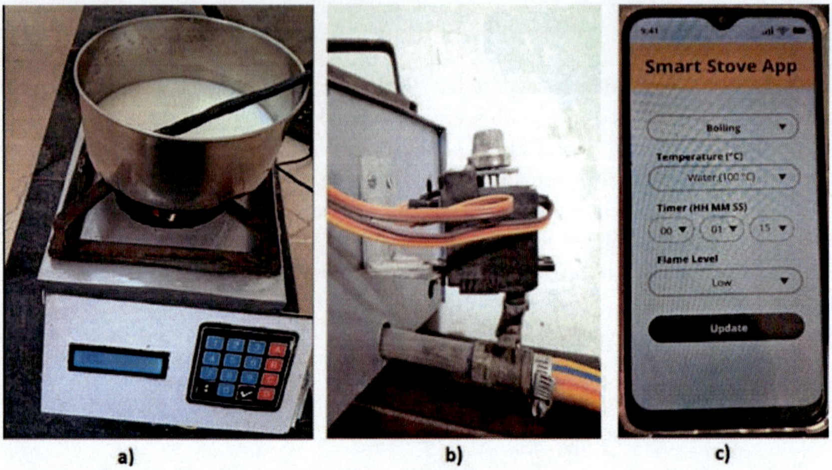


**Fig. 40.5** Smart stove using mobile app and keypad. **a** Smart stove automation with all sensors and components. **b** Servo motor to turn the stove knob ON and OFF. **c** Mobile app to control the smart stove

(a) Main Menu.

(b) Flame Level.

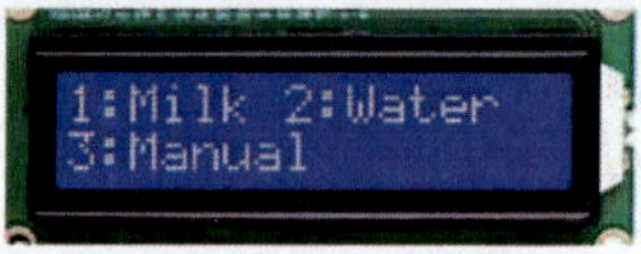


(c) Boiling Options

**Fig. 40.6** Features of proposed system

#### 40.3.3.1 Set Flame Level

In conventional gas stove, user sets flame level by turning stove knob in clock-wise or anti-clock-wise direction. However, in this research, user needs to select a number of ranges from 0 to 3. The flame level 0 means the off condition, 1 means low flame, 2 means the medium flame, and 3 means the high level of the flame. Figure 40.6b shows current level of flame, and flame levels show number of levels user can change.

**Table 40.2** Names of common liquids and their boiling point [22]

| Liquid Name | Boiling Point (°C) |
|---|---|
| Water | 100 |
| Milk | 100.5 |

#### 40.3.3.2 Liquid Boiling

The purpose for liquid boiling is to boil liquid for specified time and turn off stove knob when time elapsed. User needs to select the liquid from the list of options in the menu. When the liquid's temperature reaches to its boiling point, stove knob automatically turns off and buzzer produces sound. If the name of the liquid is not present in the options of menu, user can set temperature of that liquid manually. Table 40.2 shows temperature of common liquids such as milk. The boiling temperature of common liquid needs in daily routine such as milk and water that are already stored in Arduino.

The three boiling options are displayed as milk, water and manual shown in Fig. 40.6c. Milk option is to boil milk, water option is for boiling water, and manual option is to set temperature of liquid manually.

#### 40.3.3.3 Timer-Based Cooking

The purpose of timer-based cooking is to cook food for a specific time. User needs to enter time to cook food. Cooking time of foods shown in Table 40.3 are already stored in the system. User just needs to select from the menu. When the specified amount of time has elapsed, stove knob automatically turns off and buzzer produce sound.

#### 40.3.3.4 Gas Leakage Protection

The purpose of gas leakage protection is to protect lives from gas leakage. MQ-5 sensor senses the presence of methane or LPG gas in the air. If gas leakage detects, stove knob automatically turned off. In case, if gas leakage detected but stove knob is already off, the buzzer on the stove produced sound to aware the leakage of gas. Methane gas is present in air, but it cannot catch fire because its concentration in the air is too low. The range between lower explosive limit (LEL) and upper explosive limit (UEL) is referred to as the flammable range in which the LEL for methane is 5% and UEL is 15%. Concentrations of 9% are thought to be the most volatile. Concentrations above 15% will not be explosive as the air is too saturated with methane. However, this is when asphyxiation can be just as hazardous [24]. Gas leakage is the quantity of natural gas (methane gas CH4) or liquefied petroleum

**Table 40.3** Names of common cooking meat and vegetable and their cooking time [23]

| Food | Cut | Time |
|---|---|---|
| | BEEF | |
| Sirloin | Roast | 3–4 h |
| Brisket | Roast | 6± h |
| Ribeye | Steak | 5–15 h |
| | Chicekn | |
| Chicken | Whole | 3.5–4 h |
| Chicken | Breast | 25–35 min |
| Chicken | Thigh | 25–35 min |
| | FISH | |
| Salmon | Whole | 2–3 h |
| Salmon | Fillet | 30–40 min |
| Prawns | Whole | 10–15 min |
| | EGG | |
| Soft Boiled Egg | – | 6–7 min |
| Hard Boiled Egg | – | 9–10 min |

gas (LPG) present in the air greater than LEL is hazardous for human health as it causes diseases or even death. Symptoms of diseases caused by natural gas poisoning include dizziness, fatigue, nausea, headache, and decreased vision.

### *40.3.4 Case Studies*

The proposed system provides three features. Each function is tested below in each case

#### 40.3.4.1 Case 1

In first case of this system, temperature-based liquid boiling function is tested. In this case, milk is used for boiling. The boiling point of milk is 100.5 °C. In this test, we do not include time factor because the time, a liquid need to boil, depends on the climate. In winter, milk takes longer time to boil while, in summer, it takes less time. For testing purpose, 1kg milk is taken in a steel bowl and placed it on the stove. The flame level of the stove is set to medium, because we need normal flame level to boil liquid. The selected liquid boiling option from the main menu that includes the three options which are milk, water and manually set temperature. As we know, milk and water are common liquids which usually needed to boil that's why the value of boiling point is already stored in the microcontroller. The third option is "manual" which allows to set temperature of a liquid according to the temperature required for the particular purpose. In first case, the boiling liquid needs to select the "milk" option from the menu. When the option is selected, then LCD starts to show the current temperature of the milk. The temperature of the milk gradually rises. The rising temperature also displayed on the LCD. When the temperature of the milk rises above 85 °CC, a thin layer of cream appeared. When temperature of the milk reaches to 100.5 °C, stove automatically turned off. The algorithm for temperature-based boiling is as follows:

**Algorithm 1**

Temperature based Boiling

**Initialization**

Set Boiling_Point

Get Temperature_value

**if** Temperature_value $\geq$ Boiling_Point **then**

Send Turn Off signal to Servo Motor

**else**

Get Temperature_value again

break;

**end if**

In the same way, water boiling also tested on smart stove automation. When boiling point of water reaches to 100 °C, stove automatically turned off, as boiling point of water is 100 °C as shown in Fig. 40.7.

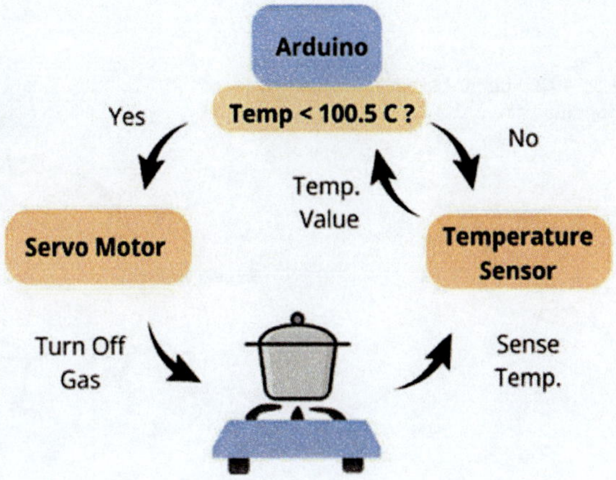


**Fig. 40.7** Temperature-based milk boiling

#### 40.3.4.2 Case 2

In second case of this system, timer-based cooking function is tested. This function is beneficial for all types of foods that needs to cook. As mentioned in Table 40.3, different foods have different cooking time. Using the timer option, user sets the cooking time of food on timer as define in the Algorithm 2. Stove automatically turned off when timer reaches to zero. This experiment is performed on the egg. Two eggs and some water are taken in a bowl. Table 40.3 shows the time for soft boil egg is 6–7 minutes and for hard boil, it is 9–10 min. For egg boiling, "Boil" option is selected from the menu and 10 min time is set on the timer. In the case of boiling eggs, the flame level is set to medium. When OK button is pressed from the keypad, time counter starts and is displayed on LCD. When the time on LCD reaches to zero, the stove automatically turns off shown in Fig. 40.8. The same experiment is performed with chicken, and result is also noted. The algorithm for timer-based cooking can be shown as follows:

**Algorithm 2**
Timer based Cooking

**Initialization**

Set Timer_Value

**if** Timer_Value == 0 **then**

Send Turn Off signal to Servo Motor

**else**

Timer_Value = Timer_Value_1

break;

**end if**

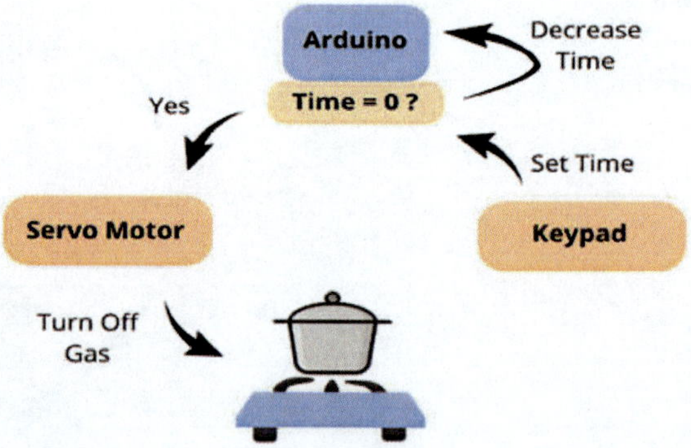


**Fig. 40.8** Timer-based cooking

#### 40.3.4.3 Case 3

In third case, we tested the gas leakage protection feature. In this case, we tested the gas leakage protection on both LPG and methane gas. We opened the stove knob but do not burnt fire. Gas started spreading all around. When gas sensor MQ-5 detected the gas leakage, stove knob automatically turned off shown in Fig. 40.9. In case, if stove knob is already at off position and the gas leakage is detected, buzzer produced sound to be aware from the gas leakage. The algorithm for gas leakage detection can be shown as follows:

**Algorithm 3**
Gas Leakage Detection

**Initialization**

Set Normal_value

Get Sensor_value

**if** Sensor_value > Normal_value **then**

Send Turn Off signal to Servo Motor

**else**

Get Sensor_value again

break;

**end if**

The implementation of this proposed system is very easy in the real world, and every type of kitchen needed this type of products. In homes, bakeries, restaurants, there are lots of cooking issues like burnt food during cooking process, hard boiled milk, and many others. To address these problems, this proposed system is a good choice.

**Fig. 40.9** Gas leakage protection

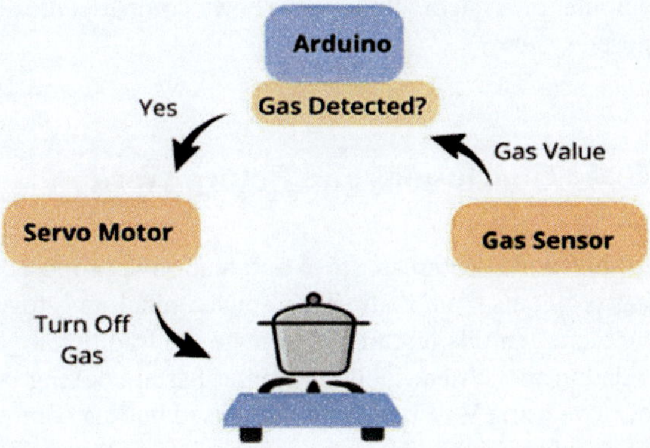

**Table 40.4** Names of common cooking meat and vegetable and their cooking time [23]

| Functionality | M. Mathew [11] | T. S. Goraya [10] | A. Sahu [13] | Proposed System |
|---|---|---|---|---|
| Temperature-Based Boiling | No | No | No | Yes |
| Timer-based Cooking | No | No | No | Yes |
| Gas Leakage Protection | Yes | Yes | No | Yes |
| Manual Control (Using Keypad or Buttons) | Yes | Yes | No | Yes |
| Wi-Fi control (Using Mobile App) | No | No | Yes | Yes |

The implementation and use of this system are similar the traditional gas stove, and there are no negative effects to the environment. On the other hand, this system protects environment by detecting and avoiding from gas leakage. In the same way, this system also saves life from gas leakage. The usage of this system is very simple. There is no need of training. Anyone who have a little knowledge to use mobile applications can easily use this system. Facilities provided by this proposed system also attract the customers.

### *40.3.5 Comparison of Existing Systems with Proposed System*

This proposed system uses Arduino UNO microcontroller to control the sensors and electronic components. Different sensors and components like ESP8266, Arduino UNO, LM35, and MG966R servo motor are used to make efficient and smart stove automation system. Table 40.4 shows comparison of existing systems and the proposed system.

## 40.4 Conclusions and Future Work

In this work, the smart stove automation is proposed to solve the daily-life cooking problems. Automating the conventional gas stove was the one of the real-life problems. In this research, automate the traditional gas stove and features are provided to the system including timer-based cooking control, remote-controlled use of stove using Wi-Fi, temperature-based boiling control, and gas leakage protection. The problem of boiling liquid is solved using the temperature sensor. Each liquid has its specific boiling point. When the temperature of liquid reaches its boiling point, the temperature sensor senses the temperature and set the flame level. Secondly, the cooking of food for a specific time is also solved using timer-based cooking. User sets up the timer value and counter start in decreasing order. When the timer's time out, stove knob automatically turns OFF. The protection of gas leakage is given using gas sensor. This sensor senses the gas in the air and sends signal to the microcontroller. Microcontroller sets the stop knob to OFF position. This system can be further analyzed in future to automate the prediction of cooking time, boiling time, temperature, and flame level. The data of experimental setup of stove in each time can be stored in database. Also, the machine learning algorithm can be applied on the stored data that help to predict of cooking time, temperature, and flame level.

## References


1. Schaefer, K., Chen, J., Szalma, J., Hancock, P.: A meta-analysis of factors influencing the development of trust in automation: implications for understanding autonomy in future systems. Human Factors J. Human Factors Ergono. Soc. **58**, 03 (2016)
2. Schütz, D., Wannagat, A., Legat, C., Vogel-Heuser, B.: Development of plc-based software for increasing the dependability of production automation systems. IEEE Trans. Ind. Informatics **9**, 2397–2406, 11 (2013)
3. Jain, S., Vaibhav, A., Goyal, L.: Raspberry pi Based Interactive Home Automation System Through e-mail, 02 2014, pp. 277–280
4. Saravanan, S.S.S.S.P., Pradeep, N., Shikhar, A., Rath, B., Paul, S.: Effect of automated burner knob for different medium on gasconsumption by local stove for household and industrialsafety. J. Ind. Pollution Control 1701–1708 (2017)
5. Nair, J.M.S.A.S., Thomas, C., Joseph, N.R.: Programmable gas stove. IJTRE, 73–78, 01 2018
6. Abdul Jamil, M.M., Ahmad, M.: A Pilot Study: Development of Home Automation System Via Raspberry pi, 03 2015, pp. 1–4
7. Singh, P., Saikia, S.: Arduino-based smart irrigation using water flow sensor, soil moisture sensor, temperature sensor and esp8266 wifi module, 01 2016, pp. 1–4
8. Kansara, K., Zaveri, V., Shah, S., Delwadkar, S., Jani, K.: Sensor Based Automated Irrigation System with IoT: A Technical Review, 12 2015
9. Schwarz, M., Milan, A., Lenz, C., Muñoz, A., Periyasamy, A.S., Schreiber, M., Schüller, S., Behnke, S.: Nimbro picking: versatile part handling for warehouse automation. IEEE Int. Conf. Robot. Autom. (ICRA) **2017**, 3032–3039 (2017)
10. Khan, A.: Gsm based automatic lpg ordering system with leakage alert. Int. J. Res. Eng. Technol. **03**, 40–42 (2014)
11. Mathew, M., Londe, D.: Automatic gas stove with advanced safety features. Int. J. Recent Contributions Eng. Sci. IT (iJES)
12. Saleque, A., Jahan, S., Talukdar, S., Islam, M., Azmir, M., Rashedin, M.: Development of smart Cooking Stove: Harvesting Energy from the Heat, Gas Leakage Detection and IoT Based Notification System, 01 2019
13. Sahu, P.S.A., Verma, N.: Automatic gas stove control by servo motor and wi-fi. IJTRE, 3377–3379, 01 2018
14. Karim, K.E., Hossain, J.A., Hossain, F.: Gas Stove with Embedded Controlling System to Stop Misuse of Natural Gas in Bangladesh, 12 2014
15. Kamelia, L., Noorhassan, S., Sanjaya, W.S.M., Mulyana, W.: Door-Automation System Using Bluetooth-Based Android for Mobile Phone, vol. 9, pp. 1759–1762, 01 2014

16. Panganiban, E.B., Dela Cruz, J.C.: Rfid-based vehicle monitoring system. In: 2017IEEE 9th International Conference on Humanoid, Nanotechnology, Information Technology, Communication and Control, Environment and Management (HNICEM), pp. 1–6 (2017)
17. Amin, M., Azel, M., Nugratama, A., Maseleno, A., Huda, M., Jasmi, K.A.: Design of cigarette disposal blower and automatic freshner using mq-5 sensor based on atmega 8535 microcontroller. Int. J. Eng. Technol. **7**, 06 (2018)
18. Auwal, A.M., Morufat, A.T., Yusuf, M.A., Muhammad, A., Sani, M., Abubakar, S.B.: Design and Implementation of a Microcontroller Based Digital Thermometer (2017)
19. Huboyo, H., Tohno, S., Cao, R.: Indoor pm 2.5 characteristics and co concentration related to water-based and oil-based cooking emissions using a gas stove. Aerosol Air Qual. Res. **11**, 08 (2011)
20. Chin, C.: Project 17 - interface with 4×4 keypad and 2×16 lcd. https://tutorial.cytron.io/2012/08/15/project-17-interface-with-4x4-keypad-and-2x16-lcd/ (2012). online: accessed 22 April 2020
21. Vergara, A.L., Villaruz, H.M.: Development of an arduino-based automated household utility power monitoring system. In: 2014 International Conference on Humanoid, Nanotechnology, Information Technology, Communication and Control, Environment and Management (HNICEM), pp. 1–6 (2014)
22. Marie, H.A.: What is the Boiling Point of Milk? https://www.thoughtco.com/boiling-point-of-milk-607369 (2012). Online: accessed 22 April 2020
23. Dataset: Adminchild, prime rib cooking time chart awesome time to take your temperature. http://childandfamily.info/prime-rib-cooking-time-chart/prime-rib-cooking-time-chart-awesome-time-to-take-your-temperature-receipes/ (2019). Online: accessed 22 April 2020
24. Chris, D.: Methane Gas Detection—Know the Risks and Detect the Danger. https://www.heatingandprocess.com/methane-gas-detection/ (2016). Online: accessed 22 April 2020

# Chapter 41
# Research on Intelligent Predictive Control of Roasting Furnace Temperature

**Cuiping Pu, Jie Ren, and Bin Xue**

**Abstract** A predictive control strategy based on the fuzzy neural network (FNN) for baking furnace is proposed. This method combines T-S fuzzy model with the RBF neural network to form a FNN. The temperature control system of anode baking furnace is taken as the object of simulation and predictive modeling. The T-S fuzzy RBF neural network (T-SFRBFNN) is used to model the off-line predictive control of the controlled system to provide online learning opportunities for the parameters of T-SFRBFNN. The self-compensating feedback correction method is used to adjust the output of the predictive model directly online to achieve the purpose of real-time control. The predictive control algorithm based on the FNN can establish an accurate multi-step predictive control model in the off-line state when the information of the controlled process is not fully understood and the precision of the model of the controlled object is not high. In online state, the predictive control algorithm can be calibrated by on-line feedback with self-compensation function. Meanwhile, the gradient descent method is used to adaptively adjust the parameters in the FNN model of the controller to realize the intelligent control of the controlled process. Simulation results show that the control method is simple, real-time and effective. It has good robustness and adaptability. It provides a theoretical basis for the producing high-quality carbon anode.

## 41.1 Introduction

The so-called roasting, that is, the process of the coal tar pitch carbonization in green carbon products, is one of the most important processes in the production process of carbon products. The purpose of the anode roasting is to improve the strength of the anode and reduce the resistance of the anode. Therefore, the roasting has a great

C. Pu (✉) · J. Ren
Kunming University, Kunming 650214, China
e-mail: pucuiping@126.com

B. Xue
Shanghai Feixun Co., Ltd., Shanghai 201100, China

 
V. Chang et al. (eds.), *Modern Industrial IoT, Big Data and Supply Chain*,
Smart Innovation, Systems and Technologies 218,
https://doi.org/10.1007/978-981-33-6141-6_41

impact on the property of the anode. In view of this, how to make baking parameters is very important.

The heating curve and many technological parameters of carbon anode and the baking furnace need to be selected strictly according to the actual requirements. There are many disturbances in the baking chamber, and most of them are unmeasurable. There is a serious coupling between the chambers. Therefore, it is a complex system [1–4] with large time delay, nonlinearity and no accurate mathematical model.

At present, in order to improve the anode quality, the traditional mathematical model and PID control are adopted to control the temperature of the roasting furnace [5]. However, because the anode baking furnace is an open furnace, it has the characteristics of model uncertainty and high nonlinearity, and the control task is complex, the traditional control method is not ideal for some indexes in practical application, and its anti-interference ability and robustness of the control system are poor.

The FNN is a new direction of intelligent control. The FNN with expert knowledge and self-learning function of the artificial neural network [6] is used to model the FNN, which can get better control effect. The T-S FNN combines the fuzzy system and the neural network, has strong fuzzy reasoning ability and the ability of the parallel process, and has stronger data processing ability than BP, RBF and other simple neural networks [7–9].

Generalized predictive control (GPC) is based on the generalized minimum variance control method. It introduces the multi-step prediction method into the optimization process and then uses the prediction model to predict the deviation between the output of the system at future time and the set value of the system. The GPC theory has been widely used as a new type of NN, which has an important impact on the development of industrial process control in China [10–13].

A predictive control system is proposed in this paper. It is based on FNN. This method combines the T-S RBF FNN structure with the multi-step predictive control. The T-SFRBFNN is used to model the off-line predictive control of the controlled system to provide online learning opportunities for the parameters of T-SFRBFNN control. The self-compensating feedback correction method is used to adjust the output of the predictive model [14] directly online to achieve the purpose of real-time control. The predictive control algorithm based on the FNN can establish an accurate multi-step predictive control model in the off-line state when the information of the controlled process is not fully understood and the precision of the model of the controlled object is not high. In online state, the predictive control algorithm can be calibrated by online feedback with self-compensation function. In the meantime, the gradient descent method is used to adaptively adjust the parameters in the FNN model of the controller to realize the intelligent control of the controlled process [15]. The simulation results show that the control method is simple, real-time, good control effect, and has excellent robustness and adaptability.

## 41.2 Intelligent Prediction Model

### *41.2.1 The T-SFRBFNN Predictive Control Modeling*

The research shows that to improve the real-time tracking and robustness of the system, it is very helpful to include sufficient information redundancy in the prediction model. Reasonable establishment of predictive model is an important part of predictive control [16, 17]. Since the fuzzy neural network model is based on a set of input, and output sample space, and there is no special requirement for the model of the controlled system, T-SFRBFNN is used to model the off-line predictive control of the controlled system [18, 19].

A general form of the time-discrete Eq. 41.1 is used to describe the input characteristics, and output characteristics of nonlinear time-delay control systems.

$$y(l+\tau) = f[y(l+\tau-1), \ldots, y(l+\tau-n), u(l), \ldots u(l-m)] \tag{41.1}$$

In the above formula, the input and output of the controlled object are expressed by $u(l)$ and $y(l)$ respectively. $m$ and $n$ are the order of the system input and the output, respectively. $l$ means the timeslot. $u(l-m)$ and $y(l+\tau-n)$ respectively represent the control input and output of the controlled object in the past. $f(\cdot)$ is the nonlinear mapping of the unknown $R^{n+m+1} \rightarrow R$. $\tau$ is the total time delay of the system. Equation 41.1 is further expressed as:

$$y(l+\tau) = f[y(l+\tau-1), \ldots, y(l+\tau-n), u(l+\tau-1), \ldots u(l+\tau-m-1)] \tag{41.2}$$

Combining Eq. 41.2 with T-S RBF FNN mode, the predictive control model of the controlled object is established, and the predictive model of the T-S RBF FNN as shown in Fig. 41.1 can be obtained.

In the prediction model, the Gauss function (GF) is the membership function of hidden nodes. The number of fuzzy rules is represented by $p$. Set $C_i$ as the center value of the GF, $r_i$ as the base width vector of the GF, $w_j$ as the network output weight vector, $a_n^j$ as the weight coefficient, and $\xi_j(X_i)$ as the GF output. As shown in Fig. 41.1, the predictive model outputs $y_m(l)$ of all controlled objects can be obtained by collecting past inputs and outputs in real time through the following formulas:

$$\xi_j(X_i) = \exp\left[-\frac{\|X_i - C_i\|^2}{2 \cdot r_i^2}\right] \tag{41.3}$$

$$w_j = a_0^j + a_1^j x_1 + \cdots + a_n^j x_n \tag{41.4}$$

$$\overline{\xi}_j = \frac{\xi_j}{\sum_{j=1}^{m} \xi_j} \tag{41.5}$$

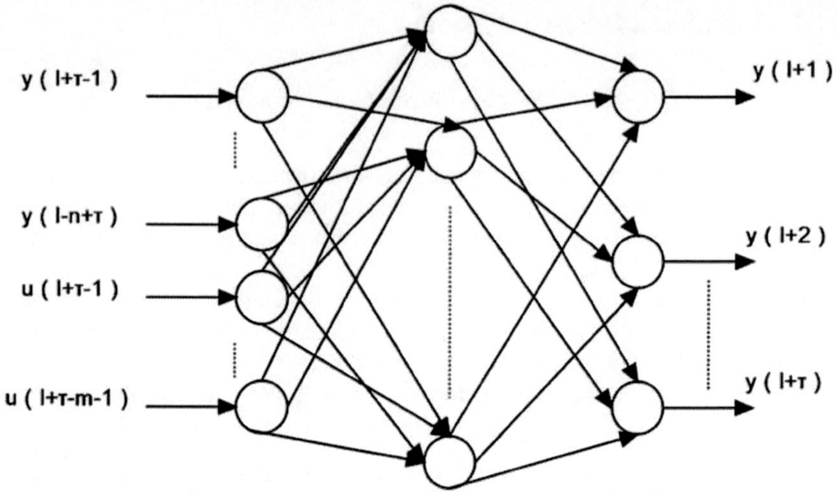


**Fig. 41.1** T-S RBF FNN prediction model

$$y_m(l) = \frac{\sum_{j=1}^{p} \xi_j w_j}{\sum_{j=1}^{p} \xi_j} = \sum_{j=1}^{p} \xi_j \overline{w}_j \tag{41.6}$$

As can be seen from Fig. 41.1, the multiple mapping of T-S RBF FNN prediction model from input to output constitutes a multi-step prediction.

In conclusion, the prediction model output of the network can be obtained as follows:

$$y_m(l+1) = f^1[y(l), y(l-1), L, y(l-n+1), u(l), L, u(l-m)]$$

......

$$y_m(l+g) = f^g[y(l+g-1), L, y(l+g-n), u(l+g-1), L, u(l+g-m-1)]$$

(41.7, 41.8)

where $f^g(\cdot)$ represents the $g$th mapping of input to the output through the FNN prediction model.

It should be pointed out that in Eqs. 41.7 and 41.8, all the time prediction models are impractical and the actual output values in the future cannot be detected now. Therefore, the predicted output values of the corresponding time of the model are used instead of the actual output values to realize the multi-step prediction of the FNN prediction model.

Through a large number of input and output data offline learning, according to the above formulas for multi-step predictive control modeling of the system, the accurate mathematical model of the time-delay nonlinear system is obtained, and the network structure is relatively simple, the number of hidden nodes is relatively small [20–22]. After off-line training, the model can learn online.

### 41.2.2 The Design of the T-S RBF FNN Controller

For the design of controller part, the structure of T-S RBF FNN is shown in Fig. 41.2.

The input is the error value and its change after the correction between the given value and the prediction model, and output is $u$. The number of fuzzy rules is equal to the number of nodes in the hidden layer of the network. Error values and their variations can be obtained from the following formulas:

$$e(l+\tau) = y_h(l+\tau) - y_m(l+\tau) \tag{41.9}$$

$l_h$: the real output.

$$ec(k+\tau) = e(k+\tau) - e(k+\tau-1) \tag{41.10}$$

As can be seen from Fig. 41.2, the input and output formulas for all levels in this network are as follows:

Input layer (IL): output $X_i = [x_{1i}, x_{2i}]^{\mathrm{T}}$ $x_{1i}$ : $e$ and $x_{2i}$ : $ec$,

Hidden layer (HL): input the output of the IL,

$$\text{output} \quad \mu_i(X_i) = \exp\left[-\frac{\|X_i - C_i\|^2}{2 \cdot r_i^2}\right],$$

Output layer (OL): input the output of the HL,

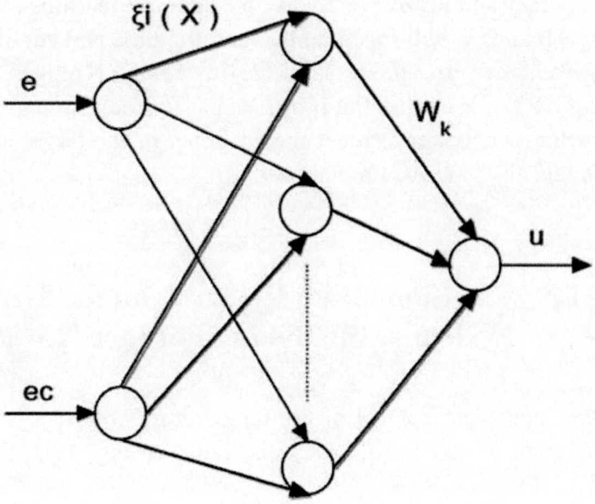


**Fig. 41.2** Structure of the T-S type RBF FNN

$$\text{output}\quad y = \frac{\sum_{k=1}^{l} w_k \mu_k(X_i)}{\sum_{k=1}^{l} \mu_k(X_i)},$$

where $w_k = a_0^k + a_1^k x_1 + a_2^k x_2$.

In the adaptive adjustment process of the T-S type RBF FNN, the center value and weight coefficient of its GF are obtained by the gradient descent method.

$$c_k(l+1) = c_k(l) + \eta_c \frac{\partial J}{\partial c_k} + \alpha_c[c_k(l) - c_k(l-1)] \tag{41.11}$$

$c$ means the central value, $k$ means the number of the weight for input;

$$a^k(l+1) = a^k(l) + \eta_a \frac{\partial J}{\partial a^k} + \alpha_a\left[a^k(l) - a^k(l-1)\right] \tag{41.12}$$

where

$$\frac{\partial J}{\partial c_k} = (y_m - y_h) \cdot \frac{\partial y_m}{\partial u} \cdot \frac{\partial u}{\partial \mu_i(X_i)} \cdot \mu_i(X_i) \cdot \frac{(x_i - c_i)}{r_k^2} \tag{41.13}$$

$$\frac{\partial J}{\partial a^k} = -(y_h - y_m) \cdot \frac{\partial y_m}{\partial u} \cdot \frac{\mu_k(X_k)}{\sum_{i=1}^{l} \mu_i(X_i)} \cdot \begin{bmatrix} 1 & X \end{bmatrix} \tag{41.14}$$

$$\frac{\partial u}{\partial \mu_i(X_i)} = \frac{w_k \cdot \sum_{k=1}^{p} \mu_k(X_i) - \sum_{k=1}^{p} \mu_k(X_i) \cdot w_k}{\left[\sum_{k=1}^{p} \mu_k(X_i)\right]^2} \tag{41.15}$$

where, $\eta$ represents the adaptive learning rate, $\alpha$ represents the momentum factor.

In the equations above from Eqs. 41.9–41.15, firstly, the error for the $d$ steps is fuzzy. Then, set the central value properly. To try to get the target central value in accuracy and faster, we involve the adaptive learning rate and momentum factor. If $\eta$ sets bigger, it will speed up the learning time and get the partial value of it and show the adaptive error process is more trusted. If $\alpha$ is bigger, the weight variability is taken seriously. To choose the proper value for the adaptive learning rate and momentum factor, it is better to meet the accuracy of the target and find the most appropriate value, also, to save time at best.

## 41.3 The Simulation Research on the Temperature Control System of the Baking Furnace [23–25]

For intelligent control of the temperature control system of the roaster, the block diagram of its control system is shown in Fig. 41.3. In Fig. 41.3, T-SFRBFNNPM

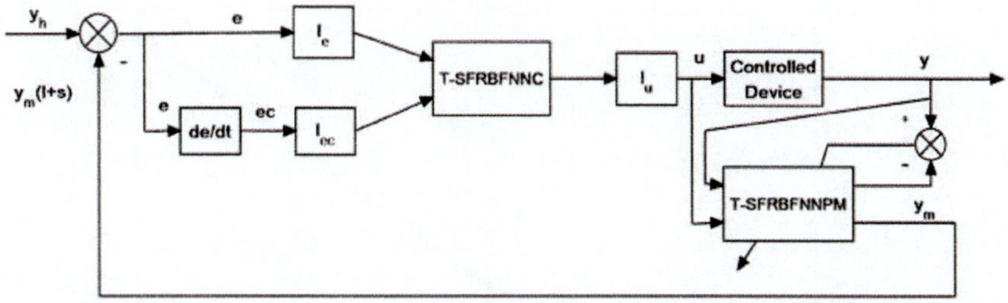


**Fig. 41.3** Block diagram of the predictive control system based on the T-S RBFFNN

is based on T-SFRBFNN, which provides online learning opportunities for T-SFRBFNNC parameters. *le*, *lec* and *lu* express the quantization factor of input variables and output variables respectively. $y_h$ rep the desired output of the system, $y_m(1+s)$ expresses the predictive correction output value of advance $s$ step for the prediction model, *e*, *ce* indicates the change of error and error, respectively. By combining off-line learning with online learning, the intelligent predictive control of the controlled object can be realized. Based on T-S type RBF FNN as the controller of the controlled object, the controller has two inputs: temperature error and temperature error change; single output: control the number of heavy fuel injection pulses. Data are provided by measurement. The principle of the control system is to establish an intelligent multi-step predictive model for the controlled object. By comparing the output of the intelligent multi-step predictive model with the given value of the output temperature of the actual controlled object, the error and the error change are produced. The FNN and intelligent controller can be adjusted adaptively online, and then the system is aligned. The system is controlled effectively so that the actual output of the system can reach a given value.

Setting the performance index of the system is

$$J = \frac{1}{2}[y_h - y_m(1+d)]^2, \tag{41.16}$$

According to processed input and output data, when the temperature of the controlled system reaches 785 °C, the control quantity is added. The FNN model and the performance index (Eq. 41.16) are used to control the theoretical heating curve (Fig. 41.4).

### *41.3.1 Simulation Analysis [26–29]*

The traditional PID control algorithm based on bias elimination is a short-sighted approach. For the calcine temperature system with great inertia and delay, it is difficult to achieve a pleasant regulating quality. T-S fuzzy RBF neural network generalized predictive control algorithm is based on the predictive model, which is provided

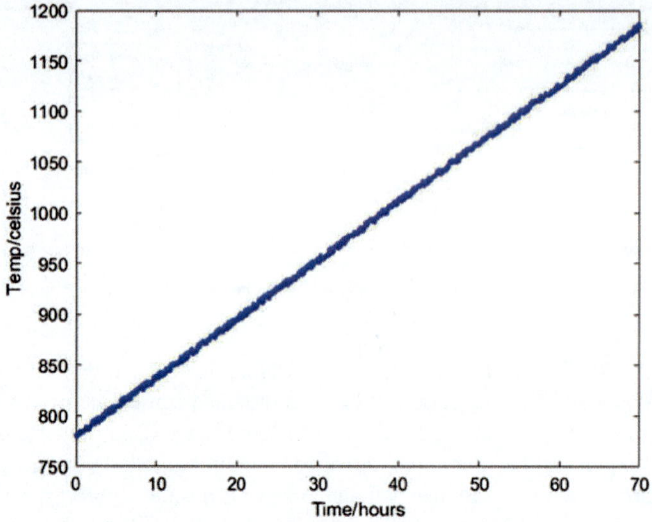


**Fig. 41.4** Theoretical heating curve

with the distinguishing features of multi-step prediction, dynamic optimization and feedback correction. Thus, optimal control can be achieved according to the dynamic characteristics of the baking furnace temperature system under different operating conditions. This method can not only obtain the optimal overshoot and regulation time, but also achieve optimal control. The adaptability of the system is improved.

The simulation results are shown in Fig. 41.5. It can be clearly seen from Fig. 41.5 that the output waveform of the system fluctuates with respect to the theoretical value in the first 13 steps. After the thirteenth step, the system can well keep up with the given value, and the error value does not exceed 0.5%. Overall, the control effect is better. It can be seen from [30] Fig. 41.5 that in the first 40 steps of traditional PID control, the system has a certain fluctuation, and the amplitude of fluctuation is large, the error reaches as much as 15 °C. Although the control system can basically keep up with the given value after 40 steps, the error value is much larger than the predicted output, reaching plus or minus 3 °C. Figure 41.5 compares the control effects between the two. The predictive intelligent control system has the fast adjusting speed and a small overshoot.

## 41.4 Conclusion

The simulation control effect of this paper is only aimed at the temperature rising process between 785 and 1150 °C, and the whole temperature rising process is not completely controlled. The input and output data are relatively limited, and the fuzzy neural network prediction model based on it has certain limitations.

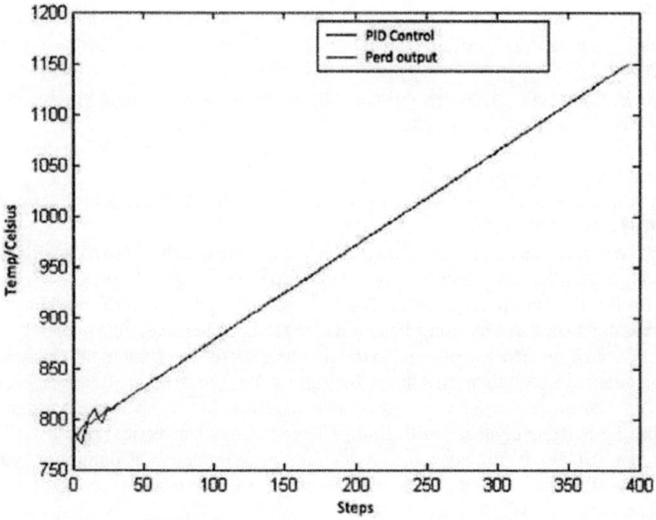


**Fig. 41.5** Comparison of control effects

In addition, the FNN controller based on the T-S RBF is also at the stage of theoretical research, and there are still many problems to be further studied and solved by the vast number of scholars. It includes a better scheme for selecting the number of hidden layer nodes of the fuzzy neural network, whether the initial center point and the width constant have certain rules and so on. At present, these can only rely on a large number of practical experience to analyze the choice, whether it is reasonable, whether it affects the control effect is still unknown. Therefore, with the deepening of research, solutions to these problems will make the predictive control system based on the fuzzy neural network proposed in this paper more widely and practical in modern industry.

**Acknowledgements** The authors acknowledge the Joint special project of Yunnan Province (202001BA070001-055), Yunnan Education Department Project (2019J0560), Innovative Training Program for College Students (201911393010).

## References

1. Li, C.: Study on the prediction and optimization control strategy of alumina roasting quality based on chaotic Grey Wolf algorithm. Guangxi University (2017)
2. Wang, Y.: Research on intelligent modeling and control system of alumina roasting process based on ELM genetic algorithm. Guangxi University (2017)
3. Xu, C., Wang, Y., Lin, X., Huang, Q.: Intelligent optimized control on temperature of alumina roaster inside. Comput. Simul. **34**(09), 319–324 (2017)

4. Xu, C., Li, C., Wang, Y., Lin, X.: Prediction model of alumina quality index based on chaotic grey wolf optimization algorithm. J. Guangxi Univ. (Natural Science Edition) **41**(06), 1869–1878 (2016)
5. Dong, S., Hao, P., Tighe, S.: A diagnostic method for evaluating the condition index of cement-stabilised base using T-S fuzzy neural network. Int. J. Pavement Eng. (2019)
6. Hou, Y., Zhao, H.: Handwritten digit recognition based on improved BP neural network. In: Chinese Intelligent Systems Conference (2018)
7. Qiao, J., Cai, J., Han, H.: Prediction of PM (2.5) based on T-S fuzzy neural network. Control Eng. **25**(03), 391–395 (2018)
8. Song, G., Wang, S., Yang, X.: Fuzzy predictive control algorithm of partial ventilator based on T-S model. J. Nanjing Univ. Sci. Technol. **41**(5) (2017)
9. Yu, J., Li, X.: Prediction of the annealing furnace temperature control based on T-S model fuzzy neural network. J. Shenyang Jianzhu Univ. (Natural Science) **30**(1) (2014)
10. Fu, B.: Research on fuzzy control of extraction temperature of traditional Chinese medicine based on generalized prediction. Autom. Instrum. **4**, 14–17 (2018)
11. Wang, Y.: Design and application of boiler reheat steam temperature system based on generalized predictive control. North China Electric Power University (2017)
12. Liu, D., Yin, J., Ding, F.: Energy consumption estimation modeling of aluminum hydroxide gas suspension calcinations based on least squares support vector machine and genetic algorithm. J. Comput. Appl. **34**(04),1217–1221 (2014)
13. Li, S.: Application of predictive PID stove temperature control technology in ladle baking system. First Heavy Technol. **5**, 64–66+71 (2015)
14. Liu, X., Guo, M., Yang, C., Zhang, J., Zhou, Y.: Design and simulation of vacuum wood carbonization equipment control system based on fuzzy neural network. J. Northeast Forest. Univ. **45**(08), 87–92 (2017)
15. Zhang, Q.: Active noise control using a simplified fuzzy neural network. J. Sound Vib. **4**(22) (2004)
16. Pu, C., Ren, J., Xue, B.: Study of predictive control model of fuzzy neural network. J. Comput. Methods Sci. Eng. Preprint **1**, 1–9 (2018)
17. Pu, C., Ren, J., Xue, B.: Dynamic self-adaptive learning algorithm research based on T-S RBF fuzzy neutral network. J. Comput. Methods Sci. Eng. **1**, 209–218 (2019)
18. Lu, Y., Nan, J., Gao, M.: Improved parallel particle swarm optimization for RBF neural network modeling. Comput. Eng. Appl. **53**(14), 45–50 (2017)
19. Zhang, X., Guo, X., Zhu, L.: Generalized predictive control for time-delay nonlinear systems based on neural networks. Meas. Control Technol. **36**(02), 54–57 (2017)
20. Shen, C., Yang, J.: RBF neural network of greenhouse temperature control system PID control. Control Project **24**(02), 361–364 (2017)
21. Song, R.: Research on sensor temperature compensation system based on improved RBF neural network. Mech. Strength **38**(06), 1225–1228 (2016)
22. Zhu, Z., Wu, H., Zheng, J.: Temperature prediction of disc solar collector based on RBF. J. Nanjing Inst. Eng. (Natural Science Edition) **15**(02), 67–71 (2017)
23. Ge, C., Wang, L., Cao, X.: Complete Collection of MATLAB Technology. People's Post and Telecommunications Publishing House, Beijing (2014)
24. Xie, Z.: Statistical Analysis and Application of MATLAB in 40 Cases. University of Aeronautics and Astronautics Press, Beijing (2012)
25. Zhou, R., Zhang, L.: Fuzzy and Neural Network Design Based on MATLAB and Fuzzy TECH. Electronic Industry Press, Beijing (2010)
26. Cheng, D.: Application of predictive control in main steam temperature system. North China Electric Power University (2016)
27. Xiu, G., Wang, L.: Simulation analysis of nonlinear model predictive control of guided projectile attitude. J. Nanjing Univ. Technol. (Natural Science Edition) **35**(01), 66–71 (2011)
28. Zhao, S.: MATLAB Modeling and Simulation. Tsinghua University Press, Beijing (2013)

29. Zhao, J., Dai, J.: Predictive tracking control simulation analysis based on MATLAB fuzzy controller. Comput. Simul. **9**, 28–30 (2004)
30. Wang, J., Zhu, Y., Qi, R., Zheng, X., Li, W.: Adaptive PID control of multi DOF industrial robot based on neural network. J. Ambient Intell. Humanized Comput. (2020)

# Chapter 42
# Investigating the Clarity Dimension of Social Norms: How Normative Ambiguity on Digital Piracy Intentions

**Chang Xiong and Victor Chang**

**Abstract** Digital piracy is widely recognized as a global concern, and it may arise due to different causes. The Chinese music market is holding unique characteristics as a result of China's distinctive institutional context and technological development. However, the extant research in explaining this unethical consumer behavior in the digital era only examined the impact of the strength of social norms and ignored the effect imposed by the clarity dimension of social norms. The purpose of this paper is to introduce the clarity dimension of social norms into explaining consumers' digital piracy intentions through a deductive approach. The current research contributes to the existing literature on digital consumer behavior by investigating the new perspective of clarity of social norms. Also, by bringing in data from an under-studied context of an emerging economy, this study offers a more comprehensive understanding of the piracy issues from a social–psychological perspective. Theoretical and managerial contributions, as well as the limitations and suggestions on future research, are also discussed.

## 42.1 Introduction

"Copyright-intensive industries, including film, television and music, support over 11 million jobs…., and piracy, with the associated loss of revenue it brings, represents a direct threat to those industries," notes Christian Archambeau, Executive Director of the EUIPO. The government reports of several countries (e.g., Australia, China, European countries, USA) highlighted that there is much work to do to tackle pirated consumption problem to help decision makers as they develop policies and solutions. The State Intellectual Property Office of China highlighted a key area where the need for more investigation is consumer's awareness. This is because

C. Xiong
HeXie Management Research Centre, Xi'an Jiaotong-Liverpool University, Suzhou, China

V. Chang (✉)
Artificial Intelligence and Information Systems Research Group, School of Computing, Engineering and Digital Technologies, Teesside University, Middlesbrough, UK

 
V. Chang et al. (eds.), *Modern Industrial IoT, Big Data and Supply Chain*,
Smart Innovation, Systems and Technologies 218,
https://doi.org/10.1007/978-981-33-6141-6_42

such studies can provide more detailed information regarding consumer's awareness with the decision markers from both governments and the industries to facilitate the sustainable development of the society and industries.

Although there are many studies that have been investigating consumer's awareness across various contexts (e.g., environment, food), there is still a little focus on how consumer's social norms impact their unethical behavior, particularly in the context of digital domain [1, 2]. Some of those studies in digital piracy behaviors have predominately applied theory of planned behavior to predict consumer behavior based on the intention to perform the behavior and levels of perceived behavioral control [3–8]. Some others have applied psychological theories to predict consumer behavior, such as egocentric heuristics and social learning perspectives [9], social cognitive theory [10], uses and gratifications approach [11] and exchange theory [12].

The literature has identified subjective norm as one the main factors that influence consumer behavior. Subjective norms are derived from the way an individual perceives that other significant referants (people or groups) would behave. The term "subjective norm" captures both social norms and the individual's susceptibility to social pressure to perform in a certain way. The literature further conceptually expanded subjective norms into descriptive subjective norms and injunctive subjective norms [13]. Descriptive subjective norms refer to an individual's observations on others' behavior, while injunctive subjective norms are formed through the perception on what others think an individual should do.

However, those existing studies have largely investigated the impact of the strength of social norms, such as injunctive norms and descriptive norms, on behavioral intentions [14, 15]. Indeed, the strength of social norms provides its significant value on our understanding of how social norms affect customer's unethical behavior. According to the psychological research [16], when we are investigating a person's social awareness, we should not only pay attention to its strength, but also its clarity. Therefore, we propose that the clarity of social norms is also an important dimension to be considered, as the clarity of norm information is a crucial assumption to enable social norms to impact an individual's behavior. We believe that, in addition to the strength of social norms, the exploration of the clarity of social norms can provide a comprehensive insight into how social norms influence unethical consumer behavior toward digital domain.

This paper makes several contributions to the literature. First, we extend the academic community's understanding of social norms by investigating a new dimension of social norms in the form of which have a direct influence on individuals' unethical consumer behavior in the digital era. Prior research in the field of unethical consumer behaviors has emphasized largely on the strength of social norms [15, 17–19]. We introduce the clarity of social norms. We provide a new perspective on how the strength and clarity of social norms work together to influence unethical consumer behavior.

## 42.2 Theoretical Background

### *42.2.1 The Strength of Social Norms: Subjective Norms Descriptive and Injunctive Norms*

Perhaps the most influential development of the normative component in TPB is the further conceptualization of subjective norms proposed by Cialdini et al. [20]. Inspired by the mixed and sometimes contradictory findings on the effect subjective norms have on behavioral intentions, Cialdini and his colleagues have suggested that "…norms do have substantial impact on human action; however, the impact can only be properly recognized when researchers separate two types of norms" [21]. These two types of norms are descriptive norms and injunctive norms, with descriptive norms referring to the perceptions of what most others do and injunctive norms referring to the perceptions of what most others approve or disapprove of doing [22].

Descriptive norms predict or explain human behavior. They have an informational value in terms of the behavior in question, and they are argued to provide an information-processing advantage and a decisional shortcut when an individual is making a choice on what to do. This type of norms is confirmed to act as reliable predictors or antecedents of a number of social behaviors even if the behaviors are in a morally neutral position [20]. Another type of social norms is the injunctive norms which involve an explicit *request* on what an individual should do in terms of a certain behavior. In differentiating these two types of social norms, Cialdini stated that "…rather than simply informing one's actions (i.e. descriptive norms), these norms (injunctive norms) enjoin it through the promise of social sanctions." Descriptive and injunctive norms are conceptually different, and they are derived from different sources of human motivation. However, these two types of norms are connected to a certain extent, as what an individual should do is always what most others do.

The theoretical expansion of subjective norms into descriptive and injunctive norms gives scholars a new perspective on the normative and influences the social context imposed on human social behaviors. Descriptive norms were argued to explain or predict human behaviors through the informational value they could provide. This kind of social norm could provide information to the decision-makers on the behaviors of others. Injunctive norms, which are conceptualized as a different source of normative motivation of human behavior, are argued to contain an explicit request on what should be done to be considered socially acceptable. The extant researches have examined the different effect of descriptive and injunctive norms on human behaviors which significantly enhanced our understandings of these normative motivations and their functioning mechanisms. These studies have relied on a crucial assumption that the descriptive norms and injunctive norms toward a certain human behavior are clear to the decision-makers. However, in some cases or under certain circumstances, an individual's perceptions of these social norms could be ambiguous.

## 42.3 The Clarity of Social Norms

### 42.3.1 Normative Ambiguity

In previous research, normative ambiguity has been identified in sociological research into conflict and violent contexts [23]. In their research, normative ambiguity was defined as the "absence of higher-order rules for reconciling contradictions between conflicting codes for conduct in situations ..." [23]. However, this definition was given in the context of violence. Following this line of reasoning, the definition of normative ambiguity could be extended to cover a broader range of human behaviors. In this research, we define ***normative ambiguity***, which was identified and conceptualized from the interviews, as the situation where the individual is unable to possess precise normative beliefs toward the behavior of interest from their social referents. In the specific context of music piracy behavior in China, normative ambiguity might be observed since the concept of copyright was not originated in China and Chinese music consumers had been enjoying the "free" music services for years, and as a result, the referents chosen by the music consumers may not be able to deliver negative descriptive and injunctive messages against music piracy behaviors to the music consumer.

According to [24–27], subjective norm captures the social influence imposed by individuals' social referents. These social referents include your family members and close friends. We argue that one crucial assumption for these social referents to be valid is that they must possess explicit normative beliefs toward the behavior of interest (i.e., whether one should conduct digital piracy or not). In another word, for the family members and close friends functioning as valid social referents, they must be able to easily decide how to respond in the context of digital piracy since there exists a higher-order rule in dealing with music acquisition behaviors. While in China and many other less-developed markets, intellectual property protection issues involved in the music acquisition process might be ambiguous, given the fact that most Chinese music consumers are not paying for the online music services in China. In this case, these potential social referents, as well as the social influence imposed by them, might be invalidated by the presence of normative ambiguity toward digital piracy.

## 42.4 Results

### 42.4.1 Construct Reliability and Model Fit

**Internal Consistency Reliability**. The composite reliability of each construct should be greater than 0.70, and composite reliability of all constructs in our study exceeds 0.85, which shows satisfactory internal consistency reliability [28].

| Variable | Items | Factor loadings | Cronbach's alpha | CR | AVE |
|---|---|---|---|---|---|
| INJ | INJ1 | 0.853 | 0.862 | 0.906 | 0.707 |
| | INJ2 | 0.796 | | | |
| | INJ3 | 0.879 | | | |
| | INJ4 | 0.835 | | | |
| DES | DES1 | 0.813 | 0.839 | 0.892 | 0.675 |
| | DES2 | 0.873 | | | |
| | DES3 | 0.799 | | | |
| | DES4 | 0.798 | | | |
| NA | NA1 | 0.899 | 0.825 | 0.895 | 0.739 |
| | NA2 | 0.798 | | | |
| | NA3 | 0.879 | | | |
| PR | PR1 | 0.887 | 0.866 | 0.918 | 0.789 |
| | PR2 | 0.884 | | | |
| | PR3 | 0.895 | | | |
| INT | INT1 | 0.902 | 0.758 | 0.892 | 0.805 |
| | INT2 | 0.892 | | | |
| DPB | DPB1 | 0.902 | 0.781 | 0.901 | 0.821 |
| | DPB2 | 0.910 | | | |
| PBC | PBC1 | 0.824 | 0.846 | 0.896 | 0.683 |
| | PBC2 | 0.823 | | | |
| | PBC3 | 0.842 | | | |
| | PBC4 | 0.817 | | | |
| ATT | ATT1 | 0.866 | 0.881 | 0.918 | 0.737 |
| | ATT2 | 0.837 | | | |
| | ATT3 | 0.856 | | | |
| | ATT4 | 0.875 | | | |

### *42.4.2 Discriminant Validity*

| | INJ | DES | NA | PR | INT | DPB | PBC | ATT |
|---|---|---|---|---|---|---|---|---|
| INJ | **0.841** | | | | | | | |
| DES | −0.427 | **0.821** | | | | | | |
| NA | 0.165 | −0.287 | **0.860** | | | | | |
| PR | 0.284 | −0.369 | 0.523 | **0.888** | | | | |
| INT | −0.543 | 0.560 | −0.348 | −0.534 | **0.897** | | | |

(continued)

(continued)

| | INJ | DES | NA | PR | INT | DPB | PBC | ATT |
|---|---|---|---|---|---|---|---|---|
| DPB | −0.194 | 0.371 | −0.175 | −0.271 | 0.545 | **0.906** | | |
| PBC | −0.384 | 0.414 | −0.239 | −0.302 | 0.550 | 0.399 | **0.827** | |
| ATT | −0.360 | 0.452 | −0.243 | −0.302 | 0.548 | 0.386 | 0.407 | **0.858** |

| | INJ | DES | NA | PR | INT | DPB | PBC | ATT |
|---|---|---|---|---|---|---|---|---|
| INJ1 | **0.853** | −0.335 | 0.081 | 0.262 | −0.447 | −0.132 | −0.330 | −0.294 |
| INJ2 | **0.796** | −0.405 | 0.207 | 0.211 | −0.416 | −0.179 | −0.293 | −0.297 |
| INJ3 | **0.879** | −0.364 | 0.139 | 0.255 | −0.500 | −0.178 | −0.342 | −0.341 |
| INJ4 | **0.835** | −0.337 | 0.134 | 0.224 | −0.458 | −0.165 | −0.323 | −0.275 |
| DES1 | −0.397 | **0.813** | −0.238 | −0.318 | 0.474 | 0.313 | 0.375 | 0.366 |
| DES2 | −0.343 | **0.873** | −0.252 | −0.300 | 0.467 | 0.333 | 0.308 | 0.386 |
| DES3 | −0.329 | **0.799** | −0.218 | −0.288 | 0.409 | 0.255 | 0.313 | 0.331 |
| DES4 | −0.329 | **0.798** | −0.234 | −0.305 | 0.481 | 0.312 | 0.361 | 0.397 |
| NA1 | 0.155 | −0.281 | **0.899** | 0.474 | −0.328 | −0.158 | −0.229 | −0.218 |
| NA2 | 0.151 | −0.209 | **0.798** | 0.424 | −0.237 | −0.095 | −0.151 | −0.183 |
| NA3 | 0.124 | −0.244 | **0.879** | 0.452 | −0.320 | −0.184 | −0.225 | −0.222 |
| PR1 | 0.240 | −0.305 | 0.456 | **0.887** | −0.471 | −0.208 | −0.272 | −0.296 |
| PR2 | 0.247 | −0.349 | 0.446 | **0.884** | −0.491 | −0.277 | −0.267 | −0.280 |
| PR3 | 0.270 | −0.329 | 0.494 | **0.895** | −0.459 | −0.236 | −0.266 | −0.228 |
| INT1 | −0.496 | 0.522 | −0.324 | −0.512 | **0.902** | 0.494 | 0.521 | 0.476 |
| INT2 | −0.477 | 0.481 | −0.300 | −0.445 | **0.892** | 0.484 | 0.465 | 0.509 |
| DPB1 | −0.194 | 0.322 | −0.146 | −0.249 | 0.487 | **0.902** | 0.368 | 0.328 |
| DPB2 | −0.159 | 0.350 | −0.170 | −0.243 | 0.501 | **0.910** | 0.355 | 0.371 |
| PBC1 | −0.349 | 0.293 | −0.197 | −0.247 | 0.447 | 0.283 | **0.824** | 0.347 |
| PBC2 | −0.329 | 0.382 | −0.199 | −0.238 | 0.471 | 0.330 | **0.823** | 0.335 |
| PBC3 | −0.336 | 0.374 | −0.172 | −0.259 | 0.468 | 0.371 | **0.842** | 0.335 |
| PBC4 | −0.250 | 0.317 | −0.225 | −0.256 | 0.430 | 0.334 | **0.817** | 0.330 |
| ATT1 | −0.313 | 0.425 | −0.246 | −0.290 | 0.500 | 0.370 | 0.395 | **0.866** |
| ATT2 | −0.246 | 0.311 | −0.190 | −0.221 | 0.417 | 0.268 | 0.322 | **0.837** |
| ATT3 | −0.326 | 0.394 | −0.215 | −0.255 | 0.452 | 0.286 | 0.331 | **0.856** |
| ATT4 | −0.342 | 0.413 | −0.182 | −0.266 | 0.504 | 0.388 | 0.346 | **0.875** |

The model fit was assessed through $R^2$ and by conducting a blindfolding test, and the $Q^2$ value (construct cross-validated redundancy) was calculated to assess the predictive relevance of the model. In this study, the $R^2$ value of the proposed structural model is 0.687 for INT and 0.311 for DPB, which indicates satisfactory explaining power. In addition, the $Q^2$ values calculated from blindfolding procedure for all constructs are greater than zero, which is indicative of predictive relevance of the proposed structural model [29, 30].

| | *R* square | *Q* square |
|---|---|---|
| INT | 0.687 | 0.513 |
| DPB | 0.311 | 0.237 |

### 42.4.3 Path Coefficients and Test of Hypotheses

The path coefficients of each path in the proposed framework as well as the corresponding *P* values are presented in table.

| Paths | $\beta$ | STDEV | *t* | *p* |
|---|---|---|---|---|
| INJ → INT | −0.192 | 0.036 | 5.330 | 0.000 |
| DES → INT | 0.168 | 0.058 | 2.907 | 0.004 |
| INT → DPB | 0.529 | 0.044 | 12.148 | 0.000 |
| PBC → INT | 0.189 | 0.043 | 4.368 | 0.000 |
| ATT → INT | 0.208 | 0.043 | 4.887 | 0.000 |
| INJ * NA → INT | −0.086 | 0.041 | 2.070 | 0.039 |
| INJ * PR → INT | −0.221 | 0.048 | 4.607 | 0.000 |
| DES * NA → INT | 0.069 | 0.054 | 1.275 | **0.202** |
| DES * PR → INT | −0.178 | 0.085 | 2.101 | 0.036 |
| Age → DPB | 0.018 | 0.039 | 0.463 | 0.643 |
| Gender → DPB | 0.042 | 0.039 | 1.063 | 0.288 |
| Income → DPB | −0.036 | 0.040 | 0.905 | 0.366 |
| Education → DPB | 0.103 | 0.039 | 2.639 | 0.008 |

## References

1. Chatterjee, S., Sarker, S., Valacich, J.S.: The behavioral roots of information systems security: exploring key factors related to unethical IT use. J. Manag. Inf. Syst. **31**(4), 49–87 (2015)
2. Chiou, J.S., Huang, C.Y., Lee, H.H.: The antecedents of music piracy attitudes and intentions. J. Bus. Ethics **57**(2), 161–174 (2005)
3. Sinha, R.K., Mandel, N.: Preventing digital music piracy: the carrot or the stick? J. Mark. **72**(1), 1–15 (2008)
4. Tjiptono, F., Arli, D., Viviea: Gender and digital privacy: examining determinants of attitude toward digital piracy among youths in an emerging market. Int. J. Consum. Stud. **40**(2), 168–178 (2016)

5. Cronan, T.P., Al-Rafee, S.: Factors that influence the intention to pirate software and media. J. Bus. Ethics **78**(4), 527–545 (2008)
6. Kwong, S.W., Park, J.: Digital music services: consumer intention and adoption. Serv. Ind. J. **28**(10), 1463–1481 (2008)
7. Wang, X., McClung, S.R.: The immorality of illegal downloading: the role of anticipated guilt and general emotions. Comput. Human Behav. **28**(1), 153–159 (2012)
8. Yoon, C.: Theory of planned behavior and ethics theory in digital piracy: an integrated model. J. Bus. Ethics **100**(3), 405–417 (2011)
9. Cho, H., Chung, S., Filippova, A.: Perceptions of social norms surrounding digital piracy: the effect of social projection and communication exposure on injunctive and descriptive social norms. Comput. Human Behav. **48**, 506–515 (2015)
10. Jacobs, R.S., Heuvelman, A., Tan, M., Peters, O.: Digital movie piracy: a perspective on downloading behavior through social cognitive theory. Comput. Human Behav. **28**(3), 958–967 (2012)
11. Culiberg, B., Koklic, M.K., Vida, I., Bajde, D.: Examining the effects of utilities and involvement on intentions to engage in digital piracy. Comput. Human Behav. **61**, 146–154 (2016)
12. Coyle, J.R., Gould, S.J., Gupta, P., Gupta, R.: 'To buy or to pirate': the matrix of music consumers' acquisition-mode decision-making. J. Bus. Res. **62**(10), 1031–1037 (2009)
13. Yoon, C.: Extending the TAM for green IT: a normative perspective. Comput. Human Behav. **83**, 129–139 (2018)
14. Cronan, T.P., Mullins, J.K., Douglas, D.E.: Further understanding factors that explain freshman business students' academic integrity intention and behavior: plagiarism and sharing homework. J. Bus. Ethics **147**(1), 197–220 (2018)
15. Fleming, P., Watson, S.J., Patouris, E., Bartholomew, K.J., Zizzo, D.J.: Why do people share file unlawfully? A systematic review, meta-analysis and panel study. Comput. Human Behav. **72**, 535–548 (2017)
16. Miller, D.T., Prentice, D.A.: Changing norms to change behavior. Annu. Rev. Psychol. **67**(1), 339–361 (2016)
17. Gunawan, D.D., Huarng, K.H.: Viral effects of social network and media on consumers' purchase intention. J. Bus. Res. **68**(11), 2237–2241 (2015)
18. Hashim, M.J., Kannan, K.N., Maximiano, S., Ulmer, J.R.: Digital piracy, teens, and the source of advice: an experimental study. J. Manag. Inf. Syst. **31**(2), 211–244 (2014)
19. Sukhu, A., Paul, P.T., Scharff, R.: Will 'doing right' lead to 'doing well'? An examination of green behavior. J. Consum. Mark. **35**(2), 169–182 (2018)
20. Cialdini, R.B., Reno, R.R., Kallgren, C.A.: A focus theory of normative conduct: recycling the concept of norms to reduce littering in public places. J. Pers. Soc. Psychol. **58**(6), 1015–1026 (1990)
21. Kallgren, C.A., Reno, R.R., Cialdini, R.B.: A focus theory of normative conduct: when norms do and do not affect behavior. Personal. Soc. Psychol. Bull. **26**(8), 1002–1012 (2000)
22. Cialdini, R.B., Trost, M.R.: Social influence: social norms, conformity and compliance. In: The Handbook of Social Psychology, pp. 151–192 (1998)
23. Horowitz, R., Schwartz, G.: Honor, normative ambiguity and gang violence. Am. Sociol. Rev. **39**(2), 238–251 (1974)
24. Ajzen, I., Fishbein, M.: Understanding Attitudes and Predicting Social Behavior, vol. 278 (1980)
25. Ajzen, I., Fishbein, M.: Questions raised by a reasoned action approach: comment on Ogden (2003). Heal. Psychol. **23**(4), 431–434 (2004)
26. Armitage, C.J.: Can the theory of planned behavior predict the maintenance of physical activity? Heal. Psychol. **24**(3), 235–245 (2005)
27. Ajzen, I.: The theory of planned behaviour is alive and well, and not ready to retire: a commentary on Sniehotta, Presseau, and Araújo-Soares. Health Psychol. Rev. **7199**, 1–7 (2014)
28. Bagozzi, R., Yi, Y.: On the evaluation of structural equation models. J. Acad. Mark. Sci. **16**(1), 74–94 (1988)

29. Hair, J.F., Hult, G.T.M., Ringle, C.M., Sarstedt, M.: A Primer on Partial Least Squares Structural Equation Modeling (PLS-SEM). Sage Publications, Inc, UK (2017)
30. Henseler, J., Ringle, C.M., Sinkovics, R.R.: The use of partial least squares path modeling in international marketing. Adv. Int. Mark. **20**, 277–320 (2009)

# Chapter 43
# Research on the Status Quo of Cultivating Innovative Talents of Enterprises in Hubei Province

**Pinghao Ye**

**Abstract** With the development of knowledge economy, innovative talents have become an important resource in the development of enterprises and are a key factor for enterprises to enhance their competitiveness. To give full play to the important role of innovative talents, enterprises must invest funds for employee training. Accordingly, this paper investigated the status quo of employee training in Hubei Province, analysed 315 valid questionnaires and summarised the satisfaction degree, training form, training measures and training status of enterprise employees in Hubei Province, with some strategies proposed to improve the training efficiency of enterprise employees.

## 43.1 Introduction

The technological innovation capability of an enterprise affects its future development and the level of the country's overall innovation level. Training is an important part of human resource development. A growing number of enterprises have realised the crucial role of training in achieving strategic business goals. As an important content of human resource development, training is highly necessary to enhance the competitiveness of enterprises in economic transformation. Training is the process of updating knowledge, developing skills, changing attitudes and behaviours and enhancing the ability of the trainees, thereby enabling them to complete tasks efficiently and effectively [1]. An increasing number of organisations invest a huge amount of time and money in training to improve the work-related competency of employees [2]. Emphasis on employee training and human resource development is not only the means to improve the efficiency and competitiveness of enterprises but also the motivation for enterprise technological, system and management innovations. Several empirical studies have shown a positive correlation between

---

P. Ye (✉)
Wuhan Business University, Wuhan, China
e-mail: yezigege1977@163.com

 
V. Chang et al. (eds.), *Modern Industrial IoT, Big Data and Supply Chain*,
Smart Innovation, Systems and Technologies 218,
https://doi.org/10.1007/978-981-33-6141-6_43

training and corporate performance. Training helps improve the company's competitive advantage, enhance the quality of output [3] and allow the company to grow and increase profitability. Niu [4] holds that training can improve all aspects of business performance, such as product quality, product development, employee relations, sales growth, profit and market share [4, 5].

Although the importance of training has attracted the attention of many companies, some companies have not fully established the training system [6]. Moreover, it has not even attracted the attention of enterprise managers. How to carry out a training systematically remains a difficult problem for enterprises. To further understand the status quo of enterprise training in Hubei Province and improve the pertinence and effectiveness of training, this paper examined the status quo of enterprise training, statistics on training initiatives, training forms, training status and employee satisfaction in Hubei Province. At the same time, it made a cross-analysis between the employees and the nature of the enterprises and put forward specific suggestions to improve the training efficiency of the enterprises.

## 43.2 Subject and Methods

### *43.2.1 Respondents*

This paper conducted a questionnaire survey for employees in Hubei area, used an electronic questionnaire to perform a secret survey and distributed questionnaires via the Wenjuanxing survey platform. A total of 315 valid questionnaires from 16 prefectures, cities and prefectures were collected.

### *43.2.2 Research Methods*

This study used the convenient sampling method. The content of the questionnaire includes two parts: the demographic characteristics of employees, including age, education, job position, working years and so on; the status quo of enterprise innovation talents training, including enterprise nature; employee satisfaction; innovation atmosphere; enterprise organisation training mode, its training frequency, its incentive measures and so on. Before starting the formal survey, 15 enterprise employees were invited for a dry run of the questionnaire, and the answers to the questionnaire were then modified to form a formal questionnaire.

### *43.2.3 Statistical Analysis Methods*

SPSS24.0 and Excel were used for statistical analysis. General information, training organisation status and training emphasis were described by frequency and percentage. The status quo of training was combined with the nature of the enterprise for cross-analysis.

## 43.3 Data Analysis

### *43.3.1 Overall Situation of the Samples*

A total of 315 valid questionnaires were collected in this survey, covering enterprises in all prefecture-level cities in Hubei Province. The survey samples include the types of industries such as machinery, chemicals, building materials and light industry. Table 43.1 shows that among the employees who participated in the survey, male accounted for 54.92%, and female accounted for 45.08%. In terms of age distribution, those aged 25 and under accounted for 19.37%, those aged 26–30 accounted for 34.6%, those aged 31–35 accounted for 26.35% and those aged 36–40 accounted for 19.68%. The academic qualifications of the investigators were undergraduates, accounting for 62.86%, undergraduates and below accounted for 19.37%, masters accounted for 16.19% and doctors accounted for 1.59%. The enterprises of the employees participating in the survey were mainly private enterprises accounting for 50.16%, followed by state-owned enterprises accounting for 29.52%, foreign enterprises accounting for 8.25% and collective enterprises accounting for 5.08%. The employees participating in the survey mainly held management and technical positions accounting for 38.1% and 35.24%, respectively, and R&D and design positions accounted for 10.16%.

### *43.3.2 Satisfaction Degree of Staff Training*

Satisfaction evaluation of training effect is a very important link in the entire training process [7]. Respondents who were satisfied with the training status accounted for 47.94%, 14.29% were very satisfied and 11.43% were not so satisfied. From the survey results, the employees have recognised the training carried out by the company. This paper made a cross-analysis on employee education and training results and found that employees with higher education have higher satisfaction with enterprise training. Conversely, employees with lower education had lower satisfaction with enterprise training. Thus, the investment of enterprises in the cultivation of innovative talents was unbalanced (Fig. 43.1), and they paid extra attention to

**Table 43.1** Demographic characteristics of effective samples

| Variable | Category | Frequency | Percentage (%) |
|---|---|---|---|
| Gender | Male | 173 | 54.92 |
| | Female | 142 | 45.08 |
| Age | 25 years old and below | 61 | 19.37 |
| | 26–30 years old | 109 | 34.6 |
| | 31–35 years old | 83 | 26.35 |
| | 36–40 years old | 62 | 19.68 |
| Years of working | Below 0–5 years | 134 | 42.54 |
| | 6–10 years | 118 | 37.46 |
| | 11–15 years | 34 | 10.79 |
| | Above 16 years | 29 | 9.21 |
| Education level | Below undergraduate | 61 | 19.37 |
| | Undergraduate | 198 | 62.86 |
| | Master | 51 | 16.19 |
| | Doctor | 5 | 1.59 |
| Job position | Production | 28 | 8.89 |
| | Technical | 111 | 35.24 |
| | Management | 120 | 38.1 |
| | R&D and design | 32 | 10.16 |
| | Others | 24 | 7.62 |
| Nature of enterprise | State-owned | 93 | 29.52 |
| | Collective | 16 | 5.08 |
| | Private | 158 | 50.16 |
| | Foreign | 26 | 8.25 |
| | Mixed ownership | 12 | 3.81 |
| | Others | 10 | 3.17 |

employees with higher educational background and lower investment in training for employees with lower educational background.

### 43.3.3 *Importance of Enterprise Training*

The importance enterprises attach to training will seriously affect the cultivation of innovative talents [8]. Survey respondents who reported that the enterprises paid extra attention to training and carried out regular annual training projects accounted for 50.16%. Employees who reported that the enterprises attached great importance to training accounted for 25.08%. Moreover, special funds were annually invested in

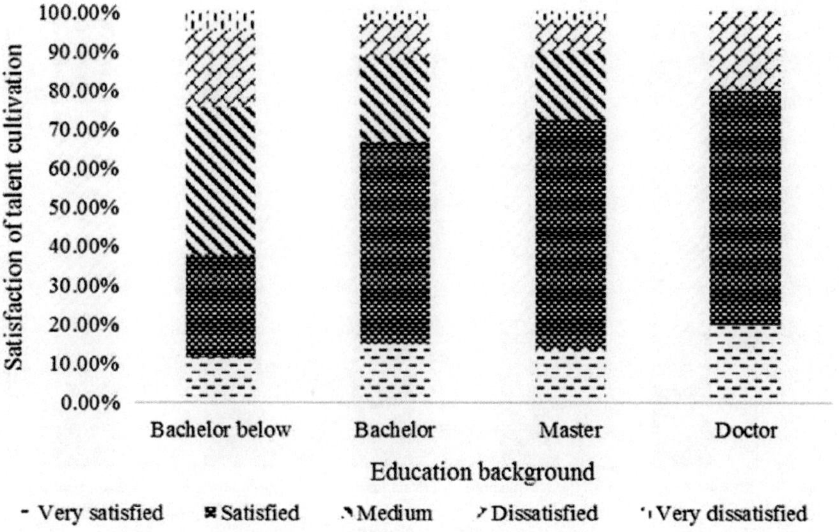


**Fig. 43.1** Satisfaction degree of staff training

training for its careful implementation. Employees who said that enterprises never carry out talent training projects accounted for 1.59%. This paper conducted a cross-analysis on the nature of enterprises and the importance of training, finding that state-owned (Fig. 43.2), collective and foreign enterprises pay more attention to training, whereas private enterprises pay less attention to training.

### *43.3.4 Training Methods*

Conducting new skills training, academic guidance of apprentices with apprentices, series of lectures by experts and establishing external academic connections are common training methods used by enterprises. Among them, 64.76% of employees said that enterprises adopted new skills training for employee training, and 46.67% of employees believed that enterprises adopted methods of masters and apprentices for training. The cross-analysis of the nature of the enterprise and the training methods found that foreign and state-owned enterprises can provide other opportunities for overseas exchanges (Fig. 43.3). Given that the academic and technical guidance of master and apprentice is one of the most feasible inexpensive methods, private enterprises often adopt this way to carry out employee training.

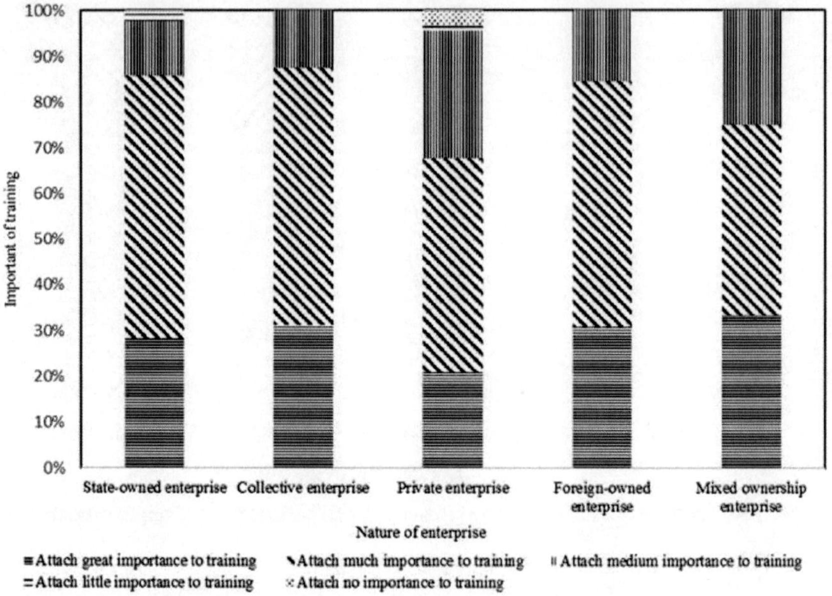


Fig. 43.2 Importance of enterprise training

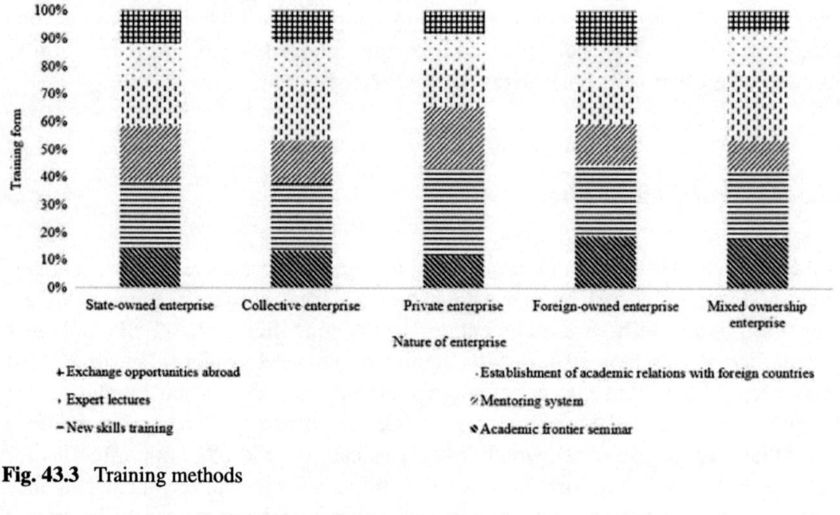


Fig. 43.3 Training methods

## *43.3.5 Incentive Measures for Innovative Talent Cultivation*

The measures that enterprises use to demonstrate the important role of training innovative talents included material compensation, training and learning activities, job promotion activities, honorary rewards and funding for innovation activities [9].

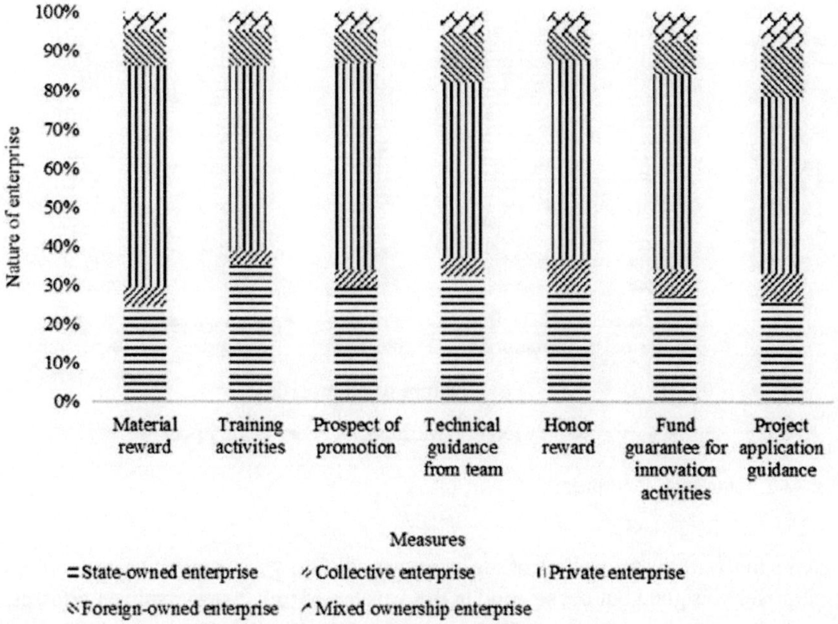


**Fig. 43.4** Incentive measures for innovative talent cultivation

Employees who believed that training and learning activities were conducive to training innovative talents accounted for 65.71%, 63.81% of them said that providing job promotion opportunities can encourage the cultivation of innovative talents and 49.21% believed that honorary rewards can also promote the cultivation of innovative talents. The cross-analysis on the nature of enterprises and incentives for the cultivation of innovative talents found that private enterprises adopt multiple incentives to promote the cultivation of innovative talents (Fig. 43.4), followed by state-owned enterprises that also invest further in the cultivation of innovative talents.

### *43.3.6 Innovation Atmosphere*

The innovation atmosphere is an organisational atmosphere with 'innovation' as the specific goal and is the subjective perception and description of whether the organisational members are innovative in the organisational environment. This perception will affect individual's attitude, belief, motivation, values and innovative behaviour [10]. It is conducive to cultivating the innovative behaviour of employees. Employees who believed that the enterprises had a good atmosphere of innovation accounted for 47.94%, and 28.89% said the innovation atmosphere was average. The results of cross-analysis on the nature of the enterprise and the innovation atmosphere showed

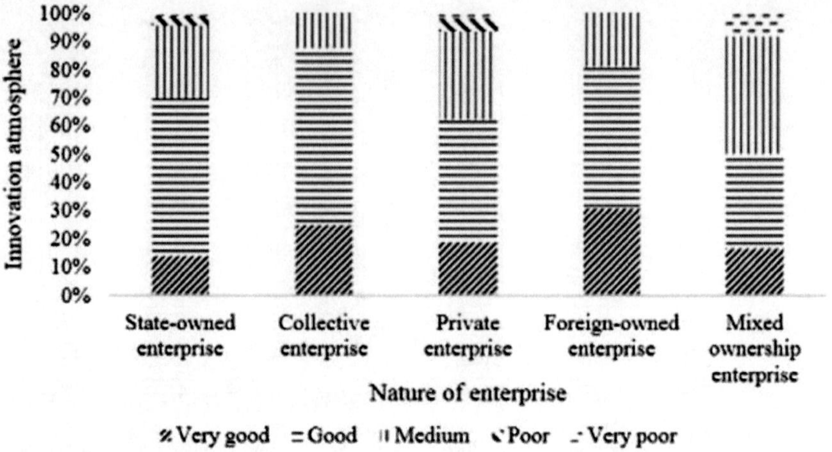


**Fig. 43.5** Innovation atmosphere

that the innovation atmosphere of the state-owned (Fig. 43.5), collective and foreign enterprises was good but not so good in the private and mixed ownership enterprises.

## 43.4 Problems in Enterprise Training

The lack of scientific and meticulous and comprehensive analysis of employee training renders the work quality of the training to have high blindness and randomness, which is disconnected from the long-term development goals of the enterprises and loses the strategic position of the training [11]. Correspondingly, employees also thought that training was a short break from work and thus did not pay adequate attention to it, which generated wrong training views. They were unwilling to participate in or did not take the training seriously, which rendered the training less effective whilst consuming substantial manpower, material and financial resources. From the survey results, both the total amount of training and the status of the training effect remained far from the actual needs of employees.

Given that different positions have different skill requirements, the training contents of many enterprises were general and less targeted. In addition, the enterprises adopted other traditional training methods, which led to susceptible learning effects and reduced the enthusiasm for employee training [12]. The training content was irrelevant to the actual work, and the training failed to achieve the expected results, consequently wasting time and money.

Some enterprises did not provide adequate support for the learning and training of innovative talents, which caused difficulty in the effective development and guaranteed sustainable development of the innovative talents in the enterprise [13]. At the same time, a good environment to fully incentivise innovative talents had not yet been formed, and no effective incentive system for talent cultivation had been formulated. Absence of such motivational strategies affected the initiative of innovative talents and restricted the healthy development of enterprises.

## 43.5 Countermeasures and Suggestions

Through the above analysis, remarkable progress has been found in the cultivation of innovative talents in enterprises, with many persistent problems and deficiencies. In response to these problems and deficiencies, this paper proposed the following suggestions from training design, training methods and training effects.

### *43.5.1 Strengthening Training Needs Analysis and Reasonably Allocating Training Resources*

Enterprises must increase the intensity and breadth of training needs analysis, with consideration of the actual needs of employees at all levels and levels that combine development goals, job skill requirements and employee career design; they must also conduct detailed and in-depth needs analysis to attain a reasonable design of course content [14]. Training needs should be clarified to obtain a reasonable allocation of training resources for employees; avoid over-concentration of training resources allocated to individual groups and carry out purposeful, step-by-step and systematic training to improve the overall quality of employees.

### *43.5.2 Enriching the Training Content*

Enterprises should not only provide training on job skills and old and new business knowledge but also provide training on employees' attitudes. Employee attitudes directly affect the effectiveness of training and work performance. Through training, employees can maintain a positive attitude towards work, which is helpful to their job [15]. The enterprise should also provide training on the personal career development of employees and combine the personal development of employees with the enterprise development, thereby enabling both employees and the enterprises to achieve progress. In addition, the training content should be targeted, and different contents should be trained for employees with different positions to match the training content with the actual work needs of employees.

### 43.5.3 *Combine Multiple Training Methods with Training Forms*

Classroom lectures, group discussions, case analysis, role-playing and other training methods have their own advantages and disadvantages. In the actual training design, various training methods can be selected for a reasonable combination according to the characteristics of training content and training conditions [16]. The internal training is highly targeted, with low training cost and flexible training time; the external training is conducive to learning new knowledge, new technologies and new concepts. Thus, the two complement each other. Whilst performing well in internal training, enterprises also carry out external training. The latter is helpful in acquiring foreign excellent enterprise management experience and production technology and in improving the competitiveness of enterprises.

**Acknowledgements** This paper was funded by the Hubei Provincial Technology Innovation Project (Soft Science Research), number: 2019ADD160.

## References

1. Palo, S., Padhi, N.: Measuring effectiveness of TQM training: an Indian study. Int. J. Train. Dev. **7**(3), 203–216 (2003)
2. Zhao, J., Qi, Z., de Pablos, P.O.: Enhancing enterprise training performance: perspectives from knowledge transfer and integration. Comput. Hum. Behav. **30**(1), 567–573 (2014)
3. Harry, J., Richard, H., Marcus, B., Jack, C.: Are training subsidies for firms effective? The Michigan experience. Ind. Labor Relat. Rev. **46**(4), 625–636 (1993)
4. Niu, C.: Study of enterprise management training based on cluster computing. Clust. Comput. **22**(6), 14889–14895 (2019)
5. Pham, N.T., Vo-Thanh, T., Shahbaz, M., Duc Huynh, T.L., Usman, M.: Managing environmental challenges: training as a solution to improve employee green performance. J. Environ. Manage. **269**(1), 770–781 (2020)
6. Arasanmi Chris, N.: Training effectiveness in an enterprise resource planning system environment. Eur. J. Training Dev. **43**(5/6), 476–489 (2019)
7. Nicod, L., Llosa, S., Bowen, D.: Customer proactive training vs customer reactive training in retail store settings: effects on script proficiency, customer satisfaction, and sales volume. J. Retail. Consum. Serv. **55**(2), 102069 (2020)
8. Bluff, E.: How SMEs respond to legal requirements to provide information, training, instruction and supervision to workers about work health and safety matters. Saf. Sci. **116**(3), 45–57 (2019)
9. Børing, P.: The relationship between training and innovation activities in enterprises. Int. J. Train. Dev. **21**(2), 113–129 (2017)
10. Bharadwaj, S., Menon, A.: Making innovation happen in organizations: individual creativity mechanisms, organizational creativity mechanisms or both? J. Prod. Innov. Manag. **17**(6), 424–434 (2000)
11. Al Mamun, A., Muniady, R., Fazal Syed, A., Malarvizhi, C.A.: Micro-enterprise development training and entrepreneurial competencies among low-income households in Malaysia. Asia Pac. J. Innovation Entrepreneurship **13**(3), 354–366 (2019)
12. Barrios, A., Reficco, E., Taborda, R.: Training effects on subsistence entrepreneurs' hope and goal attainment. Education + Training **61**(7/8), 895–917 (2019)

13. Smith, A., Oczkowski, E.: Modelling the reasons for the use of vocational training in Australian enterprises. Asia Pac. J. Human Resour. **53**(3), 370–385 (2015)
14. .Jones, J.: Training and development, and business growth: a study of Australian manufacturing small–medium sized enterprises. Asia Pac. J. Human Resour. **42**(1), 96–121 (2004)
15. Hart, S.L., Steinheider, B., Hoffmeister, V.E.: Team-based learning and training transfer: a case study of training for the implementation of enterprise resources planning software. Int. J. Train. Dev. **23**(2), 135–152 (2019)
16. Blunch, N.-H., Castro, P.: Enterprise-level training in developing countries: do international standards matter? Int. J. Train. Dev. **11**(4), 314–324 (2007)

# Chapter 44
# Collaborative Cultivation of Innovative Talents Between Enterprises and Colleges in Hubei Province

**Pinghao Ye**

**Abstract** Innovation is crucial for the development of a country and a nation, especially in the era of 'widespread entrepreneurship and innovation'. The promotion of a sustainable and healthy development of the national economy requires a large number of high-quality innovative talents. The collaboration between enterprises and universities provides new ideas and opportunities for cultivating innovative talents. This study investigated the current situation of innovation talent training in enterprises in Hubei Province, analysed the factors affecting the collaborative training of innovative talents in enterprises and universities and proposed specific measures for collaborative training of innovative talents.

## 44.1 Introduction

With the development of economy and society and the advent of the era of 'widespread entrepreneurship and innovation', collaboration has become a brand-new concept, emphasising the deep and organic integration of various elements, and collaborative development has provided new ideas for the cultivation of innovative talents [1]. As an important place for the cultivation of innovative talents, enterprises and universities should adapt to the new requirements of the development of the times, take the opportunity of collaborative innovation, accelerate the pace of reform and development of higher education [2] and explore innovative collaborative training mechanisms for talents. They can also integrate all resources and improve the quality of innovative talent training to ultimately generate a large number of high-quality innovative talents for the development of national economy and society.

Enterprises are an important subject to implement innovation-driven development strategies [3]. With the in-depth implementation of the innovation-driven development strategy, the willingness of enterprises to transform and strengthen innovation

P. Ye (✉)
Wuhan Business University, Wuhan, China
e-mail: yezigege1977@163.com

 
V. Chang et al. (eds.), *Modern Industrial IoT, Big Data and Supply Chain*,
Smart Innovation, Systems and Technologies 218,
https://doi.org/10.1007/978-981-33-6141-6_44

has increased significantly. Numerous innovative enterprises with independent intellectual property rights, independent brands and continuous innovation capabilities have emerged. China's innovative enterprises have boomed, and their economic benefits have been significantly improved, but a gap exists between the overall innovation ability of the enterprises and the international advanced level [4]. Studies have shown that the relative lack of innovative talents in enterprises or the collaborative innovation collaboration between innovative talents has not yet been formed, seriously restricting the improvement of enterprises' innovation capabilities [5]. Enterprises possess various resources such as talents, information, materials and technology. Among them, human resources play an important role in providing support and services for enterprise innovation. Therefore, to improve the independent innovation capability of enterprises and promote their sustainable and healthy development, the key is to have a certain quality of innovative talent resources [6]. Only by developing human resources, optimising the allocation of human resources and cultivating high-quality and highly intelligent innovative talents can China comply with the trend of the development of the innovation era and promote the substantial improvement of innovation capabilities.

The elite unity, collaborative research and common growth of innovative talents in different departments of an enterprise are important means to solve the shortage of innovative talents and insufficient collaboration of enterprises under the innovation-driven development strategy [7]. Innovative talents with a collaborative spirit, innovative awareness, technical understanding and management skills can be cultivated through various channels such as schools, enterprises or school–enterprise alliances [8]. Internal training is the most reliable and effective way as enterprises are the direct users of innovative talents.

## 44.2 Research Methods

### *44.2.1 Respondents*

This study conducted a questionnaire survey for employees in Hubei Province, used an electronic questionnaire to perform a secret survey and distributed questionnaires through the Wenjuanxing survey platform. A total of 315 valid questionnaires from 16 cities and prefectures were collected.

### *44.2.2 Research Methods*

This study used the convenient sampling method. The content of the questionnaire includes two parts: the demographic characteristics of employees and the status quo of collaborative training of innovative talents in enterprises. The former includes age, education, job position and working years, among others. The status quo includes the nature of the enterprise, innovation atmosphere, factors affecting the collaborative training of talents by the colleges and enterprises and main ways of collaboration between colleges and enterprises. Before starting the formal survey, 15 enterprise employees were invited to try out the questionnaire, and the answers to the questionnaire were modified to form a formal questionnaire. At the same time, the cross-analysis method was used to analyse two different variables.

## 44.3 Data Analysis

### *44.3.1 The Main Ways of College–Enterprise Collaborative Training*

The college–enterprise collaborative education mechanism is a new measure and method for innovating the integration of industry and education in local colleges under the background of the state's efforts to promote and deepen the reform of higher education. This measure reflects that the application-oriented colleges and universities are closely connected with the development of regional industries, and their traditional mode has been transformed into the initiative of facing the society and seeking social support [9]. A survey of enterprise employees found that the main ways for enterprises and colleges to develop innovative talents collaboratively include employment base training, technical consultation and entrusted training of colleges in enterprises. Enterprises also have great enthusiasm to participate in the construction of college majors and courses (Fig. 44.1).

Enterprises stated that the main factors affecting the collaborative training of innovative talents between the colleges and enterprises include the position of colleges and enterprises, difference in value pursuit, high cost of talent training and colleges' education ideas. Other factors are insufficient enterprise funding and lacking government support (Fig. 44.2).

### *44.3.2 Innovation Atmosphere and Training Input*

Innovation requires organisations to use a variety of resources to create an atmosphere that can stimulate employees' innovative ideas, encourage innovation and prompt employees to work and integrate all ideas in a unique way, thereby improving innovation performance. The cross-analysis of the enterprise's innovation atmosphere and training input found that as the enterprise's input in employee innovation training increased (Fig. 44.3), the innovation atmosphere was significantly improved.

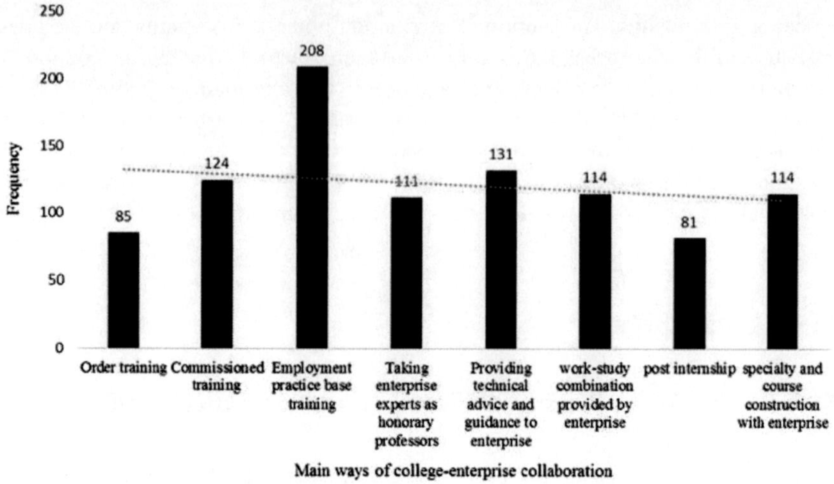


**Fig. 44.1** Main ways of college–enterprise collaborative training

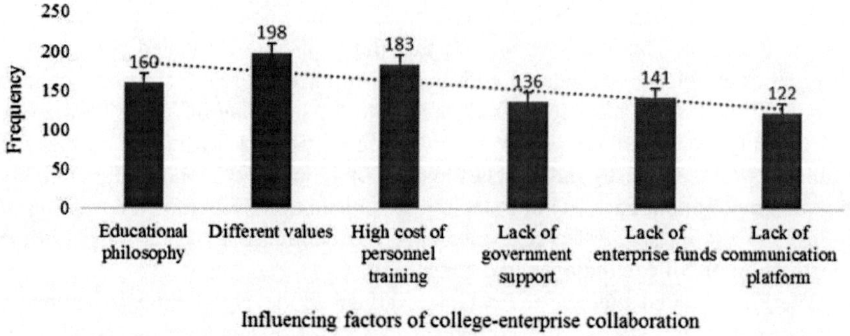


**Fig. 44.2** Influence factors of college–enterprise collaboration

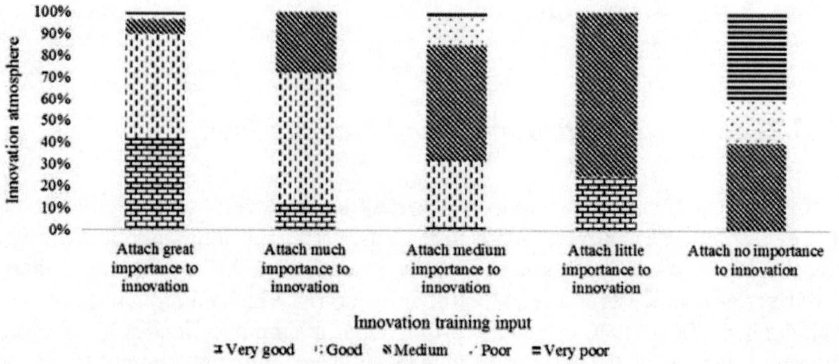


**Fig. 44.3** Innovation atmosphere and training input

### 44.3.3 *Innovative Atmosphere and Factors Affecting Collaborative Education*

Enterprise innovation activities are not merely the work of the technology R&D department. Enterprises should pay much attention to the integration of R&D activities with other management activities and market activities to create various favourable conditions to enhance the collaboration between the different departments [10]. The collaborative training of enterprise innovative talents aims to build a multi-functional department joint technical innovation team, give full play to the collaboration of 'innovative talent community' and realise the innovative collaborative value among enterprise innovative talents. The factors that affect the collaborative education of the enterprises include insufficient thought, lack of policy support, lack of economic motivation, lack of cooperation platform and lack of contact channels. The cross-analysis of the innovation atmosphere and the factors that affect collaborative education found that in enterprises with a good innovation atmosphere (Fig. 44.4), the lack of policy support and lack of economic motivation are the main factors that affect the development of collaborative education. Among enterprises with a poor innovation atmosphere, the lack of a cooperation platform is the main influencing factor.

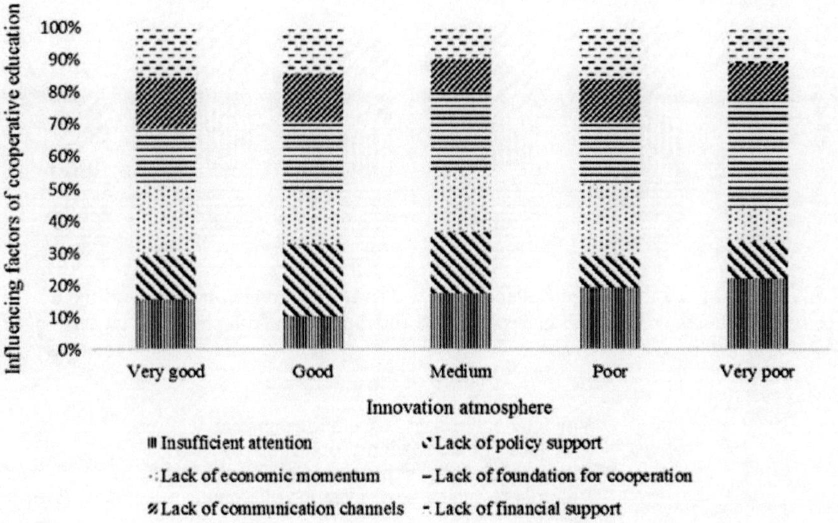


**Fig. 44.4** Innovative atmosphere and factors affecting collaborative education

### *44.3.4 Nature of Enterprise and the Type of College–Enterprise Collaboration*

The generation of collaborative effect is a superposition of the overall non-simple benefits established in an open system environment. Enterprise innovative talents can be regarded as a large talent system which needs to obtain various resources and information from outside to ensure the normal operation of enterprises [11]. At present, the main ways for enterprises and college to develop collaborative education are to establish a college–enterprise cooperation platform, a government–enterprise cooperation platform, an industry association cooperation platform and an enterprise cooperation platform. The cross-analysis of the nature of enterprises and collaborative education found that most enterprises typically establish of a college–enterprise cooperation platform to cultivate innovative talents (Fig. 44.5). Foreign and state-owned enterprises mainly develop innovative talents by establishing a government–enterprise cooperation platform. Mixed ownership enterprises mostly establish cooperation platforms between enterprises.

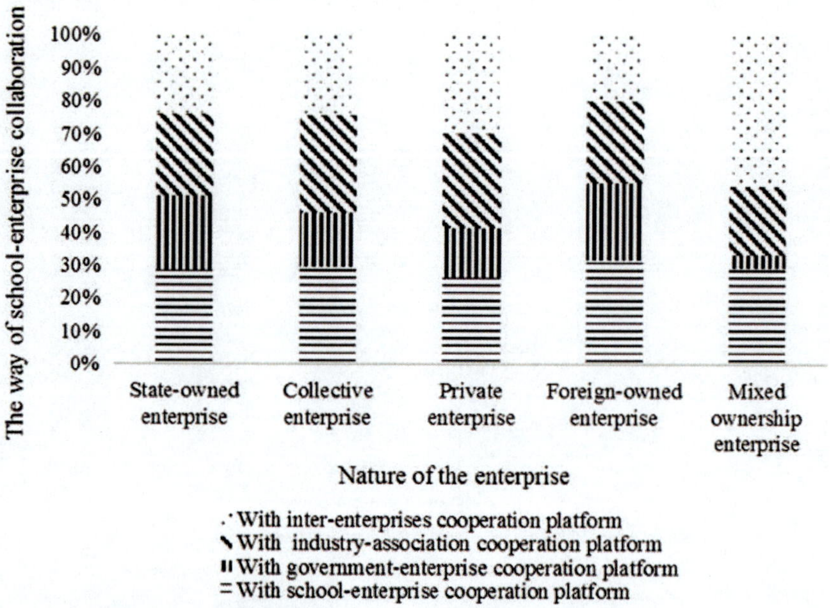


**Fig. 44.5** Nature of enterprise and the type of college–enterprise collaboration

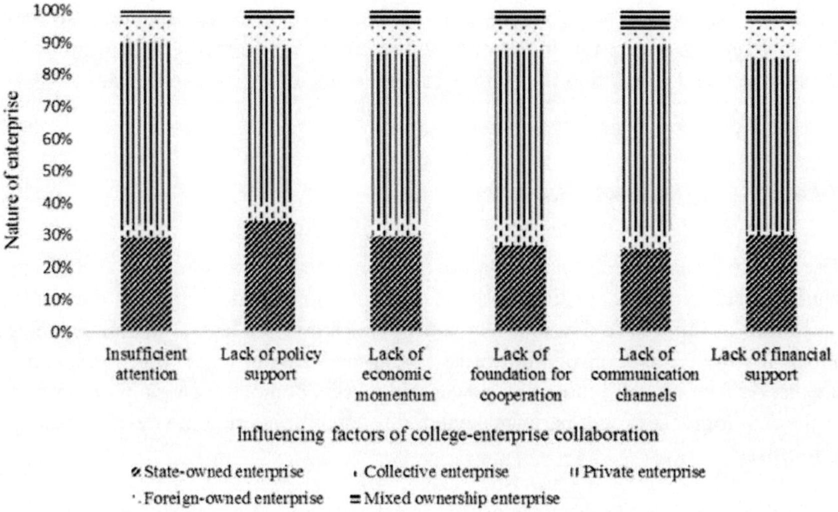


**Fig. 44.6** Nature of the enterprise and factors affecting the collaborative education of colleges and enterprises

### *44.3.5 Nature of the Enterprise and Factors Affecting the Collaborative Education of Colleges and Enterprises*

The collaborative training of innovative talents in enterprises is a social activity carried out under the condition of multiple factors, and many factors hinder enterprises from developing collaborative education [12]. The factors that affect the collaborative education of the colleges and enterprises include insufficient thought, lack of policy support, lack of economic motivation, lack of cooperation foundation, lack of contact channels and lack of financial support. The cross-analysis on the nature of the enterprise and factors affecting the collaborative education of colleges and enterprises found that private enterprises are affected by various factors in the process of collaborative education (Fig. 44.6), followed by state-owned enterprises.

## 44.4 Problems

### *44.4.1 Conceptual Barriers*

The backwardness of some managers in enterprises is the main factor restricting the collaborative training of innovative talents in enterprises [13]. Many managers of enterprises still hold the old traditional concept that training innovative talents has investment risk. They believe doing so will cause waste of resources, so they are unwilling to invest too much energy and capital to strengthen the training of innovative talents, resulting in a single training mode of innovative talents.

### 44.4.2 *Institutional Barriers*

The system is the external manifestation and concretisation of ideology. The system regulates and encourages talent activities and promotes and guarantees the implementation of innovative development concepts. Enterprises lack a guarantee mechanism for the cultivation of innovative talents and cannot retain more innovative talents. As a result, the innovation power and market competitiveness of enterprises in the development process are weakened, ultimately hindering the development of enterprises.

### 44.4.3 *Cultural Barriers*

Culture is the spiritual power of innovative talents in enterprises. Enterprises lacking spiritual power are not conducive to the full play of creative talents [14]. During technological innovation and development, enterprises achieve many important scientific and technological breakthroughs, but many cultural factors are not conducive to scientific and technological innovation. Enterprises emphasise group development, attach importance to the common character of people, ignore the development of people's personality, easily strangle the innovation and lose the exploration spirit given by the innovation core.

## 44.5 Suggestions

The essential feature of the era of knowledge economy is innovation, and the innovation of science and technology lies in the innovation of concepts, that is, the re-understanding of the value of talents by enterprises [15]. Enterprises need to change the concept of hiring and educating people. They should cultivate and use innovative talents, guide innovative talents to establish a sense of innovation, respect talents and personality and give play to the initiative of innovative talents.

The collaborative training of innovative talents in enterprises needs to accelerate the mechanism construction, update the concept of talent development, eliminate the obstacles in the mechanism of talent development, explore new training methods for system innovation and improve the mechanism of innovative talents training driven by innovation. All these actions can stimulate the enthusiasm of different subjects of collaborative training, strengthen communication, enhance mutual trust and remove institutional obstacles. A long-term mechanism of talent training can then be established.

Enterprises should embrace an innovative culture to encourage, condense, guide and promote the sustainable development of enterprises [16]. An innovative culture with rich connotation and unique personality is conducive to the innovation of enterprises. It is also helpful to promote the training of talents. The collaborative training of enterprise innovative talents must strengthen the construction of the innovative and collaborative culture of enterprises. Then, enterprises can establish innovative and collaborative culture comprehensively in an objective manner.

**Acknowledgements** This paper was funded by the Hubei Provincial Technology Innovation Project (Soft Science Research), number: 2019ADD160.

## References

1. Huo, W., Zhang, W., Chen, P.S.-L.: Recent development of Chinese port cooperation strategies. Res. Transp. Bus. Manag. **26**(3), 67–75 (2018)
2. Ferrer-Martí, L., Garfí, M., Ferrer, I.: Cooperation and human development projects as bachelor, master and Ph.D. thesis: evaluating an internship program. Procedia Soc. Behav. Sci. **196**(7), 63–68 (2015)
3. Wei, S., Zhang, Z., Ke, G.Y., Chen, X.: The more cooperation, the better? Optimizing enterprise cooperative strategy in collaborative innovation networks. Phys. A Stat. Mech. Appl. **534**(2), 120810 (2019)
4. Johnson, M.P.: Knowledge acquisition and development in sustainability-oriented small and medium-sized enterprises: exploring the practices, capabilities and cooperation. J. Cleaner Prod. **142**(2), 3769–3781 (2017)
5. Pietraszewski, D.: The evolution of leadership: leadership and followership as a solution to the problem of creating and executing successful coordination and cooperation enterprises. Leadersh. Quart. **31**(2), 101299 (2020)
6. Lisowska, R., Stanisławski, R.: The cooperation of small and medium-sized enterprises with business institutions in the context of open innovation. Procedia Econ. Financ. **23**(6), 1273–1278 (2015)
7. Rzepka, A.: Innovation, inter-organizational relation, and co-operation between enterprises in Podkarpacie region in Poland. Procedia Manuf. **30**(3), 642–649 (2019)
8. Meng, G., Arunkumar, N.: Construction of employee training program evaluation system of three exponential forecast based on sliding window. Cluster Comput. **22**(3), 6865–6870 (2019)
9. Almeida, R.K., Aterido, R.: Investing in formal on-the-job training: are SMEs lagging much behind? IZA J. Labor Dev. **4**(1), 8 (2015)
10. Jakobsone, A., Cakula, S., Florea, M.: Modelling of knowledge sharing for the provision of sustainable cooperation between adult educational institutions and enterprises. Procedia Comput. Sci. **104**(6), 160–165 (2017)
11. Gao, Y.-F., Meng, J.-Z., Guo, C.-J., Li, J.: "Construction of the practice teaching system for school-enterprise cooperation" in agricultural specialty of college under new situation. Procedia Environ. Sci. **12**(2), 1274–1279 (2012)
12. Zhao, J., Qi, Z., de Pablos, P.O.: Enhancing enterprise training performance: perspectives from knowledge transfer and integration. Comput. Hum. Behav. **30**(6), 567–573 (2014)
13. Suh, Y., Woo, C., Koh, J., Jeon, J.: Analysing the satisfaction of university–industry cooperation efforts based on the Kano model: a Korean case. Technol. Forecast. Soc. Change **148**(2), 119740 (2019)

14. Ajupov, A.A., Mikhailov, R.V., Mullanurov, A.T.: Implementation of new educational technologies through cooperation of economic clusters of enterprises with scientific and educational centers. Procedia Soc. Behav. Sci. **191**(2), 1288–1292 (2015)
15. Bliesner, A., Liedtke, C., Rohn, H.: Resource efficiency and culture—workplace training for small and medium-sized enterprises. Sci. Total Environ. **481**(6), 645–648 (2014)
16. Tsuruya, N., Kawashima, T., Shiozuka, M., Nakanishi, Y., Sugiyama, D.: Academia–industry cooperation in the medical field: matching opportunities in Japan. Clin. Ther. **40**(11), 1807–1812 (2018)

# Chapter 45
# Risk Analysis on Quality and Safety of Plastic Assembled Cabinets Based on Statistics of Injuries Inflicted by Hazardous Sources

**Qianwen Chen, Shudong Sun, Xia Liu, and Feng Qiao**

**Abstract** In recent years, the quality and safety of cabinet products have been more and more widely concerned by society, of which plastic assembled cabinets are extremely popular with the tenants in big cities, the users of small-sized houses or the households with children due to the characteristics such as convenience and lower price. Based on the use cases from consumers, from the influencing factors of the safety to use the products and the personal health of consumers, the main risk sources of the products of the plastic assembled cabinet in daily life have been described and analyzed in particular in this paper, with a view to guiding manufacturers to improve the production process during design and production to avoid potential risks.

## 45.1 Introduction

With the increasing demand for cost-effective, lightweight, and durable plastic furniture in various market areas, more and more people have adopted plastic furniture instead of traditional wooden or metal furniture. The uppermost reason for consumers to buy plastic furniture products is that plastic furniture is exquisitely refined, light and convenient, followed by natural colors, easeful comfort, and better practicality [1]. Currently, there are more than 100 professional plastic furniture enterprises in China, with an annual output of about 80 million pieces, nearly half of which are located in the provinces of Zhejiang, Guangdong, and Fujian. In 2019, the plastic furniture export volume was 62.332 million pieces from China, with an export value of 7.477 billion Yuan, and an import volume of 1.423 million pieces, with an import value of 180 million Yuan. As an important member of plastic furniture, quite a few bands of the plastic assembled cabinets have been emerging today.

Q. Chen (✉) · X. Liu · F. Qiao
China National Institute of Standardization, Beijing, China
e-mail: mis1303@163.com

S. Sun
National Center for Quality Supervision and Inspection of Furniture and Indoor Environment, Beijing, China



V. Chang et al. (eds.), *Modern Industrial IoT, Big Data and Supply Chain*,
Smart Innovation, Systems and Technologies 218,
https://doi.org/10.1007/978-981-33-6141-6_45

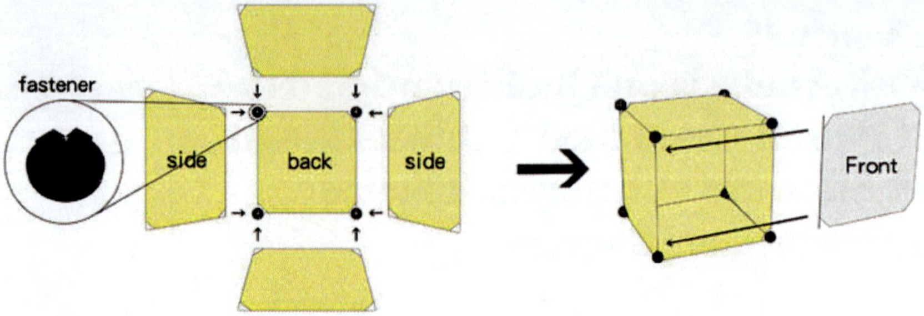


**Fig. 45.1** Schematic diagram of product assembly of plastic assembled cabinet

The plastic assembled cabinet refers to a type of furniture product made up of the individual plastic assembling boards formed by the plastic material that attaches to the metal frame and then connected by snap-fit joints with each other. Common products are mainly the wardrobes and shoe cabinets for adults or children, which are extremely popular with the tenants in big cities, the users of small-sized houses or the households with children due to the characteristics such as easy assembly and disassembly, space-saving, and lower price, etc. The sales channels of plastic assembled cabinets in China are mostly online, having nearly 40 major brands on the domestic online shopping platform, with product prices ranging from 30 to 200 Yuan. While plastic assembled cabinets are selling well, the risk incidents that threaten physical health or personal safety during consumer use also occur from time to time. The schematic diagram of product assembly of plastic assembled cabinet is shown in Fig. 45.1.

## 45.2 Hazardous Sources of Product Safety and Analysis on Injury Mechanism

Since plastic assembled cabinets are often used to store children's articles or are actually used as children's furniture, the safety risks of their products should be more considered in the scenarios of contact to and use by infants and children. Combining with the reports on quality and safety incidents, it is believed in this paper that the quality and safety risks of plastic assembled cabinets are mainly concentrated in three aspects: structural safety, snap-fit joints, and chemical hazards.

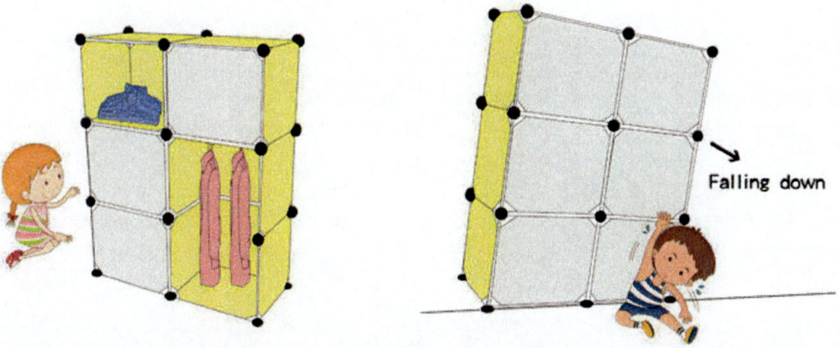


**Fig. 45.2** Schematic diagram of product dumping to hurt people

## *45.2.1 Risk of Structural Safety*

### 45.2.1.1 Tip-Over Risk

The height of the plastic assembled cabinets purchased by consumers generally exceeds 110 cm. Consumers often choose higher cabinets in order to have more storage space. When fully filled with articles, the cabinet will be deformed and collapsed if the product structure and intensity do not meet the requirements since the plastic assembled cabinet itself is lighter in the material [2]. In the process of daily use, children under the age of 3 are in a sensitive period to explore the space. One of the manifestations is to climb into the small spaces such as boxes and cabinets or climb and play in the plastic assembled cabinet being attracted by the cartoon patterns on the cabinet body. On the other hand, the plastic assembled cabinets have lower product stability than ordinary wooden furniture because of insufficient self-weight and assembly characteristics. Once the product has insufficient structural strength or is not equipped with additional fixed connectors, the latent risk of overturning and hurting people may be greater. The schematic diagram of product dumping to hurt people is shown in Fig. 45.2.

### 45.2.1.2 Sharp Points and Sharp Edges

The frame of the plastic assembled cabinet is made of thin metal rods, the metal frame of the cabinet body is a closed square, and the left side of the metal frame of the cabinet door is not completely closed, leaving a metal rod with sharp ends exposed at both ends. When inserting the metal rods into the snap-fit joints, the resin plates are assembled into a cabinet body. In daily use, since consumers often open and close the door, the surface coating of the metal rod end is cracked or thinned due to long-term wear, and the fixation of the snap-fit joints is reduced, resulting in the cabinet door often falling off from the cabinet body, which may expose the sharp

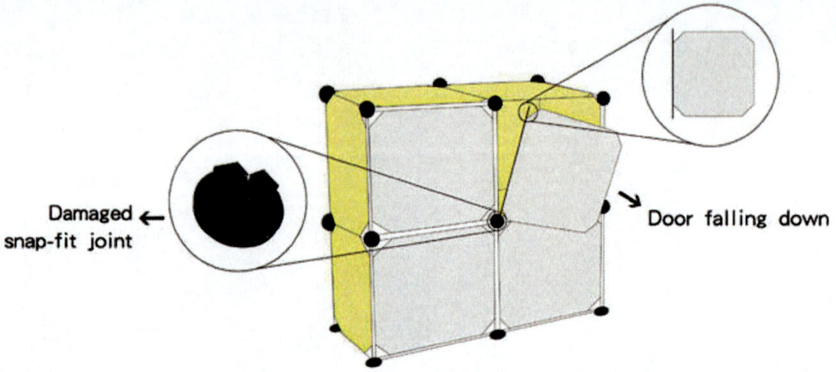


**Fig. 45.3** Sharp points and sharp edges

metal rod tip within the reach of humans. When used for a period of time, the snap-fit joints may also be damaged to form a sharp section, where the beveling angle near the snap-fit joint of the resin plate may occur the sharp edges that may be touched. Children are easily cut by sharp edges sharp points during activities near the cabinet or during daily use. The sharp points and sharp edges are shown in Fig. 45.3.

#### 45.2.1.3 Spaces Among the Components

After purchasing the products online, consumers need to assembly them at home. Since the groove of snap-fit joint clamps the thin metal rod through the raised grain design, it is easy to form gaps due to the connection not in place during the assembly, while the four corners of the cabinet assembly board may have some gaps due to the design. Owing to the unconsidered characteristics for children to use such products, there exist some unthoughtful designs such as the gaps in some products or the depth and size of the holes punching the product, which may cause the crushing risk when children extend their extremities (fingers or toes) into the product. The gaps between the cabinet door are shown in Fig. 45.4.

### *45.2.2 Accessories of Snap-Fit Joints*

An important accessory of the plastic assembled cabinet is the snap-fit joint that serves as an important supporting point of the cabinet body and plays a role of connecting components and supporting the strength of the cabinet body. There is a groove with the raised grain design on the snap-fit joint to enhance the tightness of the metal rod after insertion. If the snap-fit joint breaks after being used for a period of time, it may drop off small fragments, which may block the throat or trachea and

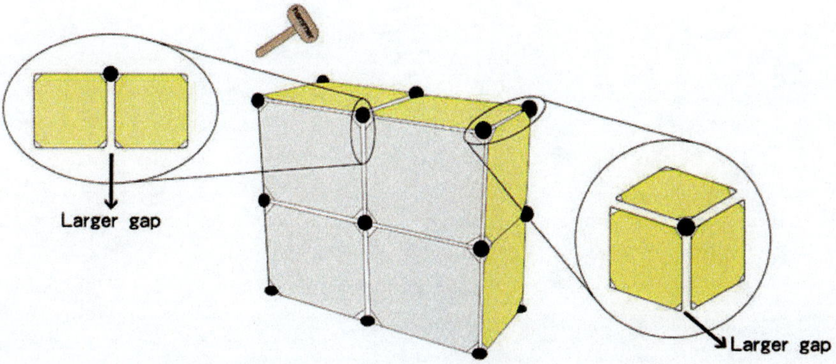


**Fig. 45.4** Large gaps between the cabinet door and the snap-fit joints to pinch the children's fingers

**Fig. 45.5** The small piece and the sharp section

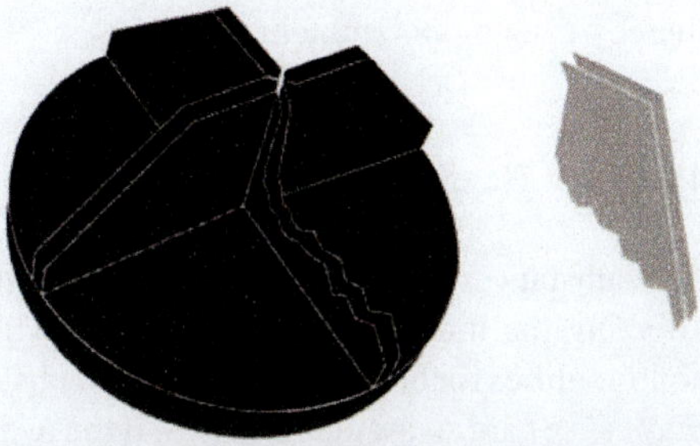

cause suffocation when the small pieces are swallowed by children or stuck in their noses. In addition, the sharp section formed after the snap-fit joint is damaged can also cause people to be scratched. The small piece and the sharp section formed by the damaged snap-fit joint are shown in Fig. 45.5.

## 45.2.3 Chemical Hazards

### 45.2.3.1 Hazards from Migratable Elements

Most plastic assembled cabinets are colorful and painted with images of cartoon animations and animals and are often used as the storage cabinets for infants and children due to better space utilization and multi partitions, but the surface coating of the resin board may have migratable elements. When the quality control of the production enterprise is not up to standard, it may cause the transferable elements contained in the dyes and pigments used in the production process to exceed the safety level. Once the infants or children bite or their skin touches the resin board,

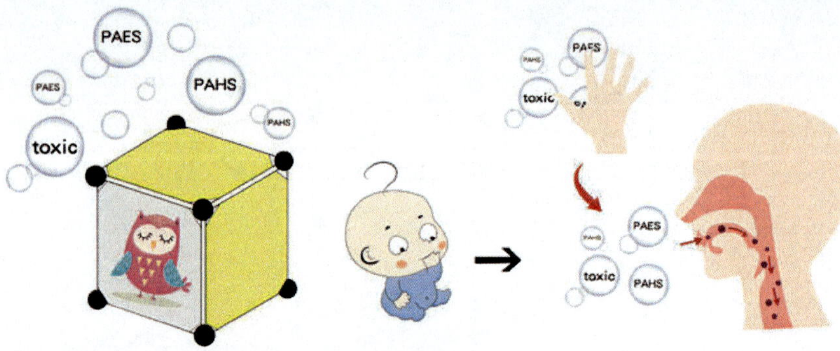


**Fig. 45.6** Schematic diagram of the chemical hazards in the products

this may cause the transferable elements to migrate the body of the infants or children, thereby causing potential harms [3].

#### 45.2.3.2 Hazards Caused by Phthalate

The substances containing phthalates are toxic to humans and organisms, mainly affecting the male reproductive ability, which may cause the reproductive system abnormalities for boys. In addition, this substance has potential carcinogenic risks and may cause cardiovascular diseases. If the resin boards in the plastic assembled cabinet are made of the materials that do not meet the safety requirements, especially the unqualified secondary recycled materials, the phthalates in the product may exceed the safety level. Phthalates can also migrate into the body of the infants and children through biting or skin contact, thereby causing potential harms [4]. The chemical hazards are shown in Fig. 45.6.

## 45.3 Data Analysis on the Use Cases by Consumers

125 pieces of case data for consumers to use the products of the plastic assembled cabinet are collected in this paper, of which the elements related to product quality and safety in the process for the consumers to use are refined. According to the statistical analysis on the injuries caused by the hazardous sources, the cases of broken or damaged products during the usage include six kinds of conditions such as the exposed sharp wire, damaged snap-fit joints, broken or damaged door panels, excessive door gap, cabinet deformation and falling into pieces, odor, etc., of which the frequency of breakage and damage to the door panel of the cabinet is 43, accounting for 34.4%. The structural strength of the door panel itself is critical to the overall structural strength of the cabinet. See Table 45.1 for the types of product damages.

**Table 45.1** Distribution table of product damage types

| S. No. | Damage type | Frequency | Ratio (%) |
|---|---|---|---|
| 1 | Exposed sharp wire | 28 | 22.4 |
| 2 | Cabinet deformation and falling into pieces | 33 | 26.4 |
| 3 | Damaged snap-fit joints | 8 | 6.4 |
| 4 | Excessive door gap | 28 | 22.4 |
| 5 | Broken or damaged door panels | 43 | 34.4 |
| 6 | Odor | 18 | 14.4 |

The main causes of the damage to the above-mentioned products are seven types of hazards: sharp points, sharp edges, snap-fit joints, intensity, chemicals, stability, and component gaps, of which the quality of the snap-fit joints that the consumers report is prominent, with a total of 63 times, accounting for 50.4%. During the assembly, it is more difficult for consumers to insert the metal rod into the deepest part of the snap-fit joint, requiring external force to insert the metal rod, which is also prone to the possibility that the snap-fit joint is broken or damaged due to improper force; during the daily use, since the door is opened and closed for a long time, and the metal rod is worn or the position changes in the snap-fit joint, it is likely to cause the damage to the snap-fit joint, resulting in that the door panel cannot be fixed. In addition, there are cases showing that children are easily scratched to the skin by the snap-fit joints when they are moving around the cabinet and because the snap-fit joint is exposed outside the cabinet. The second highlighted problem is that the gap between the door panels is larger, with a total of 64 times, accounting for 51.2%. The case shows that the linking up between the door panels fails to be perfectly connected, often leaving gaps, where the width of the larger gap exceeds 1 cm, which affects the stability of the cabinet, and easily pinches the fingers of children, and the cabinet body is dusty. The third relatively serious problem is that the intensity of the door panel is not enough, with a total of 47 times, accounting for 37.6%. According to the consumers, many of the door panels received are broken or damaged, the plastic part of the door panel cracks when used for some time, and the sharp edges at the crack have the risk of cuts. The distribution of types of product safety hazards is shown in Table 45.2.

**Table 45.2** Distribution table of the types of product safety hazards

| S. No. | Hazardous source | Frequency | Ratio (%) |
|---|---|---|---|
| 1 | Sharp points | 11 | 8.8 |
| 2 | Sharp edges | 35 | 28.0 |
| 3 | Snap-fit joints | 63 | 50.4 |
| 4 | Intensity | 47 | 37.6 |
| 5 | Chemicals | 19 | 15.2 |
| 6 | Stability | 61 | 48.8 |
| 7 | Component gaps | 64 | 51.2 |

## 45.4 Suggestions

1. **Pay attention to the safety to use and the needs of the children as a special group**. When designing and producing the products of the plastic assembled cabinet, manufacturers should consider the possibility for children as a special group to use, with focuses on the behaviors of young children who may climb and lick in the cabinet, avoiding the risks caused by possible gaps or small parts. At the same time, production enterprises should strengthen purchase control, improve the system for purchase inspection and acceptance, and ensure that the products sold meet the requirements for personal safety.
2. **Strengthening product safety supervision**. Relevant departments should strengthen the safety supervision on such type of products, carry out supervision and inspection in the production enterprises and the circulation areas, strengthen product quality rectification, and urge the production enterprises to implement the entity responsibilities.
3. **Strengthening publicity and guiding consumers to use such a product correctly**. The relevant product consumption warning information should be released in time, and the publicity should be strengthened, indicating the focuses and the emphasis to purchase the plastic assembled cabinets, reminding that consumers should keep the children from climbing and playing in the plastic assembled cabinets during usage, and regularly check the tightness of the connection points of the plastic assembled cabinet, and the damage degree of the small parts such as the snap-fit joints, replace the broken ones, and guide consumers to purchase the plastic assembled cabinet correctly.

**Acknowledgements** This paper has been supported by the national project "Research on key technical standards for quality and safety control of consumer goods" (2016YFF02022600), and "Research on common technology for integrative services by internet plus" (2017YFF0209604), Central basic scientific research project "Research on Consumer Goods Safety Hazard Identification and risk Assessment based on scenario Simulation" (552018Y-5928).

## References

1. Yang, W.: Design research to enhance the added value of plastic furniture products. Jiangnan University, Jiangsu (2013)
2. Wang, Y.: Research on the design of children's furniture based on the safety of materials and accessories. Central South University of Forestry and Technology, Hunan (2017)
3. Xu, J., Ming, Y., Luo, J.: Research on source and variety of furniture harmful substance and relevant standardization. Stand. Sci. (04), 6–9 (2016)
4. Yu, H., Zhu, Y.: Green evaluation on plastic furniture. Furniture (03), 100–102 (2016)

# Chapter 46
# Research on Safety Psychology Evaluation Index of Construction Workers Based on Statistical Analysis and Entropy Weight Method

**Xin-gang Yang, Han-sheng Qin, and Qi-quan Wang**

**Abstract** It is of great significance to study workers' safety psychology to prevent construction accidents caused by workers' mistakes. This paper analyzes the formation process and influencing factors of the unsafe mentality of the construction workers. We conduct factor analysis through statistical software SPSS21.0, determine the weight of each index by entropy weight method, and construct a safety psychological evaluation model. In addition, according to the principle of maximum degree of membership, we get the safety psychological grade of the construction workers and provide the measures to prevent the accidents caused by the workers' mistakes.

## 46.1 Introduction

At present, the prevention of construction accidents mainly focuses on technology, environment, design, and construction operations, and there are relatively few studies on why workers produce unsafe behavior. Accidents and human factors have a direct relationship, in all security incidents, due to human factors caused by accidents accounted for nearly 80% [1]. Therefore, how to reduce accidents caused by human factors has become the focus of scholars and the government. Due to the particularity of the construction industry, construction workers need to work in harsh environments, which is a great test for the construction workers. Exploring the production process and influencing factors of construction workers' safety psychology, determining the construction workers' safety psychological evaluation indicators, will help to decompose the construction workers' psychological state into several aspects for evaluation and understanding and establish a scientific evaluation system

X. Yang · Q. Wang (✉)
Department of Safety Engineering, China University of Labor Relations, Beijing 100048, China
e-mail: wqq_100@163.com

H. Qin
College of Safety Engineering, North China Institute of Science and Technology, Langfang 065201, China

 
V. Chang et al. (eds.), *Modern Industrial IoT, Big Data and Supply Chain*,
Smart Innovation, Systems and Technologies 218,
https://doi.org/10.1007/978-981-33-6141-6_46

to help managers master the building. The psychological state of the workers changes, making management more targeted. It can be seen that studying the safety psychology of construction workers is of great significance to both individuals and managers.

At present, domestic and foreign experts have studied the safety psychology of employees. In 1919, Greenwood [2] proposed the basic theory of unsafe behavior. He believed that the occurrence of accidents is closely related to people who are prone to accidents, but not directly related to their working environment, working conditions, and working pressure. In the paper [3], people's unsafe behavior is thought to be the main cause of the accident, has a lot to do with one's own character, but can through safety education, safety training and training, working environment, working pressure, communication, work experience, etc., so as to reduce the accident caused by human factors. In 2004, Geller [4] studied the role of personality in construction accidents and found that there is a close relationship between the two. It is concluded that the psychological attitude of construction workers will directly affect the safety status of workers during the construction process. In the paper [5], miner's psychology and the relationship between the safety in production has carried on the research and exploration, in the domestic many coal mine production safety accident investigation and research, and sums up the coal mine production in the process of the seven aspects of unsafe psychological influence, proposed the 25 of the most common mental state, for the workers, the psychological and the relationship between the coal mine safety research laid the foundation. In 2007, Wang [6] formulated the psychological evaluation indicators of the blaster and developed the corresponding software. In 2010, Yang [7] analyzed the role of safety psychology in safety management, mainly from the psychological perspective of employees to analyze accidents, mainly from psychological education before accidents, psychological analysis in accidents, and psychological intervention after accidents. Three aspects of research. In 2015, Li [8] studied the core of security psychology, defined security psychology and core principles, and gave the architecture of security psychology. And the current research and more specific to coal, electric power, and other industries, the construction sector research on the psychological safety of construction workers is still quite lacking and studies focusing on the construction industry in unsafe behavior and management based.

This paper mainly studies the influencing factors of construction workers' unsafe psychology and the formation process of unsafe psychology, so as to determine the safety psychology evaluation indicators of construction workers. Questionnaire method is used to investigate the indicators affecting the safety psychology of construction workers. SPSS21.0 statistical analysis software is used to process the data of questionnaires, and the main evaluation indicators of construction workers' safety psychology are obtained, which is the safety psychological aspect of construction workers.

## 46.2 Theoretical Analysis of Construction Workers' Safety Psychology

Affected by harsh conditions and long-term overload operation, construction workers have gradually become numb to safety work, which is characterized by weak safety awareness, poor self-protection, and sometimes even disregard the life of themselves or others [9]. Safety research and practice at home and abroad have proved that human error is the main cause of accidents, and the occurrence of mistakes involves the influence of many factors such as self, internal and external environment, and management [10], but the most important is the psychological state of people. As a result, the psychological state of the person is the basis for causing unsafe behavior [5]. Therefore, the construction workers have a good safety psychology which is the internal basis for ensuring that workers do not have accidents during the construction process. Next, we analyze the formation process of construction workers' unsafe psychology from the psychological characteristics of construction workers:

1. Psychological fatigue caused by long-term tedious work. Sustained nervousness leads to a state of depression, which is manifested by a decrease in enthusiasm for work and a feeling of boredom in the heart. Most construction workers clearer division of labor, workers engaged in a certain kind of job types, plus the construction period are generally longer and changes in the environment, the situation of the depressed state of mind is prone to.
2. The level of wages and salaries did not reach the sense of loss brought about by their own expectations. Most of the construction workers are migrant workers, and choosing to work in the process is a last resort choice for obtaining a higher income level. Because the construction is a heavy physical work, and the daily working hours are very long; even if the work period needs to work overtime, the construction workers want to get the salary level that is consistent with their own efforts. But when the wage level is lower than the expected value, the workers will feel lost, and there will be unsafe mentality such as psychological imbalance and negative absenteeism.
3. Inner self-esteem. The general education level of construction workers is not high. The work they do is generally heavy or low-tech jobs. Compared with other high-tech jobs, the wage level is not very high. This situation can easily lead to inferiority of construction workers. Psychologically, I think that I am "low-minded." Therefore, construction workers hope to gain the respect and affirmation of others through their own labor. Once their work is not recognized, this kind of inferiority will be intensified and induce bad psychology.
4. Work frustration caused by anxiety. The irreplaceable work of construction workers is not very strong. When work is wrong, it is easy to be psychologically frustrated. Appropriate anxiety will help workers to form a sense of tension and help to concentrate more on work, but too high a sense of anxiety can easily lead to excessive psychological stress on construction workers, leading to mental illness.

Building construction site is a "man–machine–environment" cross-cutting system, the system of psychological safety of workers by the multiple dimensions of their own conditions, internal and external environment and enterprise management, and other co-workers determined [11].

Based on this, this paper selects the influencing factors of construction workers' safety psychology from the three aspects of workers' own conditions, internal and external environment, and management methods as the basis for designing the questionnaire. From the level of the workers' own conditions, the workers' academic qualifications, self-awareness, and occupational goals are selected as the measurement factors. From the internal and external environmental aspects, noise, vibration, illumination, temperature, and humidity are selected as the measurement factors of the external working environment, and family relationships and relationships among colleagues are selected. The relationship between the superior and the subordinate is used as the measurement factor of the internal interpersonal environment; from the management mode, the salary, reward and punishment, work arrangement, workload, and safety propaganda are selected as the measurement factors.

## 46.3 Research Design and Research Methods

### *46.3.1 Evaluation Scale Design*

According to the purpose of the research, the designed questionnaire consists of two parts: the basic information of the construction worker and the safety psychology test.

The first part of the basic personal information mainly includes age, education level, length of service, and job types.

The second part is the safety psychometric test designed for the 15 factors that affect the safety psychology of construction workers. The test scale test questions are in the form of the Likert scale, which is divided into five grades, which are very consistent, basically consistent, somewhat consistent, basically non-conforming, very non-conforming, five being very consistent, and one being very non-conforming. According to the 15 major influencing factors identified, a total of 40 psychological assessment questions were finalized after the revision of the topic design and language expression.

The survey objects selected in this paper are first-line construction workers such as steel workers and scaffolding workers in a construction enterprise in Langfang City, Hebei Province. A random survey of 100 employees at the site was conducted to recover 99 valid papers with a recovery rate of 99%.

**Table 46.1** Inspection of KMO and Bartlett's psychological assessment indicators for construction workers

| | | |
|---|---|---|
| Sampling enough Kaiser–Meyer–Olkin metrics | | 0.732 |
| Bartlett's sphericity test | Approximate chi square | 2664.495 |
| | df | 780 |
| | Sig | 0.000 |

### *46.3.2 Reliability and Validity Test of the Evaluation Scale*

Using PSS21.0 to analyze the reliability of the sample data, the results show that the **Cronbach's α** of the questionnaire is 0.840 > 0.6, which indicates that the reliability of the designed evaluation scale is higher, and the reliability of the questionnaire is better. The KMO and Bartlett spherical test were used to analyze the validity of the questionnaire. It is generally considered that the KMO value is greater than 0.7, which is suitable for factor analysis. The KMO is 0.732 > 0.6, which is suitable for factor analysis. The higher the Bartlett spherical approximate chi-square test value, the higher the independence of the variable. The analysis shows that the Bartlett spherical approximate chi-square test value is 2664.495, and the significance level is 0.000 < 0.005, indicating that there is a certain significant difference between the variables set by the questionnaire. Validity of the test results is shown in Table 46.1, the analysis results, the apparent validity questionnaire was designed to meet the requirements from Table 46.1.

### *46.3.3 Factor Analysis Process*

In the validity verification of the paper, the value of KMO is 0.732, which is in line with the value requirement and is suitable for factor analysis. Factor analysis is performed by SPSS21.0, the initial factor is extracted by principal component method, and the factor rotation is performed by the maximum variance orthogonal rotation method to obtain the rotating component that satisfies the requirements. Considering the commonality of each variable, the variables with the common degree less than 0.5 are deleted. Finally, 11 major factors were obtained. The factor analysis is shown in Table 46.2.

From the factor analysis, it can be concluded that the first common factor mainly reflects the large load on noise, vibration, illumination and temperature and humidity, indicating that the first common factor comprehensively reflects the changes in these aspects, and the first common factor is named the operating environment factor. The second common factor mainly reflects the large load on salary and rewards and punishments, which basically reflects the workers' attention to these aspects, named as compensation and reward and punishment factors; the third factor is on the workers' self-awareness and other items. There is a high load, which reflects the

**Table 46.2** Construction worker safety psychometric test questions rotation component matrix

| | | Ingredient | | | | | | | | | | |
|---|---|---|---|---|---|---|---|---|---|---|---|---|
| | | 1 | 2 | 3 | 4 | 5 | 6 | 7 | 8 | 9 | 10 | 11 |
| Factor 1 | X5 | 0.883 | −0.073 | −0.020 | 0.105 | −0.107 | 0.117 | 0.083 | −0.025 | −0.023 | −0.141 | 0.019 |
| | X26 | 0.876 | −0.061 | −0.005 | 0.087 | −0.183 | 0.066 | 0.074 | 0.032 | −0.021 | −0.117 | −0.015 |
| | X24 | 0.817 | −0.051 | 0.157 | −0.068 | 0.105 | 0.121 | −0.061 | −0.106 | 0.059 | 0.092 | 0.166 |
| | X2 | 0.779 | −0.020 | −0.029 | 0.054 | 0.354 | −0.127 | 0.175 | −0.017 | −0.094 | −0.050 | −0.202 |
| | X6 | 0.763 | −0.063 | 0.067 | 0.014 | −0.049 | 0.168 | 0.041 | 0.102 | 0.012 | 0.151 | 0.111 |
| | X30 | 0.726 | −0.159 | −0.012 | 0.115 | 0.052 | 0.179 | 0.030 | −0.089 | 0.110 | −0.192 | 0.072 |
| | X37 | 0.719 | −0.026 | 0.006 | 0.091 | 0.303 | −0.102 | 0.261 | 0.058 | −0.134 | −0.058 | −0.215 |
| | X15 | 0.710 | −0.089 | 0.019 | 0.190 | 0.164 | 0.290 | −0.139 | 0.048 | −0.102 | 0.131 | 0.187 |
| | X9 | 0.695 | −0.057 | −0.011 | 0.079 | 0.209 | 0.038 | −0.082 | −0.177 | 0.100 | 0.083 | 0.074 |
| | X14 | 0.636 | −0.162 | −0.046 | 0.285 | 0.414 | 0.062 | −0.043 | 0.067 | −0.025 | 0.120 | −0.087 |
| | X3 | 0.505 | −0.073 | −0.057 | −0.177 | 0.136 | 0.486 | 0.252 | −0.156 | 0.074 | −0.081 | −0.257 |
| | X10 | 0.423 | 0.047 | −0.037 | −0.097 | 0.124 | 0.145 | 0.351 | −0.020 | 0.069 | −0.075 | 0.141 |
| | X28 | −0.060 | 0.824 | 0.063 | 0.006 | −0.021 | −0.108 | −0.169 | −0.075 | 0.046 | −0.030 | 0.027 |
| | X4 | −0.059 | 0.781 | −0.038 | −0.066 | 0.048 | −0.098 | 0.259 | −0.095 | 0.047 | 0.065 | 0.176 |
| | X39 | −0.151 | 0.744 | 0.095 | 0.017 | 0.156 | −0.146 | −0.083 | −0.010 | 0.003 | 0.093 | −0.148 |
| Factor 2 | X35 | −0.111 | 0.733 | 0.077 | 0.095 | −0.258 | −0.007 | −0.239 | 0.042 | 0.124 | 0.120 | 0.063 |
| | X20 | 0.051 | 0.708 | −0.037 | 0.035 | −0.157 | 0.129 | 0.006 | 0.092 | 0.007 | −0.137 | −0.325 |
| | X18 | −0.097 | 0.689 | −0.144 | −0.009 | 0.095 | −0.007 | 0.133 | 0.111 | −0.099 | −0.036 | 0.306 |
| | X34 | −0.179 | 0.568 | 0.084 | 0.017 | −0.117 | 0.232 | 0.190 | 0.219 | −0.219 | −0.044 | −0.342 |
| | X31 | 0.027 | 0.079 | 0.875 | 0.081 | −0.071 | 0.023 | 0.038 | 0.035 | 0.054 | 0.219 | 0.009 |

(continued)

**Table 46.2** (continued)

| | | Ingredient | | | | | | | | | | |
|---|---|---|---|---|---|---|---|---|---|---|---|---|
| | | 1 | 2 | 3 | 4 | 5 | 6 | 7 | 8 | 9 | 10 | 11 |
| | X17 | 0.167 | 0.021 | 0.790 | 0.077 | −0.187 | −0.038 | 0.032 | 0.009 | 0.109 | 0.341 | −0.005 |
| Factor 3 | X22 | −0.237 | 0.032 | 0.714 | 0.006 | 0.219 | 0.114 | 0.190 | 0.001 | 0.055 | −0.275 | −0.057 |
| | X36 | 0.038 | −0.089 | 0.614 | 0.216 | 0.051 | −0.113 | 0.084 | 0.173 | −0.236 | −0.175 | 0.105 |
| | X21 | 0.075 | 0.035 | 0.595 | 0.507 | 0.185 | −0.094 | −0.019 | −0.018 | −0.164 | −0.030 | 0.113 |
| | X23 | 0.174 | 0.090 | 0.214 | 0.769 | −0.011 | 0.186 | 0.009 | −0.018 | −0.005 | 0.117 | −0.018 |
| Factor 4 | X19 | 0.185 | −0.030 | 0.093 | 0.710 | 0.006 | −0.121 | 0.101 | 0.017 | −0.261 | 0.130 | 0.078 |
| | X7 | 0.060 | 0.035 | 0.062 | 0.660 | 0.279 | −0.171 | 0.407 | −0.066 | 0.157 | −0.210 | −0.045 |
| | X8 | 0.586 | −0.021 | −0.024 | 0.154 | 0.672 | 0.002 | 0.033 | −0.006 | −0.023 | 0.166 | 0.071 |
| Factor 5 | X27 | 0.560 | −0.016 | 0.074 | 0.137 | 0.667 | 0.027 | −0.019 | 0.024 | −0.023 | 0.211 | 0.162 |
| | X12 | 0.303 | −0.154 | 0.087 | 0.114 | 0.496 | 0.482 | 0.100 | −0.306 | −0.018 | 0.059 | 0.089 |
| Factor 6 | X1 | 0.421 | −0.076 | −0.072 | −0.034 | −0.051 | 0.764 | −0.087 | 0.017 | −0.005 | 0.013 | 0.085 |
| | X13 | 0.556 | −0.090 | −0.001 | 0.033 | 0.050 | 0.578 | −0.017 | −0.095 | −0.118 | 0.118 | 0.285 |
| | X29 | 0.087 | −0.050 | 0.194 | 0.186 | −0.096 | −0.064 | 0.780 | −0.052 | −0.022 | 0.170 | 0.090 |
| Factor 7 | X33 | 0.069 | −0.024 | 0.133 | 0.514 | 0.141 | 0.069 | 0.596 | −0.056 | 0.087 | −0.024 | 0.054 |
| Factor 8 | X38 | 0.038 | 0.054 | 0.078 | −0.108 | −0.010 | −0.108 | 0.022 | 0.838 | 0.152 | −0.013 | 0.033 |
| | X25 | −0.122 | 0.007 | 0.047 | 0.074 | −0.017 | 0.029 | −0.103 | 0.784 | 0.232 | 0.078 | 0.054 |
| Factor 9 | X32 | 0.053 | 0.076 | −0.115 | −0.119 | −0.018 | −0.020 | 0.188 | 0.172 | 0.812 | 0.107 | −0.060 |
| | X40 | −0.032 | −0.053 | 0.103 | −0.023 | −0.021 | −0.028 | −0.138 | 0.351 | 0.759 | −0.186 | 0.026 |
| Factor 10 | X16 | −0.041 | 0.040 | 0.111 | 0.090 | 0.178 | 0.047 | 0.091 | 0.044 | −0.035 | 0.795 | −0.100 |
| Factor 11 | X11 | 0.228 | 0.025 | 0.113 | 0.085 | 0.072 | 0.176 | 0.191 | 0.130 | −0.055 | −0.160 | 0.719 |

self-awareness of the workers and is named as a self-knowledge factor; the fourth factor has a higher load in the work arrangement, reflecting the work arrangement of the construction unit, and is named as the work arrangement factor; the fifth factor has a higher load in the item of the worker's superior and subordinate relationship, reflecting the worker. The relationship with the superior is named as the upper and lower relationship factor; the sixth factor has a higher load on the worker's family, reflecting the relationship between the worker and the family, and is named the family relationship factor; the seventh factor has a higher load among workers and colleagues, reflecting the situation of workers and colleagues, named as the relationship factor of colleagues; the eighth factor has a higher load on the items such as workers' education, which reflects the education of the workers. Named worker's academic factor; the ninth factor has a higher load on the safety propaganda item, reflecting the impact of safety propaganda on workers, named as safety propaganda factor; the tenth factor is the load of the item in terms of worker workload higher, reflecting the impact of the amount of work on workers, named the workload factor, the eleventh factor on the workers' occupational target items is higher, reflecting the workers' understanding of their profession, named career target factor. After the factor rotation, a total of 11 major factors were obtained, explaining 72.95% of all variables, and the 11 main factors were classified to determine the safety psychology indicators of construction workers, including four first-level indicators. There are 11 secondary indicators, as shown in Fig. 46.1.

## 46.4 Fuzzy Evaluation Process of Construction Workers' Safety Psychology

### *46.4.1 Set of Factors*

According to the influencing factors of construction workers' safety psychology, the factor set U is constructed, as shown in Fig. 46.1.

### *46.4.2 Determination of Indicator Weight*

In the evaluation process of construction workers' safety psychology, different indicators have different influences on the construction workers' psychology, so they are assigned according to the degree of influence, which makes the worker's safety psychological evaluation process more accurate and objective. In this paper, the entropy weight method in thermodynamics is used to determine the weight value of different indicators. Under normal circumstances, the smaller the entropy value of the indicator, the greater the degree of variation of the indicator, the greater the role played by the whole, and the greater the weight value. The calculation Formulas

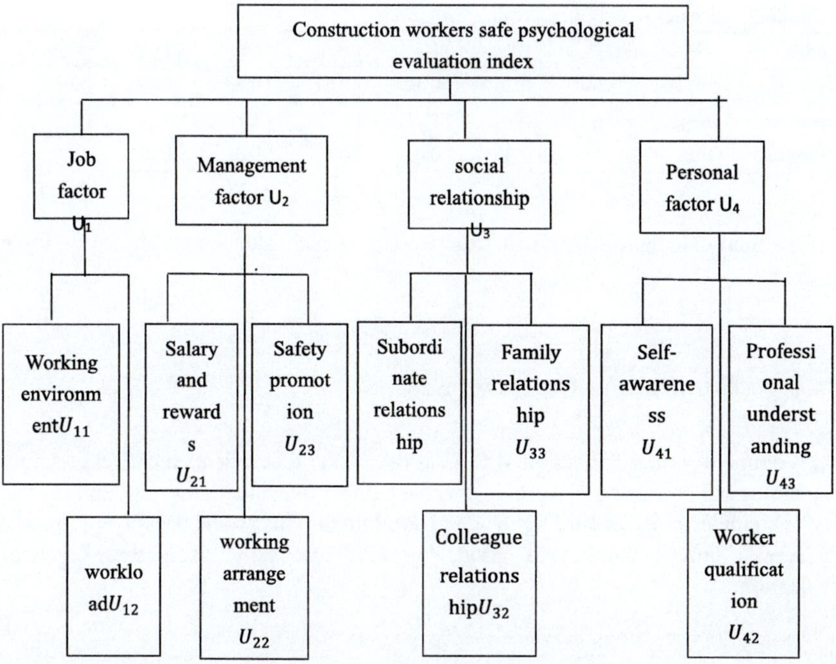


**Fig. 46.1** Construction workers safety psychology evaluation index system

(46.1) and (46.2) of the entropy value of the indicator are as follows:

$$f_{ij} = r_{ij} / \sum_{i=1}^{n} r_{ij} \tag{46.1}$$

$$H_j = -\frac{1}{\ln n} \sum_{i=1}^{n} f_{ij} \ln(f_{ij}) \tag{46.2}$$

Among them, in Formula (46.1), $f_{ij}$ is the eigenvalue of the $j$th second-level indicator in the $i$th level indicator; $r_{ij}$ is the data of the $j$ indicator of the $i$ first-level indicator, $n$ is the number of indicators; in the Formula (46.2), $H_j$ is the entropy value of the index $j$. The weight calculation method of the indicator is shown in Formula (46.3).

$$\omega_j = \frac{1 - H_j}{n - \sum_{j-1}^{n} H_j} \tag{46.3}$$

In the Formula (46.3), $\omega_j$ is the weight of the index $j$; and satisfies $0 \leq \omega_j \leq 1$, $\sum_{j-1}^{n} \omega_j = 1$.

**Table 46.3** Statistics of indicator weights at all levels

| Index | $U_1$ | $U_2$ | $U_3$ | $U_4$ | $U_{11}$ | $U_{12}$ | $U_{21}$ | $U_{22}$ |
|---|---|---|---|---|---|---|---|---|
| Weights | 0.31 | 0.34 | 0.25 | 0.10 | 0.61 | 0.39 | 0.44 | 0.35 |
| Index | $U_{23}$ | $U_{31}$ | $U_{32}$ | $U_{33}$ | $U_{41}$ | $U_{42}$ | $U_{43}$ | |
| Weights | 0.21 | 0.3 | 0.25 | 0.45 | 0.4 | 0.33 | 0.27 | |

The final calculation shows that the weight of each indicator is shown in Table 46.3.

### 46.4.3 Confirm the Comment Set

The comment set is a collection $V = \{V_1, V_2, \ldots V_k\}$ that reflects the level of safety psychology of construction workers, $k$ is the number of evaluation levels, and several levels are determined according to actual conditions. This article uses V = {"good," "general," "poor"}, that is take "good," "general" and "poor" as the three levels of evaluation.

### 46.4.4 Indicator Evaluation Process

The $i$ second-level indicators under the $j$ first-level indicators are counted and counted in $N$ surveys, respectively, where $n_1, n_2, n_3$ people consider them to be "good," "general" and "poor," and $n_1 + n_2 + n_3 = N$, then the secondary indicators. The evaluation result is ${}_2R_i^j$, that is

$$ {}_2R_i^j = ({}_2r_{i1}^j, {}_2r_{i2}^j, {}_2r_{i3}^j) = (n_1/N, n_2/N, n_3/N) \tag{46.4} $$

The comprehensive evaluation result of the primary indicator is the combined effect of all the secondary indicators it contains. Let the $j$ level one indicator include $m$ secondary indicators, and set the weight to

$$ {}_2A^j = ({}_2a_1^j, {}_2a_2^j, \ldots, {}_2a_m^j) \tag{46.5} $$

The $m$ secondary index evaluation results constitute a comprehensive matrix

$$ {}_2R^j = \begin{pmatrix} {}_2r_{11}^j & {}_2r_{12}^j & {}_2r_{13}^j \\ {}_2r_{21}^j & {}_2r_{22}^j & {}_2r_{23}^j \\ \cdots & \cdots & \cdots \\ {}_2r_{m1}^j & {}_2r_{m2}^j & {}_2r_{m3}^j \end{pmatrix} $$

Let the evaluation result of the $j$th first-level indicator be: ${}_1B^j = ({}_1b_1^j, {}_1b_2^j, {}_1b_3^j)$, then

$$ {}_1B^j = {}_2A^j \odot {}_2R^j \tag{46.6} $$

That

$$ {}_1B^j = ({}_2a_1^j, {}_2a_2^j, \ldots, {}_2a_m^j) \odot \begin{pmatrix} {}_2r_{11}^j & {}_2r_{12}^j & {}_2r_{13}^j \\ {}_2r_{21}^j & {}_2r_{22}^j & {}_2r_{23}^j \\ \cdots & \cdots & \cdots \\ {}_2r_{m1}^j & {}_2r_{m2}^j & {}_2r_{m3}^j \end{pmatrix} \tag{46.7} $$

where $\odot''$ is a modular operator, such as $M(\vee, \wedge)$, $M(+, \cdot)$, etc.

The four comprehensive evaluation of the results of a comprehensive evaluation of the composition of the matrix ${}_1R$, that is,

$$ {}_1R = \begin{pmatrix} {}_2b_1^1 & {}_2b_2^1 & {}_2b_3^1 \\ {}_2b_1^2 & {}_2b_2^2 & {}_2b_3^2 \\ {}_2b_1^3 & {}_2b_2^3 & {}_2b_3^3 \\ {}_2b_1^4 & {}_2b_2^4 & {}_2b_3^4 \end{pmatrix} \tag{46.8} $$

The weight of the first-level indicator is:

$$ {}_1A = ({}_1a_1, {}_1a_2, {}_1a_3, {}_1a_4) \tag{46.9} $$

Let the total evaluation result be: $B = (b_1, b_2, b_3)$, then

$$ B = {}_1A \odot {}_1R \tag{46.10} $$

That

$$ B = ({}_1a_1, {}_1a_2, {}_1a_3, {}_1a_4) \odot \begin{pmatrix} {}_2b_1^1 & {}_2b_2^1 & {}_2b_3^1 \\ {}_2b_1^2 & {}_2b_2^2 & {}_2b_3^2 \\ {}_2b_1^3 & {}_2b_2^3 & {}_2b_3^3 \\ {}_2b_1^4 & {}_2b_2^4 & {}_2b_3^4 \end{pmatrix} \tag{46.11} $$

where $\odot''$ is a modular operator, such as $M(\vee, \wedge)$, $M(+, \cdot)$, etc.

### 46.4.5 Application Instance Calculation

Safety assessment of the construction workers' psychology plays an important role in reducing construction accidents. In this case, the safety psychology of the first-line workers at a construction site was used as an evaluation project. After field investigation, the survey statistics of 88 people were obtained. The specific data is shown in Table 46.4.

As can be seen from Table 46.4, for the indicator "work environment," 39 people think "good," 20 people think "general," and 29 people think "poor." The evaluation result of the "work environment" calculated by Formula (46.4) is

$$ {}_2R_1^1 = (0.44, 0.23, 0.33) $$

**Table 46.4** Statistical results of the survey evaluation form

| Secondary indicators | | Statistical results | | | Evaluation results | | |
|---|---|---|---|---|---|---|---|
| Serial number | Indicator name | Great | General | Poor | Great | General | Poor |
| 1 | Working environment $U_{11}$ | 39 | 20 | 29 | 0.44 | 0.23 | 0.33 |
| 2 | Workload $U_{12}$ | 34 | 21 | 33 | 0.39 | 0.24 | 0.37 |
| 3 | Salary and rewards $U_{21}$ | 37 | 17 | 34 | 0.42 | 0.20 | 0.38 |
| 4 | working arrangement $U_{22}$ | 29 | 10 | 49 | 0.33 | 0.11 | 0.56 |
| 5 | Safety promotion $U_{23}$ | 45 | 17 | 26 | 0.51 | 0.20 | 0.29 |
| 6 | Subordinate relationship $U_{31}$ | 50 | 37 | 1 | 0.57 | 0.42 | 0.01 |
| 7 | Colleague relationship $U_{32}$ | 55 | 26 | 7 | 0.63 | 0.30 | 0.07 |
| 8 | Family relationship $U_{33}$ | 48 | 29 | 11 | 0.54 | 0.33 | 0.13 |
| 9 | Self-awareness $U_{41}$ | 53 | 27 | 8 | 0.6 | 0.31 | 0.09 |
| 10 | Worker qualification $U_{42}$ | 17 | 37 | 34 | 0.19 | 0.43 | 0.38 |
| 11 | Professional understanding $U_{43}$ | 36 | 24 | 28 | 0.40 | 0.28 | 0.32 |

Similarly, the other evaluation results of the secondary indicators are shown in Table 46.4.

Taking 1U1, the "work factor" comprehensive evaluation as an example, the data of Table 46.4 constitutes a comprehensive evaluation matrix.

$$ {}_2R^1 = \begin{pmatrix} 0.44\ 0.23\ 0.33 \\ 0.39\ 0.24\ 0.37 \end{pmatrix} $$

It can be seen from Table 46.3 that the weights of the two secondary indicators of "work environment" and "work volume" are

$$ {}_2A^1 = (0.61, 0.39) $$

Using the M(+, ··) fuzzy operator, the evaluation result of the first first-level indicator "work factor" is obtained by Formulas (46.6) and (46.7).

$$ {}_1B^1 = (0.61, 0.39) \odot \begin{pmatrix} 0.44\ 0.23\ 0.33 \\ 0.39\ 0.24\ 0.37 \end{pmatrix} = (0.42, 0.23, 0.35) $$

Similarly, the evaluation results of other first-level indicators are obtained.

$$ {}_1B^2 = (0.40, 0.17, 0.43) $$

$$ {}_1B^3 = (0.57, 0.35, 0.08) $$

$$ {}_1B^4 = (0.40, 0.35, 0.25) $$

The comprehensive evaluation matrix is composed of the first-level index evaluation results, that is,

$$ {}_1R = \begin{pmatrix} 0.42\ 0.23\ 0.35 \\ 0.40\ 0.17\ 0.43 \\ 0.57\ 0.35\ 0.08 \\ 0.40\ 0.35\ 0.25 \end{pmatrix} $$

As can be seen from Table 46.3, the weight of the primary indicator is

$$ {}_1A = (0.31, 0.34, 0.25, 0.10) $$

Using M(+, ·) fuzzy operator, a comprehensive evaluation is given by Eqs. (46.8) to (46.11) to give "psychological security" as a result of

$$ B = (0.45, 0.25, 0.30) $$

According to the principle of maximum membership degree, after comprehensive evaluation, the “safety psychology” is rated as “good.”

## 46.5 Conclusion

According to the psychological level requirements and psychological characteristics necessary for construction workers, four first-level indicators and 11 second-level indicators are finally obtained, which is helpful to construct a safety psychology evaluation index system for construction workers and adopt fuzzy mathematics evaluation method for construction. The safety psychology of workers was evaluated, which proved that the fuzzy mathematics evaluation method can effectively evaluate the safety psychological state of workers and provided a new idea for the safety psychological evaluation of employees in construction units. To classify the safety psychology of construction workers, on the one hand, it can help construction workers to grasp their own psychological state at any time, adjust their mentality in time to prevent accidents; on the other hand, the building units can grasp the safety and psychological changes of frontline workers through data and understand the workers' psychological state, avoiding the safe management of zombies, making management work more targeted and user-friendly.

**Acknowledgements** This paper is supported by National Key Technologies Research and Development Program (2017YFC0804900).

## References

1. Chen, W., Chen, R.: Study on mechanism of unhealthy occupational psychology and unsafe behaviors of construction workers. J. Saf. Sci. Technol. **12**(04), 118–123 (2016)
2. Greenwood, M., Woods, H.M.: The incidence of industrial accidents upon individuals with special reference to multiple accidents. Industrial Fatigue research Board, Medical Research Committee Report No.4., pp. 384–389. Her Britannic Majesty's stationary Office, London (1919)
3. Alaimo, B.: Unsafe behavior or unsafe condition-That is the question. Div. Chem. Health Safety Am. Chem. Soc. **2005**(2), 48–49 (2005)
4. Geller, S.: The fitting solution to respiratory hazards. Psychol. Saf. 12–14 (2004)
5. Zhang, S.: Study and exploration on the relationship between miners' psychology safety in production. In: Proceedings of the 1998 International Symposium on Safety Science and Technology, ISSST, Beijing, China (1998)
6. Wang, W.: Research on safety psychological assessment system and its application in blaster safety psychological assessment. Kunming University of Science and Technology (2008)
7. Yang, Y.: Safety psychology and its application in safety management. Central South University (2010)
8. Li, S., Wang, W., Wu, C.: Research on core principles of safety psychology. China Saf. Sci. J. (9), 8–13 (2015)

9. Pi, G.: Study on the effectiveness evaluation of safety management measures for construction projects based on the theory of safety psychology. Kunming University of Science and Technology (2017)
10. Li, N., Niu, L.: Structural model of miners' job burnout, unsafe psychology and unsafe behavior. Chin. Ment. Health J. **24**(03), 236–240 (2010)
11. Jiang, Q., Li J., Cheng, P.: Investigation of the safety awareness of the influential factors among the construction workers. J. Saf. Environ. **16**(06), 169–174 (2016)

# Author Index





V. Chang et al. (eds.), *Modern Industrial IoT, Big Data and Supply Chain*, Smart Innovation, Systems and Technologies 218,
https://doi.org/10.1007/978-981-33-6141-6